JEWS, CHRISTIANS, AND MUSLIMS IN THE MEDITERRANEAN WORLD AFTER 1492

JEWS, CHRISTIANS, AND MUSLIMS IN THE MEDITERRANEAN WORLD AFTER 1492

Edited by

Alisa Meyuhas Ginio

FRANK CASS

First published in 1992 in Great Britain
Reprinted in 2002 by
FRANK CASS PUBLISHERS
Crown House, 47 Chase Side, Southgate, London N14 BP, England

and in the United States of America by
FRANK CASS PUBLISHERS
c/o ISBS 5824 N.E. Hassalo Street, Portland, Oregon 97213-3644

Transferred to Digital Printing 2004

Website: www.frankcass.com

British Library Cataloguing in Publication Data

Jews, Christians and Muslims in the
Mediterranean World After 1492. –
(Mediterranean Historical Review Series,
ISSN 0951-8967)
I. Ginio, Alisa Meyuhas II. Series
909.08

ISBN 0 7146 3492 1 (h/back)
ISBN 0 7146 8050 8 (p/back)

Library of Congress Cataloging-in-Publication Data

Jews, Christians, and Muslims in the Mediterranean world after 1492 /
edited by Alisa Meyuhas Ginio.
p. cm.
ISBN 0-7146-3492-1 (US=.S.) : $35.00
1. Sephardim—History. 2. Jews—Spain—Castile—History.
3. Sephardim—Italy—History. 4. Jews—Italy—History 5. Muslims—
Spain—History. 6. Castile Kingdom—Ethnic relations. 7. Italy—
Ethnic relations. 8. Spain—Ethnic relations. I. Meyuhas, Ginio,
Alisa, 1937–
DS134.J48 1992
909'.04924–dc20 92-13102
CIP

This group of studies first appeared in a Special Issue on Jews, Christians and Muslims in the Mediterranean World After 1492, *Mediterranean History Review*, Vol. 6, No. 2.

Contents

Introduction

The Mediterranean has played a central role in the history of mankind's major civilizations, especially up to the early modern period; it is still the focus of universally significant events. The continuity of the Mediterranean heritage clearly emanates from the works of Fernand Braudel and Shlomo Dov Goitein. The 'Braudelian' Mediterranean history aims at wide perspectives, claiming, in terms of *longue durée*, that notwithstanding basic differences in language, religion, and culture, certain unifying elements have been apparent throughout the history of the Mediterranean basin.[1] 'Despite the many frontiers and the frequent wars' — to quote Goitein's words — 'people and goods, books and ideas travelled freely from one end of the Mediterranean to the other.'[2] This was true for the 'middle' of the Middle Ages. However, during the long centuries of a multifaceted Mediterranean history, such passage was not always of a voluntary nature. The conquest of Granada, last Muslim stronghold in the Iberian Peninsula, and the expulsion of the Jews (and later of the Muslims) from the newly unified Spanish kingdom and from other territories at the end of the fifteenth century, initiated a new era in the life of the Mediterranean world. Old communities were disrupted and new ones came into being. The impact of this change was felt throughout the Mediterranean basin, while the Age of Discovery opened up new horizons for the peoples of Europe. The following articles discuss the aftermath of the crucial historical events that took place in the Mediterranean world during the year 1492, focusing on the social, economic, and cultural consequences of those occurences, with reference to ethnic groups and religious denominations, thus integrating Jewish and world history in a wide geographical and historical context.

Maurice Kriegel wrote that 'l'expulsion des juifs était, certainement le dernier des moyens que les souverains voulaient utiliser pour briser la

1. F. Braudel, *The Mediterranean and the Mediterranean World in the Age of Philip II*, 2 vols., trans. S. Reynolds (New York, 1976).
2. S.D. Goitein, 'The Unity of the Mediterranean World in the Middle Middle Ages', *Studia Islamica*, 12 (1960), 29–41, especially 30.

progression de l'hérésie judaisante dans la société chrétienne: c'etait le plus coûteux.'[3] Haim Beinart states that, from the moment of their accession to the throne, the Catholic Monarchs had been considering the expulsion of the Jews from their newly united realms.[4] At all events, during the first two decades of their reign, prior to the promulgation of the edict ordaining the general expulsion, and despite several — sometimes harsh — anti-Jewish decrees, the policy of the Catholic Monarchs towards the Jews allowed the latter economic enterprise and community self-government. Tax assessments bear witness to the existence of a Jewish community of some size and economic standing in Medina del Campo. This Castilian town — whose economic activities were centred round its annual fairs — was, in many respects, the financial capital of the Iberian Peninsula. In her article, Maria Fuencisla García Casar discusses the bustling economic activities of the thriving Jewish community which lived in Medina del Campo until the time of the Expulsion.

The expulsion of the Jews has long been considered by many historians as a decisive cause for the decline (*decadencia*) of Spain. True enough, that concept of decline has never been free from some political prejudice: in effect, all the shortcomings of Spanish history have been traced to the expulsion of the Jews and Muslims, and the consequent inability of Spain to make profitable use of the influx of precious metals produced in the newly discovered Americas and shipped to the Castilian port of Seville. In view of the important role played by Jews and *judeoconversos*[5] in the economic activities of the Iberian Peninsula, it is clear that by expelling the Jews and the Muslims, and by instituting the 'New' Spanish Inquisition against the *conversos*, Spain turned its back on long-range economic considerations, giving priority to ideological issues. Yet, paradoxically enough, as Albert Sicroff has pointed out, even during the whole century preceding the 1492 Expulsion, the circumstances perpetuating the presence of Judaism in the Iberian Peninsula were being formed and implemented due to the very existence of the *judeoconversos*.[6] The Jews, and later the *moriscos* — even 100 years after their forced conversion to Christianity —

3. M. Kriegel, 'La prise d'une décision: l'expulsion des juifs d'Espagne en 1492', *Revue historique*, 260 (1978), 49–90, especially 87.
4. H. Beinart, *The Expulsion of the Jews from Spain* (Jerusalem, forthcoming).
5. H. Kamen, *La Inquisición Española*, trans. G. Zayas, 3rd edn. (Barcelona, 1988), Ch. III, pp. 34–89; id., 'The Decline of Spain: A Historical Myth?', *Past and Present*, 81 (1978), 24–81. Kamen claims that the migration of the *conversos* was even more damaging to the economic development of Spain than the expulsion of the Jews.
6. A.A. Sicroff, *Los estatutos de Limpieza de Sangre*, trans. M. Armiño (Madrid, 1985), Ch. I, pp. 43–4.

were expelled from the Iberian Peninsula.[7] The *conversos* lingered on, and were joined by Jews who, under the obligation of converting to Christianity, returned to Spain after 1492 in order to resettle there. The life of the *conversos* was marked by a tragic duality; for they found it very hard to be fully integrated into Catholic society, yet they had already crossed the lines and abandoned their former creed. Carlos Carrete Parrondo expounds historical documentation and literary sources reflecting the nostalgia of Castilian *judeoconversos* — living far from the cultural centres, and in direct contact with a closed, rarefied, rural society — for their former faith; their inability to part with a thousand-year-old tradition; their Messianic fervour; and their criticism, and sometimes ridicule of the Christian faith. Parenthetically, it is both noteworthy and touching to weigh the nostalgia expressed by those Castilian *judeoconversos* against the wistful memories of their former coreligionists who disdained conversion and chose to go into exile. Some of these Castilian Jews took refuge on the far shores of the Mediterranean, in Palestine and Egypt. A moving *aperçu* of those Iberian refugees' nostalgia for the land of Spain and its ways of life, and, above all, their loyalty to the Castilian language,[8] is offered in Joel L. Kraemer's description of some Spanish ladies from the Cairo Geniza.

The lot of the *judeoconversos* was dramatized by Felipe Godínez, the *Siglo de Oro* dramatist who expressed, both in the theatrical pieces he wrote and in his own life story, the duality which marked the lives of *judeoconversos*. He presented his heroine, Queen Esther, as a sort of *conversa*, for she concealed her Jewish origin upon entering the king's palace. Sara Turel points out how — against the background of the tragic crisis that obscured *conversos'* lives — Godínez wished to raise them from their inferior status in the ranks of Spanish society, and even grant them the elevated title of the chosen people of God.

The principles of the policy adopted by the Catholic Monarchs towards the Jews and *judeoconversos* were applied with the same vigilance towards the longtime converts from Islam and to the *mudéjares* of Granada. In a paper written for this volume, Miguel Angel Ladero Quesada examines the fate of the *mudéjares* of Granada and their relationship with the Christian *repobladores* since the capitulation treaties, and throughout the 1502 forced conversion of the Muslims — hereafter known as *moriscos* — and the latter's mid-sixteenth-century mutinies. These were the first two stages of the Christian-*morisco* conflict, to be followed by a third stage, the

7. See H. Kamen, 'The Mediterranean and the Expulsion of Spanish Jews in 1492', *Past and Present*, 119 (1989), 30–55, for a survey of the various figures given for the Jews expelled in the entire western Mediterranean.
8. Cf. Y.H. Yerushalmi, 'Castillian, Portuguese, Ladino: The Non-Hebrew Literature of Sephardi Jewry', in Z. Ankori (ed.), *Then and Now — Annual Lectures on the Jews of Greece* (Tel Aviv, 1984), pp. 32–53 (Hebrew).

1609–14 general expulsion of the *moriscos*.[9]

The conquest of Granada and the expulsion of the Jews enhanced the prestige of the Catholic Monarchs of Spain. An outcome of that process was the actual broadening of the geographical perception of their political activities. Elguja Khintibidze discusses the correspondence addressed by the king of remote Georgia to the Catholic Monarchs, trying to persuade them to undertake the long-standing medieval tradition of a crusade, this time against the rising power of the Muslim Ottoman Empire. Yet those efforts were doomed to fail, as Spain was too far away to indulge in bellicose activities in the eastern part of the Mediterranean basin, and was, moreover, actually fighting to tighten its grip in North Africa. This region, since the capture of Granada, had become the destination of those Muslims who tried to avoid living under Christian domination. Felipe Maíllo Salgado's article sheds new light on the origin and history of the *almogataces*, who — in their capacity as mounted cavalry, spies, guides, and scouts — secured the maintenance of Spanish military posts in the Orán region, thus ensuring Spanish domination there. It was in this region of the city of Orán, which, in many ways, became a sort of second capital of Spain, that a kind of *convivencia* — in which people of different religious creeds had to share a given space — existed during the later Middle Ages and the early modern period, long after it had come to an end in the Iberian Peninsula due to the triumphant Christian *Reconquista*.

The expulsion of the Jews from the Iberian Peninsula sounded the death bell of Spanish medieval *convivencia*. The plight of the expelled Jews emerges from the writings of contemporary authors. Reminicences of the hardships, the despair, the anguish and the trauma of 1492 and beyond, were recorded by Solomon Ibn Verga, author of *Shevet Yehuda*. Two Kabbalists — Yehuda Hayyat and Abraham Saba — alluded in their works to biographical events and depicted the agony of the exiled. Those authors wished to erect a monument to their past world, as well as to state an *apologia pro generatione sua*, in order to confront the resentment and criticism expressed by some Jewish leaders of the congregations where the exiled sought refuge. Michal Oron discusses their works and follows the route of the expelled Jews to Portugal, North Africa, Italy, and Salonica. In fact, Abraham Saba consecutively witnessed the expulsion of the Jews from Spain and the plight of those Jewish exiles, who at first were allowed into the neighbouring kingdom of Portugal, yet, after cruel pogroms and persecution,[10] in 1497 were finally forced to convert and forbidden to leave the country.

9. A. Dominguez Ortiz and B. Vincent, *Historia de los Moriscos. Vida y tragedia de una minoría* (Madrid, 1989), pp. 17–19.

10. Cf. I. Tishby, *Messianism in the Time of the Expulsion from Spain and Portugal* (Jerusalem, 1985), pp. 15–52 (Hebrew).

Italy was another refuge sought by the expelled Jews. Several waves of emigrating Iberian Jews arrived there during the century following the Expulsion. Renata Segre points out that each of the ruling powers in Italy made its own decision whether to admit the newly arrived exiles to its territories. Each of these powers likewise took upon itself to determine the Jewish identity of the exiles, and to formulate measures and pass regulations regarding coercive conversion. Church authorities in Italy demonstrated conflicting tendencies regarding the policy of tolerance towards the Marranos (one of the names applied to the converts). As a result, the situation of both Jews and Marranos in Italy was rather precarious. At their own discretion Italian rulers would encourage Jewish and Marrano merchants to settle in their lands, and at their will, they would expel or decree edicts of expulsion against Jews and persecute Marranos residing in their territories. Benjamin Ravid depicts such policy in the tale of three cities: Venice, Ancona, and Livorno (Leghorn). Yet, in their attempts to cultivate and strengthen economic and commercial growth in their states, sixteenth-century Italian authorities were indulging in competition for Jewish merchants. The economic activities of early sixteenth-century Jewish merchants stretched along the maritime and land routes leading from Italy to the Ottoman Empire, and from there overland through Poland and Lithuania, westwards as far as Germany, the Netherlands, and England.

The Republic of Venice was a most important haven sought by the Jewish exiles. It also constituted a terminal for those migrating to the Ottoman Empire. Elisabeth Crouzet-Pavan describes the wave of eschatological tension that swept Venice during the first decades of the sixteenth century. At that time, the Venetians — who until then took pride in their city as one chosen by God — regarded it as a biblical city punished and purged for its sins. This acute eschatological tension brought forth sentiments of repentance and penitence, followed by decrees of segregation against the Jews. The Venetian New Ghetto (*Ghetto Novo*) came into being: the first Jews took up residence there at the end of July 1516. Ennio Concina, who traces the history of the ghetto island's terrain and buildings, proves that, with respect to the legal procedures concerning the housing of the ghetto residents, the Republic of Venice actually acknowledged the Jews' legal concept of *ḥazakkah* (tenure) with regard to the latter's rights in the buildings and *miglioramenti* (housing improvements) they made in the New Ghetto.

The Republic of Ragusa (today Dubrovnik) was one of the major competitors of Venice and presented a serious threat to its domination of the Adriatic. Ragusa became another sheltering harbour for Iberian Jews. Jewish merchants took an active part in the commercial and diplomatic role of that republic as a mediator between Islamic and Christian states.

Ivana Burdelez describes the economic activities of those merchants, some of whom took to espionage activities alongside their commercial negotiations, making use of their idiom, the *Judezmo* (Judaeo-Spanish), as a *lingua franca*.

It was the Ottoman Empire which offered the expelled Jews a safe and secure haven for the coming generations. It was there that the exiled succeeded in the task of bringing forth a revival of Sephardic Jewry, alongside the formation of new social groupings and economic patterns in the Mediterranean world. At the end of the fifteenth century, Ottoman domination reigned over Anatolia and the Balkans. Only during the sixteenth century did the empire expand its borders to include the Levantine shores of the Mediterranean all the way to North Africa.

The Iberian Jews carried with them into exile their language, their customs, and their musical tradition, all of which they kept for many generations. Edwin Seroussi states that notwithstanding the debate as to the origin and evolution of Jewish Sephardic music, be it *piutim* of the Jewish synagogue, the medieval Spanish romances, or Moorish chants, those distant voices of the past may still be heard among Sephardic Jews throughout the Mediterranean basin. The correlation between people and location, with regard to Sephardic Jewry, attests to the continuity of Sephardic tradition and self-assertion. However, social regrouping, viewed against the background of constant movement and mobility in the Ottoman Empire and from it, mainly to Italy, might take on a somewhat different appearance in the newly settled Sephardic communities, and underwent subtle changes over the course of the centuries.

The Iberian Jews were far superior in their education, culture, and social organization to the members of oriental Jewish communities with which they mingled and integrated. This superior position caused the Iberian Jews to look down at the indigenous Jewish communities with disdain and haughtiness, which were heightened by the despair brought about by the traumatic experience of the Expulsion.[11] Despite their plight, the Sephardic Jews gained the upper hand in the Jewish communities,[12] as is clearly demonstrated by the precedence given to Sephardic Jewish ritual over the oriental tradition. Leah Makovetzky suggests that the case of Arta in the Epirus region of Greece is an exception to this rule. There — far from the main route of commercial activities — the Romaniote congregation

11. J.R. Hacker, 'Pride and Depression — Polarity of the Spiritual and Social Experience of the Iberian Exiles in the Ottoman Empire', in *Culture and Society in Medieval Jewry: Studies Dedicated to the Memory of Haim Hillel Ben-Sasson* (Jerusalem, 1989), pp. 541–86 (Hebrew).
12. See, for example, A. Cohen and B. Lewis, *Population and Revenue in the Towns of Palestine in the Sixteenth Century* (Princeton, 1978), pp. 155–61. Cf. A. Cohen, *Jewish Life under Islam: Jerusalem in the Sixteenth Century* (Cambridge, MA—London, 1984).

managed to hold its ground and adhere to the old ways of life.

The newly arrived Jews, physicians and courtiers, tradesmen and merchants, changed the face of society in their havens of refuge. One of the most important aspects of the 1492 Expulsion was the resulting transfer of western technology from the Iberian Peninsula to the lands that offered shelter to the exiled Jews. The latter brought with them expertise in glass-blowing, printing, and textile manufacture. Benjamin Braude discusses the longest lasting of those industries — the manufacture of woollen cloth in Salonica. Thanks to the development of that industry, Salonica — once devastated by the conquering Turks — was transformed into a thriving Jewish centre for many generations.

The victory of the Ottoman Turks made them masters of Palestine (in 1516) and of Egypt (in 1517). The Jewish population of these regions was constantly increasing, due to the influx of Iberian refugees settling there during the fourteenth, fifteenth, and sixteenth centuries. Yet it was only after the Ottoman conquest of Palestine that Jewish settlement there expanded significantly. A number of fifteenth- and sixteenth-century letters included in the Cairo Geniza depict Sephardic Jewish women living in Palestine and Egypt. Joel L. Kraemer states that evidently the passage of men and goods between those neighbouring countries was an everyday practice, undertaken for educational or economic reasons or due to marriage settlements. Kraemer concludes that the Expulsion from Spain had a decisive impact on the social status of Jewish women living in Palestine and Egypt in the period following the arrival of the Iberian refugees.

The articles selected by the editor and included in this volume are mainly based on papers originally read at the international conference on 'The Mediterranean World after 1492: Jews, Muslims, and Christians', held on 2-6 June 1991, at the School of History, Tel Aviv University, under the auspices of the *Mediterranean Historical Review* and the International Jewish Committee Sepharad '92, headed by Mr. Mauricio Hatchwell Toledano. The Elias Sourasky Chair for Iberian and Latin-American Studies; the Chair of the History and Studies of the Jews of Salonica and Greece; the Istituto Italiano di Cultura (Tel Aviv), and the Institut Français (Tel Aviv) co-operated with Tel Aviv University in sponsoring the conference. The Editorial Board of the *MHR* expresses its gratitude for their support. Thanks are also due to the translators, whose names appear in the relevant articles, to Marva Roth who undertook the typing, and to Yael Kaplan Typesetting Ltd. The editor fully acknowledges her debt to Ann Ussishkin, the assistant editor of the *MHR*, for her extensive and assiduous work on the entire volume.

Alisa Meyuhas Ginio

Jewish Participation in Castilian Fairs: The Example of Medina del Campo in the Fifteenth Century

MARÍA FUENCISLA GARCÍA CASAR

In the south-east of what is now the province of Valladolid, on the banks of the Zapardiel river, lies the city of Medina del Campo, one of the 42 *comunidades de villa y tierra* (town and surrounding region) that make up the Castilian Extremadura, land which was won from the Moors during the eleventh and twelfth centuries.[1]

The westernmost of all the *comunidades* of the Castilian Extremadura, with 70 villages and stretching over 1,100 square kilometres, bounded on the north by the principality of Valladolid, on the east by the *comunidad de villa y tierra* of Olmedo, to the south by the *comunidad* of Arévalo, by the *villa* (small town) of Cantalapiedra, belonging to the bishopric of Salamanca, and by the *encomienda* of Fresno el Viejo, of the Order of Saint John, belonging to the Kingdom of León; the Kingdom of León itself, and the territories owned by the Order of Saint John, today part of the province of Zamora, marked its boundaries to the west.

Medina del Campo was the main town of a large *comunidad*, with many population centres, some of which, such as Alaejos and Nava del Rey were quite considerable in themselves, spreading out through the valleys of the Zapardiel and the Trabancos rivers to the south-west, and between the Zapardiel and the Adaja rivers to the south-east.[2]

Situated in the centre of the Duero Basin, by the late fourteenth and early fifteenth centuries, that is from the time that it was granted the right to hold its free fairs, Medina del Campo grew to be a meeting point attracting people and animals travelling along the most important Roman

This article was translated by Judith Roumani

1. Another model of administrative organization in late medieval Castile was the *Castilla de las merindades* (district castle), which consisted of lands held by Christians in the eighth century, together with repopulation dating from the ninth and tenth centuries, prior to the Almanzor campaigns.
2. See G. Martín Díez, 'La comunidad de villa y tierra de Medina', in E. Lorenzo Sanz (ed.), *Historia de Medina del Campo y su tierra* (Valladolid, 1986), Vol. 1, p. 161.

roads and the great cattle trails belonging to the Mesta (association of sheep breeders) that crossed its territory.[3] The town stands at the crux of the road network crisscrossing the heart of Castile, from Santiago de Compostela to Barcelona, and from Laredo and Irún in the north of the peninsula to Lisbon, Seville, and Alicante.[4]

Originally a Berber settlement,[5] the earliest mention of Medina del Campo is the document by Alfonso VI, king of Castile and León, confirming to the bishopric of Salamanca the contents of the dowry of his daughter Doña Urraca and his son-in-law Raymond of Burgundy: 'Et ultra fluvium vero Durium Metina per suos terminos determinata'.[6]

The definitive repopulation took place immediately after the conquest of Toledo by this king in 1085, and thus happened at the same time as the conquest of the great *concejos* (regional authorities) of Ávila, Segovia, and Salamanca. Lucas de Túy, the twelfth-century historian and bishop, writes: 'King Alfonso also populated all of Extremadura, all of Castile, and the cities and towns with the following names: Salamanca, Ávila, Coca, Olmedo, Medina, Segovia, Iscar, and Cuéllar'.[7]

Half a century later we find another mention: in 1144 King Alfonso VII — who in 1170 would give it *en arras* (gift in recompense for the dowry) to his wife Doña Leónor — donated to Salamanca 'a royal town called Zufraga[8] in the territory of Medina, situated on the river Zapardiel[9] ... and exempted it from the payment of tolls in the whole kingdom, including the toll for Medina'.[10]

Thus, by the twelfth century, the city that would have on its coat-of-arms the motto 'Ni el rey oficio ni el Papa beneficio' (We depend on neither king

3. See J. Klein, *La Mesta: Estudio de la historia económica española (1273–1836)* (Madrid, 1979), pp. 38–9.
4. See H. Pascual Gete, 'Las condiciones fisicas de Medina del Campo y su tierra', in Lorenzo Sanz, *Historia*, Vol. 1, p. 22.
5. The report on the expedition that 'Abd al-Raḥmān III conducted against Simancas tells how after crossing the sierra of Guadarrama and entering Christian territory, the troops followed the ancient Roman road of Somosierra, and after four days of marching reached a city called M.d.m.h. that had been abandoned by its inhabitants. Some historians identify these consonants with Olmedo rather than with Medina del Campo (see M.J. Viguera and F. Corriente, *Crónica del califa Abdarrahman III an-Nasir entre los años 912 y 924* [Saragossa, 1981], pp. 323–35).
6. See J.L. Martín *et al.*, *Documentos de los archivos catedralicio y diocesano de Salamanca (siglos XII–XIII)* (Salamanca, 1977), No. 4, pp. 85–7.
7. *Chronicon Mundi*, ed. A. Scott, *Hispaniae Illustratae*, Vol. 4, pp. 100–101, quoted by J.M. Ruiz Asencio, in Lorenzo Sanz, *Historia*, Vol. 1, p. 141.
8. Sofraga, or Sufraga, an uninhabited place in the province of Valladolid, in the municipal territory of Rueda, and in the municipal precinct of Medina del Campo (see A.M. Barrero García and M.L. Alonso Martín, *Textos de derecho local español en la edad media* [Madrid, 1989], p. 425).
9. See Martín *et al.*, *Documentos*, No. 12, pp. 95–7.
10. This privilege was confirmed by Alfonso VIII in 1190 and by Alfonso X in 1255 (see Martín *et al.*, *Documentos*, No. 98, pp. 187–8, and No. 256, pp. 342–5).

nor Pope) was one of the most important Castilian population centres, an energetic walled commercial centre, attracting 'large trading operations, especially those that occur annually, and take place at the end of the grain or wine harvest, or involve trade in livestock'.[11] The Latin Statutes of Sepúlveda,[12] the only text that has been preserved of the so-called *Fuero de Extremadura* (Compilation of Laws of Extremadura), regulated the juridical life of the people of Medina during the twelfth and thirteenth centuries.[13]

The year 1157 was decisive for Medina del Campo. When Alfonso VII decided on his deathbed to divide the Kingdom of Castile and León between his two sons. Sancho and Ferdinand, the town was included together with Arévalo and Avila in the territory belonging to the Kingdom of Castile, although with all its surrounding land it continued to be linked to the bishopric of Salamanca, which was assigned to Ferdinand II, the future king of León.[14]

At about that time — the second half of the thirteenth century — the town was fortified with three walls[15] and a rudimentary castle on the La Mota hill, while the outlying inhabitants sought security on another height close to it called the Castillon, which has the remains of a wall dating back to the eleventh and twelfth centuries.[16]

The process of history went on chiselling at it, and by the fourteenth and fifteenth centuries Medina del Campo had three aspects, urban, social, and commercial, as well as being a setting for political meetings, those of the Cortes, and a battlefield.[17]

(a) *The urban aspect*: From the traces of urban life in those decades, we can surmise that there was an 'old town' around the hill of La Mota, on the top of which King John II (1405–54) had ordered the present castle to be built. He used brick, the symbol of the tradition called *mudéjar* (after the Muslims living under Christian rule) as his construction material, and also built the 'new town', on the flat area at the foot of the hill.

11. See Ruiz Asencio, in Lorenzo Sanz, (ed.), *Historia*, Vol. 1, p. 154.
12. See Y. Muñoz y Romero, *Colección de fueros municipales y cartas pueblas* (Madrid, 1987), pp. 281–6.
13. The statute of Medina del Campo, predating 1177 and now lost, was granted to Sufraga (see n. 8 above) by Vidal, the bishop of Salamanca (see Barrero García and Alonsa Martín, *Textos*, p. 298).
14. See J. González, 'Fijación de la fronteras castellano-Leonesa en el siglo XII', 'En la España medieval', *Estudios en memoria del profesor D. Salvador de Moxó*, Vol. 2 (Madrid, 1982), pp. 419–23.
15. Today the remains of only the oldest one survive, due to its sturdiness and the reinforced cement of quartzite and mortar that was used.
16. See Ruiz Asencio, in Lorenzo Sanz (ed.), *Historia*, Vol. 1, pp. 150–51.
17. In support of Henry of Trastamara, in 1354 a group of nobles, among them Juan Alfonso de Albuquerque, rebelled against Pedro I, who in 1355 crushed the rebellion in Medina itself (see J. Valdeón Baruque, 'Medina del Campo en los siglos XIV y XV', in Lorenzo Sanz [ed.], *Historia*, Vol. 1, pp. 208–9).

We do not know exactly when the population shift from the hill to the plain occurred, although it could well have been at the end of the thirteenth century. The proximity of the river Zapardiel to the castle meant that, in order to continue the expansion of the urban nucleus, the town's growth had to be extended across the river. This is documented in what is perhaps the most important source, the *Ordenanzas para el aposentamiento de las ferias* (Statutes on the holding of fairs) dated April 1421. An analysis of this document shows that the most recently settled area was the left bank of the Zapardiel. Rúa Street, leading from the San Miguel Bridge to the Salamanca Gate, was the main axis of that urban neighbourhood, in which numerous streets intersected at a right angle with the Rúa Street, and this in turn at a certain point opened into the San Antolín Square, the heart of the 'villa nueva', and the nerve centre of its political and economic life.[18]

The settlement on the right bank of the river is very old, predating the period of the fairs, and included specific trades such as shoemakers or blacksmiths.

The lack of sources does not allow us to reconstruct any later date for the various phases of this process of urban growth. It is safe to say that it was both the growth of the inhabitant population from the early fourteenth century, and the seasonal increase in the floating or transitory population involved in the fairs[19] that contributed in different degrees to the town's expansion.

(b) *The social aspect*: The Jews lived side by side with the Christian inhabitants. The first documentary reference to Jews is quite late, appearing in the second half of the thirteenth century: Abrahán, a Jew of Medina, sold a donkey for 25 *marcos* (one *marco de oro* = 50 *castellanos*),[20] and Moriello, a Jew, the son of Yuçef Colodro of Medina, appears as the guarantor of an agreement.[21]

The information that we have on the location of the Jewish quarter of Medina is practically non-existent, and any mention that includes some detail, even in an indirect way, dates from the years after the Expulsion. In October 1494 Juan Pérez de Barradas, the *corregidor* (town magistrate), was ordered not to allow the *aposentador* (housing officer) to officially move the pack-saddle makers, who had abandoned their homes in San Francisco Street and had gone to live in an area between 'the first gate, where there had been a *judería* [Jewish quarter], up to the middle gate, close to the old wall, near the Capardiel'.[22]

18. Ibid., p. 217.
19. See M.I. Del Val Valdivieso, 'Medina del Campo en la época de los Reyes Católicos', in Lorenzo Sanz, *Historia*, Vol. 1, p. 237, and J. Heers, *Occidente durante los siglos XIV y XV* (Barcelona, 1948), p. 16.
20. See Martín *et al.*, *Documentos*, No. 421, pp. 529–31.
21. Ibid., No. 461, pp. 587–8.
22. Archivo General de Simancas, Registro General del Sello (hereafter AGS, RGS),

There is no mention of the Jewish community properties, such as a synagogue, butcher's shop, ritual bath, or cemetery, either before or after the Expulsion. There is, however, evidence of the Jews being kept apart in accordance with the provisions to that effect in the Cortes of Toledo of 1480. In 1490 the Catholic Monarchs reminded the *corregidor* to make sure that the prohibition against Jews working outside the Jewish quarter was being observed: 'While we order to separate and keep the Jews of Medina del Campo apart in accordance with the law of Toledo, we order that there may not and shall not be shops or workshops where they work or do any other thing, except in the said separate Jewish quarter'.[23] The following year, in view of the protest of Abraham Serrano, the *procurador* (representative) of the *aljama* (Jewish community), in which he stated that 'the Jews of this town are separated in accordance ... with the law .... made in the Cortes of Toledo, and because of this the Jewish quarter is very distant from the dealing and from communication with the town',[24] the Catholic Monarchs allowed the Jews to have shops 'in the square and outside ... the Jewish quarter' on condition that 'they neither eat nor sleep in them'.

The economic data tell us that in the *Repartimiento* of Huete of 1290, the *aljama* of Medina del Campo figures within the bishopric of Ávila. It was assessed a payment of 44,064 *marcos*, which was a considerable amount, almost the sum assessed for Ávila itself (59,591), and higher than the assessments for Olmedo (31,659) and Arévalo (12,377).[25]

The tax assessments that have come down to us, corresponding to the years 1464, 1472, 1474, 1479, 1482, 1485, 1489, 1490, and 1491, provide the following picture:

1464: 5,000 *marcos*[26]
1472: 5,000 *marcos*[27]
1474: 5,000 *marcos*[28]
1479: 5,000 *marcos*[29]
1482: 5,000 *marcos*[30]

fol. 72 (XI–3482), 27 Oct.

23. AGS, RGS, fol. 57, 23 Dec., published in L. Suárez Fernández, *Documentos acerca de la expulsión de los judíos* (Valladolid, 1964), pp. 348–9.
24. AGS, RGS, fol. 167, 19 Jan.
25. See C. Carrete Parrondo, 'El Repartimiento de Huete de 1290', *Sefarad* 336 (1976), 121–40
26. With Bobadilla and Fuente el Sol (see M.A. Ladero Quesada, 'Las juderías de Castilla según algunos "servicios" fiscales del siglo XV', *Sefarad* 31 [1971], 259).
27. Ibid.
28. Ibid. 'Medina del Campo (con Fresno de los Ajos, Bobadilla y Fuente el Sol) ... 8,500', Suárez Fernández, *Expulsión*, p. 69.
29. See n. 26 above.
30. With Fresno de los Ajos, Bobadilla and Fuente el Sol (see Suárez Fernández, *Expulsión*, p. 69).

1485: 191,5 gold *castellanos*[31]
1489: 63,15C *marcos*[32]
1490: 52,030 *marcos*[33]
1491: 64,000 *marcos*[34]

The amounts levied indicate a Jewish community of some size and strength up to the time of the Expulsion.[35] This is especially true when it is compared with the two nearest Jewish communities, those of Tordesillas and Olmedo, which in 1491, for example, paid taxes of 14,980 and 5,800 *marcos* respectively.[36] In contrast, of the 48 Jewish settlements that existed, according to documentary evidence, in what is now the province of Valladolid[37] — mostly in the northern part of the province, between the Valderaduey and Pisuerga rivers, which provide water for lands on which very bitter events happened in Spanish history — the Jewish communities of Bobadilla del Campo,[38] Cigales,[39] Fuente el Sol,[40] Mucientes,[41] Portillo,[42] and Valladolid[43] after 1489 maintained, like the community of Medina, the ability to pay high taxes.

Other statistical data go back to 1309. Ferdinand IV notified the Jewish community of Medina del Campo that he had donated the toll of Medina, together with a tax of over 2,100 *marcos* levied on the *aljama*, to the

31. Ibid., p. 60 (that year, one *castellano* = 485 *marcos*).
32. Ibid., p. 69.
33. Ibid.
34. Ibid.
35. A much higher amount than that paid by the *aljama* of León, or of Palencia (see Suárez Fernández, *Expulsión*, pp. 68–9).
36. Ibid., pp. 69, 71.
37. The list goes as follows: Aguilar de Campos, Alaejos, Bamba, Barcial de la Loma, Becilla de Valderaduey, Bobadilla del Campo, Bolanos de Campos, Cabreros del Monte, Castronuno, Cigales, Cuenca de Campos, Curiel, Encinas de Esgueva, Fresno el Viejo, Fuente el Sol, Iscar, Medina del Campo, Medina de Rioseco, Melgar de Abajo, Melgar de Arriba, Montealegre, Moral de la Reina, Mucientes, Olmedo, Palazuelo de Vedija, Peñafiel, Peñaflor de Hornija, Piñel de Abajo, Piñel de Arriba, Portillo, San Cebrián de Mazote, San Pedro de Latarce, San Roman de Hornija, Tamariz de Campos, Tordehumos, Tordesillas, Torrecilla de la Orden, Torrelobatón, Urueña, Valdenebro de los Valles, Valladolid, Villabrágima, Villafrechos, Villagarcía de Campos, Villalón de Campos, Villadefrades, Villavieja y Zaratán (I am grateful to Carlos Carrete Parrondo for providing this information).
38. AGS, RGS, fol. 280, 28 April.
39. With Peña Flor de Hornija and Torrelobatón (AGS, RGS, fol. 163, 6 April).
40. With Valderas (León) and Villalobos (Zamora); see Ladero Quesada, 'Juderías', p. 260.
41. With Bolaños de Campos, Medina de Rioseco, Moral de la Reina, and Tamariz de Campos (AGS, RGS, fol. 258, 10 Feb.)
42. With Aguilar de Campos, Becilla de Valderaduey, Medina de Rioseco, Moral de la Reina, Tamariz de Campos and Villabrágima (see F. Cantera Burgos and C. Carrete Parrondo, 'Los repartimientos de Rabí Jacó Aben Nuñes', *Sefarad*, 31 [1971], 224).
43. With San Román de Hornija and Toro (Zamora) (AGS, RGS, fol. 282, 22 April).

convent of Santo Domingo de Caleruega,[44] instead of the 7,800 *marcos* due from Salinas de Añana.[45] We also know that in 1444 the head tax paid by the Jews of Medina amounted to 8,800 *marcos*.[46]

(c) *The business aspect*: Documents that have been preserved relating to the late fifteenth century show how the business and social aspects were closely interwoven with the development of the well-known fairs. They began by lordly decree — Don Fernando de Antequera (Ferdinand I, king of Aragon, c.1379–1416) established them when he was the lord of the town[47] — and were soon mentioned in the chronicles, such as those of Don Alvaro, dating from 1450: 'And since at that time there was the fair of Medina del Campo, to which great crowds of people of diverse nations, both Castile and other kingdoms, usually come and attend it ... [the king of Castile is invited to come and visit the fair, together with the queen] to see the extent of the great assembly and many people, and also the variety of merchandise and many kinds of other things that are there';[48] the fairs also attracted the attention of travellers such as Pero Tafur, the 'Castilian knight'. He knew Europe well and, when speaking of the square in front of the church of San Marco in Venice, said: 'in front of this door there is a great square, larger than that of Medina del Campo'. Referring to Antwerp and its fair[49] he gives us this impression: 'I do not know how to put into words an event as large as the fair of this city; although I have seen others, such as that of Genoa,[50] which is in the Duchy of Savoy, and that of Frankfurt, which is in Germany, and that of Medina, which is in Castile, it seems to me that none of those can equal this one'.[51] Thus, there is no doubt that Medina del Campo, which Hernando del Pulgar, the chronicler of the Catholic Monarchs, described as 'the most opulent town in Castile', deserved comparison, with its large square and fairs, with more famous cities of Christian Europe in the fifteenth century.

Several kings issued decrees concerning its fairs. Henry IV, king of Castile (1425–74) stipulated in 1473 that the Crown was taking the fairs of Medina del Campo and the merchandise assembled there under its protection, care, and security, requiring that 'free fairs and free markets should not be held or take place in our kingdom or subject lands, except

44. In Burgos.
45. In Alava.
46. AGS, Diversos de Castilla (hereafter AGS, DC), folder 4, doc. 6, published in L. Suárez Fernández, 'Las rentas castellanas del infante Don Juan, rey de Navarra y Aragón', *Hispania*, 75 (1959), 192–203.
47. See Valdeón Baruque, in Lorenzo Sanz, *Historia*, Vol. 1, p. 226. They are prior to 1412.
48. Ibid., pp. 227–8.
49. See Heers, *Occidente*, pp. 142–3.
50. Ibid., pp. 129ff.
51. See Valdeón Baruque, in Lorenzo Sanz, *Historia*, Vol. 1, p. 205.

our fair of Medina and the other fairs that have our permission and privileges confirmed, and are entered in our books'.[52]

The fairs of Medina received maximum protection in 1491, when the Catholic Monarchs provided that merchandise originating in Seville, Córdoba, Jaén, Ubeda, Baeza, Cuenca, Ciudad Real, Segovia, Ávila, Salamanca, Zamora, Toro, the province of Extremadura, the Kingdom of Galicia, the Principality of Asturias, and other cities of the kingdom, might be included, becaue the fairs of Medina were *general fairs* for the whole kingdom.[53] The key question of this study arises unavoidably at this point: How much did Jews — both local and transitory — participate in everyday trading and in trade related to the fair, which lasted for 100 days?[54]

In addition to the agricultural activities of Medina del Campo, there was everyday trading, trade related to the fair, and trade in money, in other words, business was 'abundant, rich, and varied'.[55] The location of crafts was controlled by the *concejo*. Thus, any shoemaker, whether or not he was an inhabitant of the town, was obliged to set his stall in the street of the *Zapatería* (shoe shop);[56] 'drapers and other traders' were in Rúa, the street that led to the main square and thus to the precinct of the fair.[57] These measures were intended to restrain 'competition between the various workshops and merchants, and give all of them the same chance to reach the buyer'.[58] The pack-saddle makers constituted a large group in the town and were located in the street of San Francisco; but because their work was considered dangerous and the straw that they stored in their workshops caused many fires of which there are numerous echoes in the documents of the *Registro General del Sello* (General Registry Office) in the Simancas Archives,[59] they were transferred, as we have seen, to a location close to the Jewish quarter.

Food, household items, clothing, and luxury goods were the four main sectors into which the crafts of Medina could be grouped: first, information about mills, bread, and pastry ovens,[60] and cask-makers connected with white winemaking, the renown of which together with the wine of Alaejos, a town 24 kilometers from Medina, made Quevedo exclaim 'French tables

52. Ibid., p. 228.
53. AGS, RGS, file 110, 13 April.
54. There were actually two fairs, each of which lasted 50 days; the first — the May fair — began 30 days after Easter, and the second on 1 Oct. Inbetween there was the Lenten fair.
55. See Del Val Valdivieso, in Lorenzo Sanz, *Historia*, Vol. 1, p. 260.
56. AGS, Camara Pueblos (hereafter AGS, CP), folder 12, folio 26.
57. AGS, RGS, fol. 140, 12 July 1494.
58. See Del Val Valdivieso, in Lorenzo Sanz, *Historia*, Vol. 1, p. 260.
59. AGS, RGS, fol. 18, n.d., Nov. 1492 (regulations concerning fires, ovens, and tanneries).
60. Ibid.

do not harbour half/of what a good Alaejos wine-butt contains';[61] secondly, floor and roof tile manufacturing, and particularly carpentry, a profession engaged in largely by the Moors of the town; thirdly, tailors, tanners, hosiers, and cloth-shearers;[62] and, finally, leather-working: tanners, shoemakers, clog-makers and makers of laced shoes.[63] Mention should be made of glovemakers, hatters, jewellers,[64] and silversmiths,[65] professions that achieved great importance and were highly developed in late medieval Medina.

The sale of each of these types of merchandise was carried out in fixed stalls or shops, regulated by the statutes for fairs which, as stated previously, dated from 1421. The Jewish artisans and merchants of Medina had considerable trouble conducting business, living side by side with their neighbours, but in tension and hostility, and experiencing both rejection and tolerance.

Already in 1489 a statute had prohibited the selling of wood, coal, and baked bread in the Jewish quarter. The municipal authorities were trying to exclude the Jewish community from ambulatory or semi-ambulatory trade, which provided the Jews with raw materials that were essential to them. Shortly afterwards, the Catholic Monarchs revoked the prohibition, and the *concejo* allowed dealers who crossed through the Jewish quarter[66] to sell their products there.[67]

Two years before the Expulsion, an attempt was made to keep Jews away from doing business in the town: confined to their quarter as they were,[68] they were also prohibited from placing stalls in the square,[69] and prevented from participating in the business life of the fair. The Jews succeeded in defending their rights,[70] but some neighbouring tradespeople protested, and a compromise was only reached regarding 'day stalls'

61. Talía, Musa 6, romance 98; quoted by H. Pascual Gete, 'Medina y su tierra durante los siglos XV–XVI: Una economía agraria en el apogeo comercial de sus viñedos de calidad', in Lorenzo Sanz, *Historia*, Vol. 1, p. 322.
62. See G. Moraleja Pinilla, *Historia de Medina del Campo* (Medina del Campo, 1971), pp. 379–95.
63. See n. 56 above.
64. AGS, RGS, fol. 172, 7 May 1477, and Casa y Descargo de los Reyes Católicos, folder 6, fol. 529.
65. AGS, RGS, fol. 34, 2 July 1488; fol. 95, 10 Nov.; fol. 158, 7 March 1489, and fol. 180, 7 Nov. 1492.
66. The splendid sketch that the Flemish painter at the court of Philip II, Antón Vaden Wyngaerde, made in 1565 of Medina del Campo gives an approximate idea of where the Jewish quarter was located in the late fifteenth century, namely behind the main square (see *Ciudades del Siglo de Oro: Las vistas españoles de Antón Vanden Wyngaerde* [Madrid, 1986], pp. 382–4).
67. AGS, RGS, fol. 94, 3 June, and in Suárez Fernández, *Expulsión*, pp. 326–7.
68. AGS, RGS, fol. 57, 23 Dec., and in ibid., pp. 348–9.
69. Ibid., p. 49.
70. AGS, RGS, fol. 167, 19 Jan. 1491 and in ibid., pp. 349–50.

(removable street stands), but not at all for fixed Jewish shops in the square.

In view of the unpopularity of this measure, the Catholic Kings asked the guardian of the Convent of San Francisco to inform them of what damage the Christian population of Medina would suffer if the Jews had their shops in the square.[71] The *pesquisa* (investigation) was still going on at the time of the Expulsion.

Together with these far from insignificant problems affecting Jews living in Medina and involving discrimination, and other disputes that we could define as 'usual' — lawsuits over houses, such as the one brought by the Jew Mosé Torre against the commander of Sant Llorente regarding two houses in the street of la Especiería (the spice shop);[72] unpaid rents, regarding which the same Jew appealed to the *consejo* in 1488, in view of the delinquent behaviour of the tenant Manuel Fierro;[73] accusations of usury, such as that brought in 1489 against the Medina Jew Bienveniste. a clog- and shoemaker by trade, accused by Marina Gonçales of demanding from her 2,000 *marcos* in interest on a loan of 6,000 *marcos*,[74] there were also protests at not keeping agreements. In 1485 Mayr Abenfarax, a resident of Medina del Campo, requested payment of the sum that he had lent to the duke of Alba when the wall of the fortress of Sieteiglesias de Trabancos, close to Alaejos, was built;[75] together with these disputes were those of Jews who had come from outside the town for the fair, bringing their livestock[76] and goods.

All sorts of products changed hands at the fair: fish, meat, honey, oil, dates, clay pottery, glass, iron, dye-stuffs such as indigo and 'Berber colour', Bohemian knives (brought by merchants from Burgos), alum, jewellery, paper, and 'type books', in other words, printed books. Most activity, however, was related to three basic types of goods: textiles (wool and fabrics), spices, and money.[77]

The Jews who attended the fairs dealt in fabrics and wool. Most of the documents that have been preserved relate to them, not specifically to their economic status, like Symuel Avençar el Rico, an inhabitant of Medina del

71. AGS, RGS, fol. 206, 22 March 1491, and in ibid., pp. 355–6.
72. AGS, RGS, fol. 35, 27 Sept., and in F. Suárez Bilbao, *Judíos castellanos entre 1432 y 1492: Ensayo de una prosopografía* (Madrid, 1990), Vol. 1, p. 539.
73. AGS, RGS, fol. 65, 6 Dec., and in ibid., Vol. 2, p. 877.
74. AGS, RGS, fol. 156, 10 Jan., and in ibid., Vol. 2, p. 885.
75. AGS, RGS, fol. 190, 20 July, and in ibid., Vol. 1, p. 503.
76. All traders were obliged to pay one *maravedí* per loaded animal as a toll. Through this tax, the *concejo* covered the expenses of the *Hermandad* (fraternity) (see Del Val Valdivieso, in Lorenzo Sanz, *Historia*, Vol. 1, p. 270).
77. Ibid., p. 277.

Campo,[78] or as owners of houses like the aforementioned Mosé Torre,[79] or owners of houses with gardens as was Don Yuzaf Zalama, the son of Don Zaf Zalama, a Jewish resident of Medina del Campo,[80] or for farming the sales tax as did Yuçé Abenzón, a Jewish resident of the town of Zapardiel.[81] Rather, they are mentioned due to the incidents and vicissitudes in which they were involved in the 'town of fairs': from the loss of merchandise by theft during the journey or by fire, to all sorts of crooked business among themselves or with other traders, as the victims or perpetrators of fraud, unreturned loans, merchandise not paid for, fraudulent or overpriced sales.[82] Here are some examples:

On 30 September 1480, David de Çea, a Jew from Alburquerque, made an agreement with Pedro Texelo, a mule-driver and inhabitant of the town of Valencia de Alcantara, that

> he would bring to the fair of Medina del Campo a load of wax for a fee of 400 *marcos*, and some barley, on condition that on the return trip he would bring back the mule loaded with fabric ... when he had been paid for his work. and went off with the load of fabric, when he reached the River Jerte near Plasencia, the said Pedro Texelo plotted and intentionally did him wrong and damage, which seemed to have been done maliciously, and he says that he has witnesses to the fact that ... Pedro Texelo had previously threatened him and said that he would find a way for David de Çea to lose his possessions.[83]

On 8 August 1488, Rodrigo de Enciso, the receiver of the Inquisition in Medina del Campo, presented the following matter to the *Audiencia* (court of the Inquisition): Abrahán Biuaz and Mosé de Cantos, Jews living in Xeres, near Badajoz, 'as joint owners, were obliged to give and to pay Pero de Palacios, a merchant and resident of Medina de Rioseco, 93,000 *marcos* ... for some merchandise that he says he bought, in which the money was invested ... and they were obliged to pay him by the May fair of this year ... held in the town of Medina del Campo ... and when the fixed time had passed ... Abrahán Biuaz and Mosé de Cantos neither gave nor paid Pedro

78. Archivo Chancillería (hereafter AC), Valladolid, Ejecutorias, folder 45 (olim 23), 3 fols.
79. Also in San Francisco Street and the *rúa* (AGS, RGS, fol. 173, 20 Sept. 1485). See p. 21 above.
80. Archivo Histórico Provincial, Valladolid, Register 2/13, No. 234, fol. 905r; draft No. 171, fol. 63v; cit. in J. Castro Toledo, *Colección diplomátia de Tordesillas* (Valladolid, 1981), doc. 175, p. 121.
81. He farmed taxes around Zamora (AGS, RGS, fol. 186, 9 Oct. 1490).
82. AGS, RGS, fol. 51, 30 Sept., and in Suárez Bilbao, *Judíos*, Vol. 1, p. 262.
83. AC, Valladolid, Ejecutorias, folder 8, 2 fols.; see F. Baer, *Die Juden im christlichen Spanien* (Berlin, 1936), Vol. II, p. 428, for a different reading.

de Palacios the ... 93,000 *marcos* of their obligation'.[84] The receiver obtained a writ of detention against the Jews, 'who had fled ... towards the border with Portugal'.

16 January 1489: The protagonist is a Portuguese Jew, Salomon Carruchel, to whom Diego Sazedo owed 8 silver *marcos*. Seeing that he was a foreigner staying temporarily in Medina del Campo, the Christian thought that he could well take advantage of him, and deliberately delayed paying the debt. When the time came to go home, the Jew would leave without claiming his payment. But Carruchel was not such an easy target, and he appealed to the *consejo* to put pressure on his shrewd debtor to pay up.[85]

11 November 1489: Mosé Hayn, a Jew from Badajoz, appealed to the *consejo* because 'when he was going from the fair of Medina with some loads of merchandise, and reached the corner of Gila near Badajoz, the muleteers who were transporting the merchandise, Juan de Porras, Alfonso de Salamanca, Diego Mexía, and Rodrigo de Barrera, gave him cause to lodge a complaint against them'. The Jew from Badajoz won the suit.[86]

4 October 1490: A Portuguese Jew from Evora, Yosepe Casvana, who was staying that year in Medina, bought a roomful of tapestry for 40,000 *marcos* from Fernán López de Calatayud. He was not satisfied with his purchase, for in his opinion the quality was not up to what had been agreed upon, and he accused the seller of swindling him through Juan de Cádiz, Fernán López' agent. He appealed to the *consejo*, but information on the outcome has been lost.

Lastly, the following event occurred on 4 October 1490. Nicolás de Madrigal and Rodrigo de Dueñas, residents of Medina, requested the *consejo* to grant them a moratorium on their debts, including one to Abrahán Melamed, 'the son of the chief rabbi,[87] resident in the noble town of Medina del Campo', to whom they owed '31,000 *marcos* in connection with some wheat shipments'. The reason for the delay in payment was 'the fire and great burning that affected them disastrously and in which they suffered much damage, occurring four months before, one fire in La Rinconada and the other in Cal de la Plata where they lived ... with most of

84. AGS, RGS, fol. 273, and in Suárez Bilbao, *Judíos*, Vol. 2, p. 887.
85. AGS, RGS, fol. 43, and in ibid., p. 1001.
86. AGS, RGS, fol. 407, and in ibid., p. 1447.
87. AGS, RGS, fol. 278, and in Suárez Bilbao, *Judíos*, Vol. 2, pp. 1383–5. The son-in-law of Abrahán Seneor, Rabbi Meir Melamed, who was christened on 15 June in the monastery of Guadalupe, took the name Fernán Núñez Coronel; see C. Carrete Parrondo, 'La hacienda castellana de Rabbi Meir Melamed (Fernán Núñez Coronel)', *Sefarad*, 37 (1977), 251–6.

their possessions, which must have been worth altogether over 500,000 *marcos*.[88]

The Edict of Expulsion interrupted many things in this trading town. For some it was a permanent interruption, for others it just marked a parenthesis, after which they took up their lives again as residents of Medina and *judeoconversos*. Thus, for instance, Abrahán Melamed, who took the name of Pedro Nuñez Coronel after attending the May fair in Medina del Campo in 1492,[89] or the physicians *maestre* Enrique and *maestre* Pablo, *judeoconversos* who were assigned salaries so that they would not leave the town.[90]

The fairs of Medina would continue to flourish up to the end of the sixteenth century. Other fairs, that were spin-offs from it, were held in Medina de Pomar, Medina de Rioseco, Villalón, and Valladolid. In the fifteenth century Don Bienveniste, the Jewish merchant who also farmed the taxes of Calahorra, was selling wool in Medina del Campo for amounts ranging from one and a half million to three million *marcos*. On the Sorian branch of the river, a *judeoconverso* from Segovia, specifically Antón del Río, in 1524 sold for a price of four and a half million *marcos* to Genoese traders some 'fine, white, *merino*' wool, from his own flocks or, as he himself noted in the document of sale, 'of my own rearing and branding'.[91] The *judeoconverso* traders and stockrearers of Medina would constitute, however, another subject of study, on which I hope to embark shortly.

88. AGS, RGS, fol. 279, 3 Nov., and in ibid.
89. AGS, RGS, fol. 420 (XI–4339), 15 Dec. 1494; and fol. 32 (XI–2670), 2 Sept. 1494.
90. See Falaḥ Ḥassān 'Abd al-Ḥussein, 'Las ferias de Medina y el comercio de la lana: 1514–1573', in *Historia de Medina del Campo y su tierra* (Valladolid, 1986), Vol. 2, pp. 13–42.
91. The cloth trade of Medina took place in the four streets — Toledo, Segovia, Cuenca, and Palencia — named after the towns of Castile that were predominant in the cloth industry (see 'Abd al-Ḥussein 'El comercio de los géneros textiles: seda, paños y lienzos', in ibid., Vol. 2, pp. 44–66).

Nostalgia for the Past (and for the Future?) among Castilian *Judeoconversos*

CARLOS CARRETE PARRONDO

The title to this article may cause some confusion. The term 'nostalgia' has to be defined according to its precise semantic meaning. No definition is better than the one given by the *Diccionario de la Lengua Española*:[1] 'Loneliness or sorrow due to the absence of a person or thing', in which the temporal aspect is uncertain and even ambivalent; the present time conditions, to a large extent, the personal perspective of looking back to something that happened in the past but that hypothetically might recur, albeit with some modifications, at some time in the future. This is the frame of reference which I hope to use for this article, because it is actually the same as the one which defines the documents to be examined in terms of the level of ideology involved. After presenting some concrete illustrative examples, I hope to explain this axiom where relevant.

I would like to make it clear from the outset that these pages do not at all pretend to present a way of thinking adopted by educated Castilian *judeoconversos* (Jewish converts) of the late fifteenth and early sixteenth centuries who might have belonged to Erasmic or illuminati academic, intellectual circles. I have only been concerned here with indicating the opinions expressed by the simple *judeoconversos*. These *judeoconversos* lived far from the cultural centres, and in direct contact with a closed rarefied society, in which popular tradition prevailed, as well as the repressive system of the Inquisition and, of course, the close vigilance of one's neighbour, who sometimes turned out to be the most diligent and unsuspected informer. A rural rather than urban setting can best provide this kind of reality. In this case I am most interested in analysing the motives, which were no doubt essentially human ones and devoid of scholarly concerns, that impelled simple *judeoconversos* living on Castilian soil when, without any premeditated disloyalty on their part, they were obliged to accept a new religion and try to forget their thousand-year-old tradition.

This article was translated by Judith Roumani

1. 19th edition (Madrid, 1970), p. 98b.

Faced bv such circumstances, admittedly, the emotions aroused are incontrollable. Theological and philosophical subtleties are not useful in this case: we have to start with an essentially human postulate — which does not imply a rustic and unpolished character — a postulate which is, above all, essentially Castilian. This specific shade or trait is important enough to be considered a significant element in a broad panorama that, at times, has nothing in common with most other geographical areas, including those that are close by. In my opinion this point is fundamental. Although I am not Castilian, my longstanding contact with the land of Castile allows me to say that though from afar all sheep look the same, there are — and the distinction is essential — coarse wool and merino sheep, and their bleating sounds different; likewise, there is also a basic difference between a good wine from a well-known vintage and a grape juice that does not even qualify as cheap wine. In other words, the phenomenon of the *judeoconversos* of Castile cannot be understood adequately unless we first understand the age-old phenomenon of Castile itself. This affirmation may no doubt be risky, but it still must be stated. Again, here I am not interested in cultural exceptions, but rather in the general qualities of a social group, a largely anonymous entity. Although comprised of individual names, fortunately they will not go down in history even in the form of sophisticated graphic tables. However, such characteristics imply provincialism: as they say in Castile, each saint must put up with the candle lit in his honour. But reality speaks for itself.

Some years ago the French Hispanist Marcel Bataillon wrote a few suggestive reflections entitled 'Melancolía renacentista o melancolía judía?'[2] in which he discussed certain texts by Spanish authors of the sixteenth century. But neither the general term 'melancholy'[3] nor its strictly literary meaning is comparable to what I am suggesting here. I would like to elaborate on my earlier suggestion and describe three aspects of it, relating it to a very specific area of Castile, the former bishopric of Cuenca-Sigüenza, and confine it to a particular time, the late fifteenth and early sixteenth centuries. Examples taken from documents until now largely unpublished are the best and most loyal exponents of this theory.

2. In *Homenaje a Archer M. Huntington* (Wellesley College, 1952), pp. 39–50, repr. in his *Varia lección de clásicos españoles* (Madrid, 1964), pp. 39–54.
3. The *Diccionario de la lengua española*, p. 862b, defines *melancolía* in its primary meaning, as 'a vague sadness, deep-rooted, sustained, and permanent, originating in physical or mental causes, and preventing the person suffering from it from finding pleasure or distraction in anything'.

I. THE IMMEDIATE CONSEQUENCES OF CONVERSION

Since the traditional practices that most of the documents of the Inquisition depict generously, such as Sabbath ceremonies, dietary prescriptions, criticism of the Tribunal of the Holy Office, and so forth, are not to be found in our subject matter, some other distinguishing aspects have to be identified. The documents themselves, presented below in all their eloquence, will best illustrate the case.

1. Grief at Having Abandoned the Former Religion

- Diego de la Peña, an inhabitant of Atienza (Guadalajara) used to say that 'it grieved him that he had turned Christian'.[4]

- In 1501 Luisa, the wife of Luis de Velasco, an inhabitant of Almazán (Soria), was accused as follows: 'When I said to the aforementioned person and to her husband what a favour God had done them in making them become Christians, she said, "All this talking is a mess", shaking her head in anger; and then a little later this witness said it again, and the aforementioned person, in great anger, repeated what she had said'.[5]

- Luis de la Cerda, an inhabitant of Medinaceli (Soria), was accused in 1504 of 'saying the he was grieved that he had become Christian, and that the law of the Christians was a joke; and also that he thanked God that it was Lent, so that he did not have to partake of the food of the Christians'.[6]

- Also in 1504, the scribe Gonzalo Alonso, an inhabitant of Sigüenza (Guadalajara), who was the father of the *regidor* (magistrate) Fernán Alonso and of the abbot of the *cabildo* (chapter of the cathedral) of Sigüenza, was accused of responding (in conversation with a Jew who had said to him, 'There are many honourable *conversos* in this town'):[7] 'Yes there are, but would to God we were still under Jewish law'.[8] The Jew commented, 'Don't say anything of this',[9] and Gonzalo added, 'gesticula-

4. '... que le pesava por se aver tornado christiano', Archives of the Diocese of Cuenca, Inquisition Section (hereafter ADC), 748/7 (1504). I hope to publish this unpublished documentation in the near future, in association with Ma. Fuencisla García Casar of the University of Salamanca.
5. '... que diziendo a la susodicha e a su marido quanta merçed les avía fecho Dios en se tornar christianos, la susodicha dixo: "Comamos del caos de estas pláticas", boluiendo la cabeca con enojo; e dende a vn poco lo tornó a desyr este testigo, y la susodicha, con mucho enojo, torno a dezir lo mismo', ADC 749/14, and see C. Carrete Parrondo and C. Fraile Conde, *Fontes Iudaeorum Regni Castellae* (hereafter *FIRC*), Vol. 4, *Los judeoconversos de Almazán, 1501–1505: Origen familiar de los Lainez* (Salamanca, 1987), para. 240, p. 114.
6. '... desyr que le pesava por se aver tornado christiano, e que la ley de los christianos hera toda burla; e asymismo agradesçia a Dios, porque hera quaresma, por no comer de las cosas de los christianos', ADC, 768/7.
7. 'Muchos conversos honrados están en esta villa', ibid.
8. 'Sý ay, mas sy Dios quisiera bien nos estábamos en la ley judayca', ibid.
9. 'No digáis nada deso', ibid.

ting with his hand, "What I have, I will never abandon." '[10]

- Of even greater interest in terms of its psychological connotations is the exclamation of which Gonzalo de Vargas, an inhabitant of Almazán, was accused: 'Oh, if only there were someone here who could understand me! If evil has come upon us, we have deserved it, for we performed neither the ceremonies nor the rites as we should have, and that is why exile came upon us'.[11]

- When he was informed that the mother of another *judeoconverso* had remained in Portugal as a Jew, Antonio del Valle, an inhabitant of Almazán, answered unhesitatingly, 'Truly, if I were now in that country, I would not become Christian',[12] and in the bitterness of his words it is not difficult to trace the implications.

- Catalina, the mother of *maestre* Antonio de Alves, an inhabitant of Almazán, was accused of saying 'When I lived in the market quarter (*barrio Mercado*) I was a good Jewess'.[13]

- Ruy Díaz Lainez, an inhabitant of the same town, was accused that 'they were talking about how God had enlightened them and they had to change their customs, and then this witness saw and heard Ruy Díaz Lainez saying "I will not change the faith and the truth for Jesus Christ"; and this witness answered him that he should weigh his words more carefully and not keep on saying that, because when he had been a Jew he used to say "I will not change by *Moshe Rabenu*", which means Moses',[14] convincing proof of genuine religious continuity.

- Catalina, the wife of Pablo, an inhabitant of Almazán, used to say 'that she could not stand her husband Pablo's criticizing her because she was an Old Christian, and telling her that God had done him a really bad turn by taking away his blood and his companion, who was a Jewess, and giving him her, an Old Christian, and that if God willed her to die he would only marry a New Christian like himself',[15] an affirmation that deserves to be taken into consideration in future, more detailed studies.

10. 'Meneando la mano: "Lo que tengo nunca lo dexaré."' ibid.
11. 'O, sy oviese aquí alguno que me entendiese! ... Sy mal nos a venido a nosotros somos merescçedores dello, que no fasyemos las çerimonias nin las cosas que avíamos de faser, que por eso nos vino el destierro', *FIRC*, Vol. 4, para. 239, p. 110.
12. 'Por mi vida, sy yo estouiera agora en aquella tierra no me tornara christiano', ibid., para. 238, p. 111; accusation of 1501.
13. Ibid., para. 17, p. 30; accusation of 1505. The Jewish quarter of that town was situated near the Church of Santa María del Mercado (Saint Mary of the Market).
14. -'... platicauan cómo Dios les avía alunbrado e avian de mudar la costunbre, e a esto vio e oyo este testigo cómo Ruy Días Laynes dixo: "No me trocaré en la fee e la verdad por Ihesu Christo"; y este testigo le respondió que parase mientes, non lo dixese muchas veses, porque solía desir seyendo judio: "No, me trocaré por Mosé rabenu", que quiere desyr Muysén', ibid., para. 52, pp. 42–3; accusation of 1505.
15. '... que no podía sufrir a ... Pablo, su marido, que reñía con ella porque era christiana vieja, e le desía que farto mal le avía fecho Dios en quitarle su sangre y su conpañera, que era judía, y darle aella, que era christiana vieja, y que sy Dios

- Aldonza, the wife of the *judeoconverso* Pedro Lainez, who were inhabitants of the same town in Soria, would remember, with some nostalgia, 'how when they were Jews they had never lacked good things, and now they lacked everything',[16] and she would add that 'things used to be better for us before and we used to have more than now'.[17]

- Pedro García, an inhabitant of Matute (Soria), said in the presence of some Old Christians: 'I tell you, I am more deceived in this religion than I was before'.[18] And there can have been no mitigating circumstances that would have excused such a statement.

- Francisca, an inhabitant of Almazán, being ill during Lent and unable to eat meat, said 'This is what I deserve because I became Christian, for before I could eat anything I liked',[19] a statement which was certainly debatable.

- In 1505 Catalina la Jaena, an inhabitant of Almazán, was accused of of remarking about a young *judeoconverso*: 'Would to God he belonged to the twelve *pares*;[20] now he is mud and worth nothing',[21] a statement endowed with refined picturesqueness.

2. The Impossibility of Abandoning a Thousand-Year-Old Tradition

This aspect is both very human and usually poorly understood. Three examples should suffice to bring us closer to a reality free of academic connotations: here feelings overcome any urge to embark on scientific disquisitions.

- Juan Sánchez de Móron, an inhabitant of Almazán, was accused of having said that he wanted to die 'like his father and his ancestors before him',[22] to which someone countered, 'Juan Sánchez, this is no longer a time for dying under the law of your ancestors, for now you have a better law',[23] to which the *judeoconverso*, without hesitating, said: 'Look, good woman, if someone cut you off from your customs now, what would you want to do?' [24]

quisiese se moriese, que nunca casaría syno con christiana nueua como él', ibid., para. 203, p. 101; accusation of 1505.

16. 'Que mientra[s] fueran judíos nunca les faltara el bien, y que agora todo les faltava', ibid., para. 4, p. 20; accusation of 1505.
17. 'Mejor nos yva antes e más teníamos que agora', ibid., p. 24.
18. 'Agora os digo que estoy más engañado en esta ley que fasta aqui', ADC, 749/15; accusation of 1505.
19. *FIRC* Vol. 4, para. 59, p. 45.
20. The nobles of the old French Court.
21. 'Quando Dios quería hera de los dose pares; ahora es lodo y es nada', *FIRC*, Vol. 4, para. 179, p. 93; accusation of 1505.
22. '... como murió su padre e sus abuelos', ibid., para. 227, p. 107; accusation of 1505.
23. 'Juan Sánches, no es tienpo ahora que os muráys en la ley de vuestros abuelos, que otra ley tenéys ya mejor', ibid.
24. 'Anda, buena muger, quien vos sacase agora de vuestras costunbres, ¿qué faríades?', ibid., para. 227, p. 107; accusation of 1505.

- *Maestre* Pedro, a doctor in Almazán, who was probably a New Christian himself, accused the *judeoconverso* Pedro de Guadalajara of saying to him: 'Don't you think, Sir, that, *hay* [*sic*] *Adonai*,[25] by the law of Moses, that I'm telling you the truth?[26] And this witness then laughed and the accused responded, "Oh, Sir, what men have become greatly accustomed to, they cannot leave just like that!"'[27]

- And there is the plain statement, attributed to the *judeoconverso* Pedro López de Ayala, another inhabitant of Almazán, in vigorous Castilian language: 'I'll have you know that I will never forget the milk that I suckled'.[28] In the face of a statement like this, all commentary is superfluous.

3. Jewish Practices

Because of its resilience and the emotions it arouses, Jewish religious practice among the Castilian *judeoconversos* may perhaps be one of the most illuminating areas for study, and no doubt deserves close scrutiny.

- María González, an inhabitant of Cuenca, was accused of saying 'that her mother would appear to her in dreams and would tell her that her soul was greatly grieved when her daughter did not keep the Sabbath'.[29]

- Her neighbour Isabel García was accused in 1488 of saying to the physician don Simuel Abenxuxén, the doctor of the *cabildo* and council of Cuenca, the fact being fully documented: 'Sir, commend me in your prayers'.[30]

Francisco de Montealegre, an inhabitant of Cuenca, 'at table would say, "Pass me the wine of the *berahá*[*sic*]",'[31] and 'when he was about to die he

25. Heb. *ḥai*, alive.
26. '¿No cres, señor, hay Adonay, por la ley de Moysén, que dygo verdad a vuestra merçed?', *FIRC*, Vol. 4, para. 89, p. 59.
27. Ibid. If the accuser's Jewish ancestry can be proved — and since the information is accessible — this instance of hostility has to be added to the many others showing such an attitude among *judeoconversos* themselves.
28. 'Pues mirad que nunca olvidaré la leche que mamé', ibid., para. 166, p. 87. In 1490 Juan Sánchez de Almazán, a resident of Soria, was accused of saying, when he was reproached for praying in Hebrew, 'Get along with you, this does me as much good as seven masses do you' ('Anda, vete, que tanto me aprouecha a mi esto como a vosotros siete misas'); see C. Carrete Parrondo, *FIRC*, Vol. 2, *El Tribunal de la Inquisición en el Obispado de Soria (1486–1502)* (Salamanca, 1985), para. 23, p. 29.
29. '... que le apareçía su madre entre sueños y le dezía que reçebía grand peña su ánima quando no guardaua bien el sábado', ADC, 12/244 (see ibid. 32/555). Tried during the years 1492–93 and relaxed [i.e. released to the secular arm for execution by burning — Tr.] in effigy.
30. ADC, 1/16. Tried during the years 1489–94 and underwent formal penance. It should be noted that after 1391 there was no Jewish community in Cuenca and that the doctors who visited it belonged to the neighbouring *aljama* of the town of Huete.
31. Heb. *brakha*, blessing. 'Quando estaua a la mesa desía: "Dadme aquí del vino de la berahá",' ADC, 3/45. Tried in 1490.

instructed his wife not to sell some Jewish books but rather to sell or pawn some other object from his house'.[32] As he was dying, *circa* 1470, 'He got up, dressed himself, and walking around praying and saying "Adonai *adóname*[33] my family and all that is mine and *adona* my father and my mother"',[34] who actually were still alive. '*Adóname* my brothers and sisters, Adonai, Adonai, Adonai'.[35] [was said] three times as he lowered his head, facing a wall, and said, 'Adonai, *adóna* me and all that is mine".' Here profound religious emotion has forged a word which expresses so many facets of feeling, but which has not entered the official dictionaries of the Spanish language. That is all the more reason to consider it with respect. Perhaps, also, it should be viewed as a possible linguistic *revanche*: this word, *adóname*, constitutes part of the free, simple language of the Castilian *judeoconversos.*

- 'Pero de Cáñama, a tailor resident in Almazán ... said that it must have been about 20 years ago ... when he was living in the town of Almazán with *maestre* Juan Tornero, now dead, a Catalan resident of Almazán, though afterwards he went to live in San Esteban de Gormaz[36] where he died, with whom this witness lived for about two years, more or less; and during that time the witness saw him most evenings and after eating, when he was leaving the table, he would say without crossing himself: "In the name of God the Father who created (*crío*)[37] the heaven and the earth and the sea and the sands", and then he would take a little book with a binding, written in Hebrew letters, and he would read from it. And often during that time he would see him reading from the said book, sometimes standing and sometimes seated, nodding his head up and down. And he would sometimes say to the witness when they were alone together on Saturdays, "Son, take care of the house, I am going to hear a little of a sermon that is being given in the synagogue by a great preacher" ... Likewise, he said that one day he told Antón de Santo Antón, a resident of Almazán, about that small book, and he told him to take it, and bring it for him to see what it said, and the witness took it and gave it to him, and the said Antón showed it to the Count of Monteagudo, who showed it to a Jew, who told him that they were prayers of the Jews'.[38]

32. '... quando se quiso morir mandó a su muger que no vendiese vnos libros judaycos y antes vendiese ottra prenda de su casa o enpeñase', ibid.
33. As far as I know, the expression is new to our historical lexicography, in the sense derived from the word *Adonai*, of 'cover with your presence'. In this context, it should not be related to the verb 'donar', meaning 'to give', but rather to the Hebrew name for God.
34. '... se leuantaua vistiéndose, que andaua resando, e que desía: "Adonay adóname a mi casa e a todo lo mío e adona a mi padre e a mi madre"'.
35. 'Adóname a mis hermanos e hermanas, Adonay, Adonay, Adonay'.
36. San Esteban de Gormaz is in Soria.
37. According to the Jewish tradition, and not *creó*.
38. 'Pero de Cáñama, que es sastre, vecino de Almaçán ... dixo que puede aver veynte

Following Jewish custom when uttering an oath, generally in sudden, unexpected situations, is also understandable, coming from the mouths of people who had not been sufficiently instructed in Roman Catholic catechism. Here are some examples.

- Pedro Vizcaíno, a shoemaker living in Sigüenza, told the inquisitors 'that sometimes he heard Juan Gonçales, a baker, now dead, of the same town, swear, "By the Law of Moses it is true!" '[39]

'Fernando de Molina ... said that it might have been about eight or nine years ago ... that one day he was in Almazán playing cards with Luys Mendes, a merchant, now dead, a New Christian, living in Almazán ... And as they played they began to argue vehemently and he saw and heard how Luys Mendes said, "It is so by the Law of Moses". And the witness said to him, "You dog Sir [Don], I will tell the inquisitors". And Luys Mendes said again, "It is so by the Law of Moses." '[40]

- Luis González de Caracena, a resident of Sigüenza, was accused of swearing 'by the Law of Moses'.[41]

- Lope de Arriaga, a resident of Almazán, was accused of saying 'By God it will not be so'.[42]

More interesting is the expression used by Hernando Calatro, a resident of Sigüenza, during the course of a game: 'By God (*Dió*) and the *Sefer Torah* it is not so'.[43]

años ... que este testigo biuió en la villa de Almaçán con maestre Juan Tornero, catalán, difunto, veçino que fue de ... Almaçán, e después se fué a biuir a Santisteuan de Gormás e allí falleçió, con el qual biuio este testigo dos años poco más o menos; y en este tienpo este testigo le vido ... las más noches e a comer, en alçando la mesa, desya syn santiguar: "En el nonbre de Dios Padre, que crió el çielo e la tierra e la mar e las arenas", e luego tomaua vn librillo pequeño enquadernado con sus coverturas, escripto de letra abrayca y leya por él. E muchas veses le vido en el dicho tiempo leer porel dicho librillo algunas veses en pie e otras estando asentado, alçando e abaxando la cabeça. Y desía algunas veses a este testigo, estando solos algunos días de sábados: "Hijo, guarda la casa; yo vo a oyr vn poco de vn sermón que disen en la sinoga de vn grand sermonador" ... Yten, dixo que vn día dixo a Antón de Santo Antón, veçino de Almaçán, esto de aquel librillo, y le dixo que lo furtase e se lo truxese para ver lo que desía, y este testigo lo hurtó e se lo dió, y desquel dicho Antón lo mostró al señor conde de Monteagudo, el qual le mostró a vn judío, e que le dixo que heran oraçiones de los judíos', ADC, 749/15; accusation of 1505.

39. ADC, 748/2; statement made in 1492.
40. 'Fernando de Molina ... dixo que puede aver ocho o nueve años ... que vn día, estando este testigo en Almaçán, jugaua a los naypes con Luys Mendes, mercader, defunto, christiano nuevo, veçino de Almaçán ... Y estando jugando començaron a porfiar en le juego, e vyo e oyó este testigo cómo dixo Luys Mendes: "Por la ley de Moysén asy es". Y este testigo le dixo: "Don perro, yo lo diré los inquisidores". Y tornó a desyr luego ... Luys Mendes: "Por la ley de Moysén asy es."' ADC, 749/15; statement made in 1492.
41. '... a la ley de Moysén', ADC, 748/2; statement made in 1492.
42. 'Para el Dió no sera'. 'Dió' in the singular form, following the Jewish tradition, *FIRC*, Vol 4., para. 30, p. 34.
43. 'Para el Dió yo tome el çéfer [Torah] en braços, que no es tal cosa', ADC, 748/2; statement made in 1492.

The inevitable memory of one's Jewish ancestors and thus of the Jewish religion, a memory that survived an abrupt conversion to Christianity, gives rise to constant references.

- Graciana Lainez, a resident of Almazán, said, 'By the memory of her sons and her husband, and by their bones'.[44]

The Jewish cemetery, like the synagogue, could not be erased from the memory of the Castilian *judeoconverso*. Such precious institutions could not be ignored:

- A Jew of Sigüenza, Jacó Donates, declared to the inquisitors that his neighbour Ruy García de Lucio used to go 'to the cemetery of the Jews to pray over the *huesa* (grave) of his mother'.[45]

- Alfonso Fernández Agarejo, a resident of Uclés (Cuenca), was denounced for 'sitting with the Jews and praying and reading with them'[46] in the synagogue and for saying, 'I'm a son of a worthy father. This place was left to me by my ancestors and my father; I come here to pray for them'.[47]

- Fernán Sánchez de Castro, a *judeoconverso* of Cuenca, 'had great devotion to the law of Moses and for this reason and in order to say and hear the ceremonies he had a reserved seat in the synagogue and he would sit in it like a good Jew, honoured among his relatives'.[48]

The apparent willingness to compromise in the area of dogma and broadminded tolerance is one attitude which should in no way be interpreted as conciliatory, despite appearances.

- Gregorio Lainez, a resident of Almazán and son of the *mayordomo* Alfonso Lainez, 'confessed and asked himself how was it possible to know, or who could know, which law was better, ours or that of the Moors or Jews'[49] an attitude which, in my opinion, should not be confused with a hypothetical ecumenicalism, but should be recognized as the bitterness of nostalgia for something now out of reach.

- We should also note an expression of the now traditionally accepted

44. 'Por el siglo de sus fijos e de su marido y por los huesos dellos', ADC, 749/14; statement made in 1501, *FIRC*, Vol. 4, para. 131, pp. 65 and 70. Several examples of the document exist.
45. '... al fonsario de los judíos a resar sobre la huesa [losa sepulcral] de su madre', ADC, 748/2; statement made in 1492.
46. '... asentado con los judíos e resando e meldando con ellos', ADC, 8/165. Tried in 1491–92. See C. Carrete Parrondo, 'Los judaizantes de Uclés, 1490–1492', *Tenth World Congress of Jewish Studies*, Division B, Vol. 2 (Jerusalem, 1990), p. 170.
47. 'Fijo de buen padre, este lugar me dexaron mis avuelos e mi padre; vengo a hacer oraçión por ellos', ADC, 8/165.
48. '... tenía muy grand devoçión en la ley de Moysén, y por esto y por desir y oyr sus çerimonias tenía asentamiento aseñalado en la synoga y sentávase en él como buen judío, honrado entre los otros sus parientes', ADC, 5/103; accusation of 1491.
49. '... confesó e dezía entre sy que cómo hera posible o que quién sabía quál hera mejor ley: ésta nuestra o la ley de los moros o judíos', ADC, 749/14; statement made in 1501.

belief, expressed on this occasion by Isabel de Madrid of Cuenca, who 'would say that the Jew gained salvation through his own law just as the Christian did in his law'.[50]

II. THE REAFFIRMATION OF OLD BELIEFS

The preceding pages may have served to show that the Castilian *judeoconverso*, who was frequently anonymous — and usually went unnoticed, if he was lucky — suffered from internal yet exteriorized malaise that impelled him to show his or her real feelings: full loyalty, though we can only assume this, to the ancient religious creed. The *judeoconverso* affirms this loyalty through repeated expressions: the traditional divinity is 'the good Lord', Judaism constitutes the best religious practice, there is a predisposition towards Messianism and, lastly, as an obvious human reaction to coercion, the *judeoconverso* launches biting criticism of the majority religion of the surrounding society.

1. Adonai is the Best God

- Diego de Luján, a resident of Almazán, was accused 'when talking about how God had brought the Jews across the parted sea and given them manna and provided what they wanted, and they made the Golden Calf, Diego said, "It was the Good Lord who was in authority then";[51] and then, after a while, he added "And the Lord of these days is [also] good."'[52]

The same *judeoconverso* was accused of saying 'how God had given the Jews manna, and when they wanted him to provide chickens and quails, he had provided them too',[53] an assertion that seems steeped in local colour.

2. Judaism is the Best Religion

Valuing Judaism more than Christianity appears quite frequently. Here are some examples.

- Juan del Escuela, a resident of Cuenca, was accused of being 'very keen on and devoted to the law of Moses, maintaining that it is holy and calling it holy, entreating for it in his prayers ... and saying, "By the holy law ... he

50. '... desya que tanbién se saluava el judío en su ley como el christiano en la suya', ADC, 2/33. See C. Carrete Parrondo, 'Las judaizantes de Cuenca procesadas por la Inquisición en 1490', in *Actas del II Congreso Internacional Encuentro de las Tres Culturas* (Toledo, 1985), p. 103.
51. '... estando fablando cómo Dios avía dado el mana e cómo les sabía a lo que ellos avían gana e fisieron el vezerro [Ex. 32]', *FIRC*, Vol. 4, para. 171, p. 90; statement made in 1505.
52. '... que buen Dios hera aquél que executaua luego; e luego, dende a poco rato dixo: "E bueno el de agora."' ibid.
53. '... cómo Dios había dado a los judíos el maná, y quando querían que les supiese a gallinas e a capones aquello les sabía', ADC, 749/14; accusation of 1501.

would fast",[54] and to those who would not fast like him he would say, "Quiet, you ignorant people, for hell is full of the likes of you." '[55]

Endogamy was a general practice among these *judeoconversos*. Here is one example: Pedro García el Rubio, a resident of Matute, was accused of saying 'that the law of the Jews was true, that cousins should marry in order not to disperse property, and ... that the synagogue was a house of prayer and that he would go to the synagogue and pray; and he would also say that the Jews observed the true law and that they would keep their law well, for they would not even touch fire on the Sabbath'.[56]

- The accusation levelled against Diego de Madrid, a resident of Castillo de Garimuñoz (Cuenca), shows more pronounced intentionality, for he was accused of saying with some arrogance 'that Jews bore good and pure blood, and that they were of royal blood ... And moreover, because the Jews were of such pure blood, God had chosen Our Lady for his incarnation'.[57]

- Diego de Olivares, the son of the *bachiller* Gonzalo Sánchez de Olivares, a resident of Cuenca, was discussing questions of religion with an Old Christian, 'raising many doubts about it and affirming that the sure and true law is that of Moses, and that to defend this point he would defy many authorities'.[58]

- Lastly, here is an assertion made by a *judeoconversa* of Almazán: Graciana de Santa Cruz was accused of saying, while talking about the expulsion of the Jews and the consequent loss of property, 'Don't talk to me about that, for I have kept what our law requires and because of that I hold myself to be more fortunate than those who stayed here, for they did it in order not to lose their properties'.[59]

54. On the holiday of Yom Kippur.
55. '... muy afiçionado y devoto de la ley de Moysén, teniéndola e llamándola santa, yimando [*sic*] por ella en sus priegas y neçesydades, disiendo: "Para la ley santa que Dios, Nuestro Señor, dio a Moysén en el monte Sinay"... Ayunava, y a los que non ayunavan como él dezíales: "Callad, neçios, que de los tales se hinchan los ynfiernos."' ADC, 3/55; accusation of 1491. See. C. Carrete Parrondo, 'Dos ejemplos del primitivo judaísmo en Cuenca', *El Olivo* 13, 29–30 (1989), 67.
56. 'Que la ley de los judíos que hera verdadera, que casauan primos con primas por no desfazer las faziendas, y que ... desía que la synoga que hera casa de oración e que él yva a la synoga y hasya oración; y que asymismo desía que los judíos guardauan la ley verdadera y guardauan bien su ley, que avn los sábados no llegauan al fuego', ADC, 749/15; accusation of 1505.
57. 'Que la sangre de los judíos hera buena y linpia y que heran de sangre real ... Avn por ser tan linpia sangre la de los judíos avía Dios escogido a Nuestra Señora para su encarnación', ADC, 8/167; accusation of 1492. See Y. Moreno Koch, 'La comunidad judaizante de Castillo de Garcimuñoz: 1489–1492', *Sefarad*, 37:1–2 (1977), 370–71.
58. '... poniéndole muchas dudas çerca dello e afirmándose ser la çierta e verdadera ley la de Moysén, e que para ello reñia muchas abtoridades', ADC, 5/109; accusation of 1491.
59. 'Ya dexadme deso, que yo conplí lo que nos mandó nuestra ley y por eso me tengo

3. A Messianic Movement among Judeoconversos

A collective state of mind like the one reflected in the foregoing documented examples had to lead, logically, to a Messianic movement:

- Fernando de Uclés, a resident of Cuenca, would say that 'he wanted to go to Judaea'.[60]

- His neighbour Gonzalo, a jeweller, as far as we can understand from the *reconciliación*[61] of his daughter Leonor de Castro, would, together with other *judeoconversos* of the town, 'say that they were going to Judaea to await the Great Day, as they would call it, together'.[62]

- The accusation against the father of maestre Gabriel, a resident of Calatayud (Saragossa), is more endearing, human, and full of local colour. The father had close connections with the *judeoconversos* of Soria, and an informer says that the father 'held *maestre* Gabriel to be a page of the Messiah, and ... *maestre* Gabriel had very long hair, down to below his waist, and his father would say that he was not going to cut his son's hair until the Messiah came'.[63]

4. Criticism of the Religion of the Majority

This pent-up emotion also entailed a critical attitude towards the Catholic Church, its dogmas, rituals, and ceremonies.

- María de los Caballeros, a resident of Medinaceli, was accused of commenting: 'Look what the *hanies* (poor)[64] Christians eat'.[65]

- Fernando de Brihuega, a resident of Sigüenza, 'told his workers that they should sew on the holiday of *Santa María de la O*, telling them that it was a holiday only because on that day Our Lady had said, "Oh, now I am pregnant!"'[66]

por mejor aventurada que los que se quedaron acá, que lo fisieron por no perder sus fasiendas', *FIRC*, Vol. 4, para. 230, pp. 108–9; accusation of 1505.

60. '... se quería ir a Judea', ADC, 33/567. Tried in 1490(?). In that year Orovida, a resident of Soria, stated to the inquisitors that her brother Francisco Martínez 'was sorry that he had become a Christian, and that he was going to give all his possessions to his wife and go to his father's house because he wanted to go to Jerusalem' ('se arrepentía porque se avía tornado christiano, e que daría a su muger toda la hazienda e que se fuese a casa de su padre, porquél se quería yr a Iherusalem') (*FIRC*, Vol. 2, para. 9, p. 22).
61. Reconciliation to the Church following spontaneous or forced repentance (Tr.).
62. '... desían que se yuan a Iudea a estar todos juntos el día mayor que disen', ADC, 2/37; statement made in 1490.
63. '... cómo tenia a ... maestre Gauriel para paje del Mexías, y ... maestre Gauriel traya los cabellos muy largos, más que fasta la çinta, e desya que non se los cortaría fasta que viniese el Mexías', ADC, 749/15; accusation of 1505.
64. Heb. *'ani'im*, poor or simple people.
65. 'Mirad que comen los haníes de los christianos', ADC, 748/7; accusation of 1504.
66. '... dixo a sus obreros que cosyesen el día de Santa María de la O, disyendo que non hera fyesta, syno porque dixo Nuestra Señora aquel día: "¡O, qué preñada estó!"', ibid.

Sarcastic comments about the Mass were not lacking, nor were those about the Apostles and Christian institutions:

- An Old Christian who worked for the shoemaker Diego de Vera, a resident of Monreal de Ariza (Saragossa), 'said to him one morning, "Come Diego, let us go to Mass", because he was a married man, and Diego answered, "You go, Sir, and I will go for the trifle", and he did not go to Mass with this witness, nor did he see him there in the church at ... Mass. And that same day, after eating, the witness and the aforementioned Diego left for ... Harisa [Ariza, in the province of Saragossa], and along the way the witness asked him this question: "Diego, why did you say you were going to Mass — a mere trifle? What do you mean by trifle?" And he answered that it was when the priest finishes the Mass and turns to give the benediction'.[67]

- During the Inquisition's trial of Gabriel Castiel, a resident of Cuenca, the prosecutor brought the accusation that 'in conversation concerning the virtues of the blessed mother of God, he made the following sacrilegious statement: "Quiet, don't believe in such jokes, believe in the Creator who made heaven and earth, and everything else, and gave the law to holy Moses", and with even more pertinacity ... he would say that he would throw himself into the fire for the law of Moses and he would not burn'.[68]

- Juan de Castro, his neighbour, 'when talking with other persons would say that they would not be saved by believing in the words of the holy Gospels'.[69]

- Martín de Medina, another resident of Cuenca, used to say 'that the sayings of the prophets were all true, the more so as the Apostles could only tell lies'.[70]

- When someone reprehended Rodrigo de Torres, a resident of Almazán, for not attending the processions, he answered: 'You should go

67. '... le dixo por la mañana: "Diego, vámonos a misa", porque hera honbre casado, y a esto respondió el dicho Diego: "Id vos, señor, que yo yré a lo menudillo"; y que no fué a misa con este testigo ni lo vido allá en la yglesia en la ... misa. Y quel mismo día, después de comer, se partieron este testigo y el dicho Diego para ... Harisa, y en el camino este testigo le preguntó a dixo: "Diego, ¿por que me dezías que avias de yr a lo menudillo de la misa? ¿quán tienes tu por menudillo?" e le respondió que cuando acaba el abad la misa y buelue a dar la bendiçión', ADC, 749/15; accusation of 1505. The same attitude emerges in *FIRC*, Vol. 4, para. 196, p. 99.

68. '... hablando de las virtudes de la bendita madre de Dios dixo este deprauado sacrílego: "Callad, non trugáys tal burlería, crehed en le Criador que fizo cielo e tierra, etc. y Dió la ley al santo Muysén", e avn con grand pertinasçia ... desía que se echaría an vn fuego por aquella ley e que non se quemaría', ADC, 33/568; accusation of 1490(?).

69. '... fablando con otras personas desía que non curasen de creer en las palabras de los santos Evangelios', ADC, 5/95, and ibid. 2/29; accusation of 1490.

70. '... que los dichos de los porhetas que todos eran verdaderos, más que los apóstoles non pudieron desir sino cómo con quien andauan', ADC, 1/14; accusation of 1489.

to the processions, because you were born to go, whereas I was not'.[71]

- The resident of Cuenca, Fernando el Zarco, spoke harshly of the sacraments, exclaiming: 'I swear to God that it is all a joke, from the Pope to the cope'.[72]

- Once the mother of Alonso Díez de Cifuentes, a receiver, resident of Sigüenza, was being visited because she was ill, and her son 'said: "Mother, confess and receive communion and put your soul in order and you will then return to health", and ... his mother answered, "Leave that, son; let them bleed me and have me taken care of, for that will do me more good". And they brought her the Corpus Christi and she refused to receive it'.[73]

- García, a tanner and inhabitant of Ayllón (Segovia), was accused because 'when the witness was with the aforementioned they were tolling a bell and the witness said, It must be cracked now; the aforementioned said, They should all be cracked'.[74]

- A farm-hand accused Juan de Buendía, a resident of Serón de Nágima (Soria), that when he was 'in the house of Mosén Gil, the priest of Santa María de Ariza, talking about our holy Catholic faith, the witness heard the accused saying ... that Christians had always been mad, and when the witness asked him why they were mad he answered, "Because they left the Father and took the Son"; and the witness said to him, "Watch what you are saying, for whoever loved the Father loved also the Son and the Holy Ghost, being the one true God". And he was quiet and did not answer'.[75]

- A grave affirmation was made by María Alonso de Alcázar of Cuenca, who would proudly say that 'my ancestors killed Jesus Christ'.[76]

71. 'Vos avéys de yr a las proçesiones, que naçistes para yr a ellas, que yo no', *FIRC*, Vol. 4, para. 236, p. 110; accusation of 1505.
72. 'Juro a Dios todo es burla, desde el Papa hasta la capa', ADC, 3/51, and ibid., 3/61; accusation of 1489.
73. '... dixo a ... su madre: "Señora, confesad y comulgad y ordenad vuestra ánima y avréys luego salud", e que ... su madre le respondió: "Dexáos, hijo, deso; sángrenme y curen de mi, que me aprovechará más". E que le traxeron el Corpus Christi e que no lo quiso recibir', ADC, 748/7; accusation of 1492.
74. '... que estando con el susodicho tañían vna canpana y dixo este testigo: Quebrada estoviese agora, e respondío el susodicho: Quebradas estoviesen todas', ADC, 748/2; accusation of 1504.
75. '... en casa de Mosén Gil, cura de Santa María de Harisa, hablando sobre cosas de nuestra santa fee católica, oyó este testigo cómo dixo ... Juan de Buendya qué los christianos sienpre fueron locos, y este testigo le dixo que de que heran locos: "Que dexauan al Padre e tomauan al Fijo"; y este testigo le dixo: "Catad que dezis, que quien amo al Padre amo al Fijo y al Spiritu Santo, que es vn solo Dios verdadero". E calló e no paso más', ADC, 749/15; accusation of 1492.
76. '... se glorificaua en dezir que "mis parientes avían muerto a Ihesu Christo", ADC, 10/190; accusation of 1492.

III. THE RETURN TO JUDAISM

The almost general nostalgia generated by precipitate conversion and the wish to maintain their former creed and religious practices in some cases led the *judeoconversos* of Castile to move to the neighbouring kingdom of Portugal and to live there openly as Jews. Here are two examples:

- Gabriel Castiel of Cuenca 'was seen wearing the dress and bearing the sign of a Jew,[77] running a school for teaching Jewish children, and for that reason would say that our whole law was a joke ... He went to Portugal, and returned to being a Jew in Portugal, according to what the witness heard from a brother-in- law of his ... who saw him as a Jew in Portugal'.[78] In about 1488 they saw the accused in Viseu 'dressed in a shawl with the sign of a Jew on it; and when the said Gabriel bid farewell to the witness, and went off towards the Jewish quarter, where he was staying; and this witness asked some Jews of the town what Gabriel was doing there and why he was wearing Jewish clothes, because in the town of Cuenca here in Castile he was thought to be a Christian; and the above-mentioned Jews answered him that he [the accused] had a school for teaching the children of Jews to read and they told him the name by which Gabriel was known, something like Isaque or Jusé that he [the witness] does not remember'.[79]

- The case of the following *judeoconverso* denounced to the inquisitors by Jews themselves is typical. The accusations went like this: 'Jamila, a Jewess, the wife of Liaçar [*sic*] of Burgos, and a resident of Sigüenza ... said that in the town of Valladolid she knew a Jew called Ysaque (who is now in the prison of the Inquisition), who had a sister called Çinhá, the wife of Alifaz; and when she was in that town she heard it said that the said Isaque had turned Christian ...; and afterwards the witness went to live in Sigüenza ... and after she had been living there for about three or four months ..., it was last summer, she saw him wearing the sign of a Jew and

77. Following the Fourth Lateran Council (1216), the Council of Arles, 1235, ordered Jews to wear a yellow patch four fingers wide on the chest as an identifying sign. See the edict of the Catholic Monarchs (Burgos, 2 Dec. 1491), which includes a law punishing Moors and Jews who did not wear the signs (General Archives of Simancas, Diversos de Castilla, folder 8, No. 116).

78. '... le vieron en hábito de judío con su señal, teniendo escuela de enseñar moços judíos, e por aquesta causa desía que toda nuestra ley era burla ... Se fué a Portogal, a tornar judío a Portogal, segund que le dixo a este testigo vn cuñado suyo ..., que lo vido judío en Portogal', ADC, 33/568; accusation of 1490 (?).

79. '... vestido con vna lobilla e vna señal de judío puesta en la dicha loba; e que cómo el dicho Grauiel lo viera que se escusara deste testigo e se fuera fasia la judería, a donde posaua; e que este testigo preguntara a çiertos judíos de la dicha çibdad qué que hasia allí el dicho Grauiel o cómo estaua en ábito de judío, porque en la çibdad de Cuenca, acá en Castilla, era tenido por christiano; e que le respondieron los dichos judíos que tenía allí vna escuela de enseñar leer a los fijos de los judíos e le dixeron el nonbre dél como allá se llamaua, e que no se acuerda si le dixeron Ysaque o Yuçe', ibid.

asking for *çedacá*.[80] The witness called him to her door and asked him where he was from; he answered that he was from the town of Evora,[81] and the witness told him that this was not so, but that he was from the town of Valladolid and that she recognized him, and knew that he had turned Christian. The former answered that it was true, and the witness asked him, "What are you doing around here? If the Inquisition comes you will be burnt". And he answered ...: "I want to go to Portugal."'[82]

- 'Çimá [*sic*], a Jewess, the wife of don Salamón Acueste, a Jew in charge of the hospital, said that she knew a Jew called Ysaque, who is now in the prison of the Inquisition, and who had stayed in her house three months before ... and she saw him wearing the sign of a Jew, praying like one, and going to the synagogue to do all the other things that all Jews do; and that now, last Wednesday, the same Ysaque came to her house and she saw him coming with his sign of a Jew and he dined there, and after dining ... he made the *berahá*, and while he was there at the dinner table she saw him being taken away as a prisoner by some of the inquisitors ...'

- 'Lezar de Burgos, a resident of ... Sigüenza ... said that he knew a Jew whom they call Ysaque, who is now incarcerated in the prison of the Inquisition, and of whom he heard say in Valladolid that he had become Christian ..., and that three or four months before ... he had seen him in this city of Sigüenza dressed as a Jew, wearing his sign ... going to the synagogue and praying there and asking for *çedacá* in public, keeping the Sabbath and performing all the other rites and ceremonies like a very good Jew. And last Wednesday night he saw the said Ysaque coming to ... Sigüenza with his sign of a Jew. And that is all he knows, since he had heard say that [Ysaque] had been arrested by order of the inquisitors'.[83]

80. Heb. *çedacá*, charity.
81. In Portugal.
82. 'Jamila, /judía/, muger de Liaçar [*sic*] de Burgos, veçina de ... Sigüença ..., dixo que conosçió en la villa de Valladolid a vn judío que se llamaua Ysaque, /agora está preso en la carcel de la Inquisiçión/, el qual tenía vna hermana que se llama Çinhá, muger de Alifaz; e que estando en la dicha villa oyó desir cómo el dicho Ysaque era tornado christiano ...; e que después se vino este testigo a beuir a ... Sigüença, e beuiendo en ella, avrá tres o quatro meses ..., lo qual fue este verano pasado, al qual este testigo vió con señal, e que este testigo le dixo que non era synon de Valladolid e que ella le conosçía, que sabía cómo se auía tornado christiano, e quel susodicho le respondió cómo era verdad; e que este testigo le dixo: "Pues ¿en que andays por esta tierra? Que handa la Inquisiçión e os quemarán". E que el le respondió ...: "Quiero yr a Portogal."' ADC, 748/2: accusations of 1492. See Ma. F. García Casar, 'Judíos castellanos colaboradores con el Tribunal de la Inquisición del distrito Cuenca-Sigüenza', in *Tenth World Congress of Jewish Studies*, Vol. 2, *The History of the Jewish People* (Jerusalem, 1990), pp. 175–82.
83. 'Çimá [*sic*], judía, muger de don Salamón Acueste, judío espitalero ..., dixo que ella conosce a vn judío que llaman Ysaque, el qual agora está preso en la cárçel de la Inquisiçión, el qual ella vió posar en su casa avrá tres meses ... e que le vió traer señal de judío e rezar cómo judío e yr a la synoga a haser todas las otras obras que hazen todos los judíos; e que agora, miércoles pasado, vino asymismo el dicho

Earlier in the article I promised some final comments. However, confronted by the selections from the documents that have just been quoted, a selection that could easily be extended, the real state of affairs is obvious: after an abrupt conversion, unaccompanied by even the most basic teaching in the catechism, with a clergy that predated the Council of Trent and faced with a simple people with highly defined characteristics, as were the Castilians, the *judeoconverso* (who, let us not forget, was for these purposes just as much of a Castilian) adopted a position which cannot simply be described as challenging but — at the risk of repetition — authentically Castilian. Brother Hernando de Talavera, a Jeronomist friar, the first archbishop of Granada and the descendent of *judeoconversos*, to my mind a likeable, and profound, thinking Castilian, who in 1487, in his open and Paulinistic way of thinking, was capable of writing the following somewhat conciliatory remarks:

> It is not true that Christian people hold them [the Jews] as enemies, but rather they treat them humanely and almost as neighbours, despite the fact that Christians wish the Jews to live separately and to wear signs, and not to occupy certain offices in order to prevent the many sins that do and will result from mingling and much familiarity and from not observing all that the holy canons and civil laws order and require concerning the Jews' dealings with Christians. It is true that some Christians denounce and mistreat them, bearing them hatred and dislike, without any new cause for it, and this not only towards Jews but also towards the Christians who descend from Jews.[84]

Ysaque a casa deste testigo, e le vió venir con su señal cómo judío e cenó allí, e acabado de çenar ... le vió haser la berahá, e estando asy çenando le vió leuar traer preso a vnos onbres de los inquisidores.
Lezar de Burgos, veçino de ... Sigüença ..., dixo que el conosçió a vn judío que llaman Ysaque, que agora está preso en la cárçel de la Inquisición, al qual oyó desir en Valladolid que se auía tornado christiano ..., e que avrá tres o quatro meses ... que le uió en esta çibdad de Sigüença en ábito de judío, trayendo su señal ... e entrando en la synoga e rezando en ella e pediendo çedacá publicamente, e guardar los sábados e haser todos os otros ritos e çerimonias cómo muy buen judío. E que el miércoles en la noche que agora pasó vió venir al dicho Ysaque a ... Sigüença e traya su señal como judío. E que no sabe más de quanto oyó desir que lo auian preso por mando de los inquisidores', ADC, 748/2; accusations of 1492.

84. 'No es verdad que el pueblo cristiano les tiene enemiga, antes los trata humanamente y cuasi cómo proximós, no obstante que quiere que vivan apartados y anden señalados y que no usen de algunos oficios por evitar muchos pecados, que se siguen y seguirán de la mezcla y mucha famiaridad y de no se guardar todo lo que cerca de su conversación con los cristianos por los santos cánones y leyes civiles es ordenado y mandado. Verdad es que algunos cristianos los denuestan y maltratan y les tienen malquerencia, sin les dar a ello nueva causa y no solamente a los judíos, más aún a los cristianos que descienden de ellos' (Hernando de Talavera, *Católica impugnación*, ed. F. Martín Hernández [Barcelona, 1961], pp. 148–9). One should not forget the accurate opinion of Americo Castro, *Hacia Cervantes* (Madrid, 1967), p. 342: 'The New Christians of Semitic descent took one of three directions:

Before concluding, I would like to refer to a distinguished figure of medieval Spanish *Cancionero* (balladry). Some brief examples of the work of one of the most outstanding *judeoconverso* poets, Antón de Montoro, follow. A few years ago I studied his work in association with Francisco Cantera Burgos. Montoro once wrote:

More decorated with faith
Than a cherry tree with leaves:
Don't you know how I won
The credentials of an old Christian?[85]

— lines which contain underlying irony, bitterness and obvious nostalgia, as do these others, which are typical of the *judeoconverso* who makes fun of his own appearance:

Hood and hat
Gloves and pointed cap
The Torah like don Juza
A leather shoulder bag
A purse for money
And a belt for tightening
And a handkerchief for wiping
And a comb in a combcase.[86]

Our lighthearted poet does not conceal his Jewish past at all:

As when my ancestors
Enter the holy temple
To worship their Creator.[87]

that of excelling in zeal over Old Christians and collaborating with the Inquisition, inspired, originally by their ancestors; that of floating between Judaism and Christianity in a series of delicate shadings going from crypto-Judaism to ambivalent forms of Christian belief that, in the second or third generation, led to their not knowing exactly what they believed; and that of cultivating a spiritualized form of Christianity resting on the desire to go further into one's individual, sincere Christian faith, rather than following everything that others believe and practise — that is, choosing to combine one's belonging to the Church with individual attitudes and with criticism of some religious customs and practices'.

85. 'Muy más bordado de fee / que non de fojas el guindo; / ?no sabeís cómo gané / carta de cristiano lindo?' See F. Cantera Burgos and C. Carrete Parrondo, *Antón de Montoro: Cancionero* (Madrid, 1984), poem 82, ll. 1–4, p. 211.
86. 'Capirote y sombrero, / los gauntes y caperuza, / la Torá cómo don Juza, / la barjoleta de cuero, / la bolsa con el dinero / y la cinta de apretar / y el paño de alimpiar / y el peyne y el peynadero', ibid., poem 87, vv. 1–8, p. 219.
87. '... como quando mis parientes / entran en el templo santo / a adorar su Criador', ibid., poem 114, vv. 4–6, p. 269.

He poses an enigmatic question to Toledo, the king of arms:

> Tell me in rhyme
> Answer a doubt of mine:
> Which would you rather have lost
> The faith, or Noah's plant?[88]

Lastly, let us taste the delicate irony, not unmixed with venom, with which he addresses 'a woman who called him a Jew':[89]

> And if I told you it was greed
> By God, although it cannot be eaten,
> You would say that you have a bull
> From the prelate of Sodom.[90]

88. 'Declárame por conpás / vna dubda que non sé: / ¿quál querrádes vos más, / que se perdiese la fe / o la planta de Noé', ibid., poem 130, vv. 5–9, p. 294.
89. '... a una muger que le llamó judío'.
90. 'Y si os dixere que es gula, / por Dios, aunque no se coma, / vos diréis que tenéis bula / del perlado de Sodoma', ibid., poem 148, vv. 11–14, p. 368.

The Annunciation to Esther: Felipe Godínez' Dramatic Vision

SARA TUREL

On 30 November 1624, in San Francisco Square in the city of Seville, an *auto-da-fé* took place. Fifty penitents were punished by the Court of the Inquisition. Among them was Felipe Godínez, a priest and dramatist of Jewish descent. Godínez was sentenced for being a heretic and judaizer, and the text of the *auto-da-fé* indicated that, an adherent of the law of Moses, he had written two comedies, *La reina Ester* (Queen Esther) and *La harpa de David* (David's Harp), both of which were based on the Old Testament.

The punishment imposed on him involved walking on the platform wearing the *sambenito* (penitent's robe), confiscation of his possessions, one year's reclusion, six years of exile, and being permanently deprived of the right to practise as a priest.[1]

According to the available information, Godínez complied with all the punishments except the last one. After a year of imprisonment in Seville, he went into exile in Madrid, where he stayed until his death in December 1659, 35 years after his conviction. While in Madrid, he continued to practise as a priest and also to write plays, including some relating episodes from the lives of Christian saints and *comedias de capa y espada* (cloak and dagger dramas).

Felipe Godínez is presently known only to researchers on the Golden Age of the Spanish theatre. Until recently, his plays were studied as part of the general literary history of his age, no work being devoted to him exclusively. Interestingly, Julio Caro Baroja's book *Los judíos en la España moderna* refers to Godínez as a pure and simple Jew,[2] and when Marcelino Menéndez y Pelayo mentions him in his *Historia de los heterodoxos españoles* he states that 'Godínez, like all the poets of his race, stands out

This article was translated by Judith Roumani.

1. On the text of the *auto-da-fé*, see C. Menéndez Onrubia, 'Hacia la biografia de un iluminado judio', *Segismundo*, 25–6 (1977), 116–19.
2. J. Caro Baroja, *Los judíos en la España moderna y contemporánea* (Madrid, 1978), Vol. 2, pp. 241–2.

due to his attachment to subjects from the Old Testament'.[3] It would seem that these scholars shared the opinion, held by the Inquisition, that rather than being a Catholic priest serving the Church, Godínez was a Jew.

It is only recently that studies dedicated solely to this author, trying to reconstruct his life and define and evaluate his work, have been published. They include the work of Menéndez Onrubia (1977), Profeti (1982), Bolanos Donoso (1983), and Vega García-Luengos (1986).[4]

In this article I examine one of the works mentioned in the text of the *auto-da-fé*. Many Golden Age dramatists took the Bible as their source of inspiration. Moreover, they used the same biblical themes as Godínez employed; among them were Lope de Vega, Tirso de Molina, and Calderón de la Barca. In other words, the theme that Godínez chose is neither strange nor unusual. The question then arises of whether or not his identity as a *converso* is reflected in the vision with which he approaches the Old Testament. In this presentation and analysis of *La reina Ester* I attempt to answer this question.

A single manuscript copy of this drama, dated 1613, has been preserved. The manuscript is now in the National Library of Madrid (Cat. no. MS 17.120). It was published for the first and only time together with another drama by Godínez, *Ludovico el Piadoso*, in German Vega García-Luengo's scholarly edition. This is the version which is followed here.[5]

The play is rather long, 3,461 lines. Although the ending has been lopped off, it is obvious that only a few lines are missing. Godínez digresses freely from the biblical narrative, adding many episodes and characters and eliminating certain elements from the Book of Esther. The following synopsis of the play indicates which parts differ from the biblical original.

The play begins with a dialogue between King Ahasuerus and his counsellor, Haman. They are talking about Queen Vashti, who has refused to participate in the royal banquet. Vashti is described as a rebel with an unbearable temperament, and Ahasuerus decides to repudiate her. Haman suggests that the king should go hunting in order to forget the queen. The king and his entourage arrive at the village where Esther and Mordechai live. Ahasuerus meets Esther by a water well. He begins to court her but Esther, modest and retiring, defends her honour even when the king reveals his identity.

3. M. Menéndez y Pelayo, *Historia de los heterodoxos españoles* (Madrid, 1978), pp. 228–9.
4. See Menéndez Onrubia, 'Hacia la biografía'; M.G. Profeti, *Per una bibliografía de Felipe Godínez* (Verona, 1982); P. Bolaños Donoso, *La obra dramática de Felipe Godínez: Trayectoria de un dramaturgo marginado* (Seville, 1983).
5. G. Verga García-Luengos, *Problemas de un dramaturgo del Siglo de Oro: Estudios sobre Felipe Godínez con dos comedias inéditas: 'La reina Ester', 'Ludovico el Piadoso'* (Valladolid, 1986). Page numbers given for citations refer to this edition.

Ahasuerus decides to take a new wife, and Haman advises him to choose one from among the most beautiful virgins of Persia. Prince Egeo will be responsible for seeking them out. In the meantime, the royal hunting party lodges in the house of Mordechai, described as the richest man in the village, 'a nobleman, although he is a Jew'. Haman's hostility towards the people of Israel is already apparent at this stage of the action and continues throughout the play as the Jew refuses again and again to kneel before him.

As Esther appears on stage again, Godínez alludes to the main theme of the play: the salvation of the Jews. The maiden tells her uncle Mordechai about a dream that she has dreamed three times. In the dream she is a queen and saves her people. After relating this, she retires to pray. Egeo tells Mordechai about the king's decision to choose a new queen and suggests that Esther should also present herself as a candidate. This dialogue implies that Mordechai is Esther's adoptive father, and that he is hiding the maiden's Jewish identity. Prince Egeo looks for Esther and finds her immersed in her prayers. Egeo announces to her that she will be the new queen. In both language and form the whole scene suggests the Annunciation to Mary, by means of descriptive details recalling the countless paintings on the theme.

Ester: Balgame Dios! Quien me busca?
Egeo: (hincado de rodillas)
Dios te salbe, Hermosa Ester.
Contigo es Dios, y seras,
entre todas las demas,
La mas dichosa muger
. . . .
Dios te salbe, Ester graciosa. (p. 221.)

(Esther: Lord save me! Who seeks me?
Egeo [kneeling]:
May God save you, beautiful Esther.
God is with you, and you shall be,
Of all women
The happiest
. . . .
May God save you, gracious Esther.)

When Mordechai intervenes, the allusion to the Virgin becomes more explicit. He, too, sees in Esther a prefiguration of Mary.

Mordoqueo:
Seas, gran Dios, glorificado;

que aqui se a representado
una sonbra, una figura
de quando bengan a dar
aquella alegre enbaxada
a la donsella sagrada
en quien Dios a d'encarnar. (p. 222.)

(Mordechai:
Good Lord, may you be glorified;
For here has been represented
A shadow, a figure
Of when they shall come to give
That happy mission
To the holy maiden
In whom God will be incarnate.)

Esther is chosen by the king, and henceforth she is shown as a faithful, efficient, and energetic protector of the king's interests. She turns idlers and parasites away from the palace, and takes an active role in the unfolding of secondary action during the play.

Mordechai uncovers a plot being hatched by Seleuco and Leoncio, two princes of Vashti's family. They are planning to assassinate the king to revenge the honour of the repudiated queen. When the Jew denounces them, thereby saving Ahasuerus's life, the Temple of Mars suddenly collapses. Here Godínez again introduces an episode totally foreign to the biblical story: the king sees the fall of the temple as an evil omen that has already been announced in an ancient prophecy. The prophecy says that

> Cobrarán libertad y vida en el mismo dia en que pensaron perderla, triumpharan de sus enemigos, y quedaran libre y honrados los sujetos. Moriran a sus manos los que pensaron ensangrentar las suias. Saldra el leon a coronarse de sus angostos limites, y correra victorioso hasta la India. Empecara una nueva monarquia, y perderan los persas la suia. Hundirse a el templo de Marte, y nacera el hijo de la hebrea (p. 256). (The subjects shall win liberty and life on the day that they thought to lose it, they shall triumph over their enemies, and shall be free and honoured. Those who thought to bloody their own hands shall die at the hands of their intended victims. From his narrow confines, the lion shall emerge to be crowned, and shall run triumphantly as far as India. A new monarchy shall begin, and the Persians shall lose theirs. The Temple of Mars shall collapse, and the son of the Jewess shall be born).

Ahasuerus asks the wise men of the Court to help him decipher this enigmatic text, but they are not able to do so. In the end, Haman is the one who manages to provide an interpretation: the Jewish people will be responsible for causing the collapse of the Persian Empire, because the Messiah, the 'son of the Jewess' whom the text presages, will be born from the ranks of these people. Haman therefore proposes to exterminate the Jews, in order to prevent the prediction from coming true, and the king accepts his advice. Mordechai hears about the decision and turns to Esther to ask her to intercede with the king to revoke the edict.

Esther turns to the Holy Scriptures for help. In this scene Godínez introduces a series of biblical characters, among them a mythological one, who encourage the queen and tell her that she will succeed in her mission to save her people. The characters who appear in succession are the prophet Daniel, King David, who assures Esther that her sons will not disappear because from them the Messiah will arise, and the mythological Cumaean Sibyl, immortalized by Virgil and whom Christianity has adopted as a prophet of the coming of the Messiah. The sibyl prophesies to Esther that a new heavenly dynasty will arise from a virgin queen. In the end, the Archangel Gabriel makes an appearance. He summarizes everything that the previous figures have prophesied and adds that God will cause a maiden named Mary to be his mother, and that Gabriel himself will announce this birth to the virgin, conceived without sin. The archangel concludes with these words:

Gabriel: No tiene, pues, que temer
la muerte de tu pueblo, Ester
que túas de librarle della,
siendo figura de aquella de
quien Dios a de nacer. (p. 293.)

(Gabriel: Thus you should not fear
The death of your people, Esther
For you shall deliver them
Since you are a figure of her
From whom God will be born.)

Events in the last scene develop in a similar way to those of the biblical source: Esther invites the king and Haman to a banquet in her quarters. During this meal the queen reveals her Jewish identity, asks for clemency and accuses Haman. Angered by this revelation, the king leaves the stage. Haman rushes to Esther pleading for his life; at this moment Ahasuerus returns and seeing the situation, thinks Haman is trying to rape Esther. Haman will be hung on the very gallows that have been prepared for Mordechai, who will be given his enemy's position as the king's adviser. The Jewish people will be saved.

The secondary plots, following the tradition of Spanish comedy, also have happy endings.

THE RELATIONSHIP TO THE BIBLICAL SOURCE AND THE PERSONALITIES OF THE CHARACTERS

In order to assess Godínez' attitude to the biblical source, the first question that arises is: on which version of the Book of Esther did he base himself? The Greek translation, known as the Septuagint, contains a series of additional chapters that are non-existent in Hebrew. It is usually thought that one of the purposes of these additions was to introduce into the story certain religious elements that the original lacks and that the translators considered necessary, such as the name of God, which is absent from the Hebrew text, and also some prayers. There are six additional sections: the first relates a dream of Mordechai that predicts the victory of the Jews; the second presents the text of the edict decreeing annihilation of the Jews; the third consists of prayers by Esther and Mordechai, and in the fourth Esther puts her life in serious jeopardy by appearing before Ahasuerus without having been invited. The fifth addition expands on the king's edict annulling the previous order, and the last one provides an interpretation of Mordechai's dream, concluding that God has not forgotten his people and will always protect and love them.

In the Latin translation, the Vulgate, the additions were taken out and presented as an appendix to the sacred text. Evidently Godínez did take these additions into account when he created his own version of the Book of Esther, and elaborated on them to develop further the various personalities and dimensions that he gives his characters. In his play it is Esther, rather than Mordechai, who dreams the prophetic dream. She is the one who, impelled by a deep sense of responsibility towards her own people, decides on her own initiative to go to the king's palace, surprising her uncle with her decision. Esther, full of courage and positive initiative, is the central character in the story; she constitutes a bridge both between the Jewish people and the Gentile authorities, and between the Old and New Testaments.

Understanding this duality in Esther's configuration can help us to understand the *converso's* way of thinking and to some extent can illuminate how the author confronted the existential duality of his class.

The dominant trait in the character of Mordechai is his loyalty to both God and the king, whom he considers to be the representative of the Creator on earth. He is only willing to pay homage to kings. Bowing down before Haman implies for the Jew an act of idolatry, something which is forbidden by his faith.

Godínez' Ahasuerus is essentially different from the biblical figure on which he is based. In the Book of Esther he appears to be a little dull, grotesque, a person who acts in arbitrary fashion. Here, in contrast, his reactions are rational and just. The development of all the events in the comedy generally responds to the principle of cause and effect. The repudiation of Vashti, for example, is not based, as in the original, on her having refused to display her beauty at the royal banquet; here she is presented as an incorrigible rebel. Another example of cause and effect is that the men who plot against the king are actually princes from Vashti's family, in other words they have a motive for conspiring. The biblical source mentions two eunuchs, without providing any reason for their action. In a general sense, the king's personality is constructed according to the conventions of the age: it is a positive portrait of a supreme ruler.

Haman, the king's personal counsellor, is in the grip of pride. This fervour hastens him on to his tragic end. In addition to this character trait the play insinuates, and here it coincides with the biblical original, that Haman bears an atavistic burden: he is of the dynasty of Amalek, Israel's eternal enemy. His hatred for the Jews is inherent to his being, and this inexorably condemns him in advance.

Although one does find in this comedy some situations and characters that are comic, they are not based on the farcical and hyperbolic elements found in the original source. On the contrary, it is precisely those elements that disappear in the version by Godínez, who does not refer at all to Ahasuerus' somewhat ridiculous behaviour, nor to the banquet that lasted for six months, nor to the letters sent to all the men of the kingdom exhorting them to rule over their wives. Nor does Godínez mention the massacre of 75,000 Persians in a single day. The comic characters in the play are the fruit of the author's imagination, responding to the theatrical conventions of the age rather than being connected to the original nucleus of the plot. Godínez develops this original plot in all seriousness, to the point that it acquires the dimensions of a transcendental theological message.

Among the themes of *La reina Ester*, honour — the central motif in Golden Age Spanish drama[6] — comes into focus through various perspectives, each corresponding to one of the main characters. In the case of the king, honour is an attribute inherent to his being king. Haman demands for himself the outward signs of honour, and requires that homage be rendered to him. Mordechai will revere only his God and his king, and because of his loyalty to these principles he himself becomes the

6. On the motif of honour in the Golden Age, see, for example, A. Rastro, 'Algunas observaciones acerca del concepto del honor en los siglos XVI–XVII', *Revista de filología española*, 3 (1916), 44–73; J. A. Maravall, *Poder, honor y elites en el siglo XVII* (Madrid, 1984).

object of honour. But the greatest honour of all falls to Esther, who is shown in the play as a forerunner of the Virgin Mary — from her seed shall the Messiah be born. This reveals an aspect essential to Godínez' version: his prophetic theological vision.

A PROPHETIC THEOLOGICAL VISION

Our *converso* playwright turns to medieval typology in which Esther, together with Judith and other female biblical figures, represents a prefigurement of the Virgin Mary. Godínez weaves a plot that creates a link between the two figures, and the examples are explicit: in the first act, when Egeo comes to Esther and announces that she will become the king's wife, the scene is alluding to the Annunciation to Mary. In Act II, when the enigmatic text that so preoccupies Ahasuerus is read, and Haman interprets it as an announcement of the Messiah, Haman himself specifically states that the Jews are waiting for a king whom they will call Messiah, who will come from Nazareth and will be the son of a Jewess. In other words, Haman's explanation is the correct one according to the beliefs of the public addressed by Godínez. But Haman, at the same time, is the evil man who attempts to prevent the prophecy from becoming a reality by trying to exterminate the Jews.

The play's prophetic tone is confirmed in the third act. The overwhelming presence of characters related to Messianic prophecies emphasizes the linkage between Esther and the Virgin, affirming that Mary would be free of original sin. It should be noted that although the latter idea was already very popular in Spain, it had not yet been adopted as a dogma of the Catholic Church in the year when Godínez wrote his play.[6] The dogma of the Immaculate Conception was only proclaimed in the year 1854 by Pope Pius IX.

These examples convincingly attest to the prophetic spirit that pervades the work. In this spirit, as previously indicated, Esther serves as a bridge unifying the Old and New Testaments, a link between the people of Israel and Christianity.

CONCLUSION

Thus, the analysis of the play in relation to biblical history shows that Felipe Godínez strayed from the source in various ways. I will refer to two of these digressions.

The first expresses the author's wishes to bring the plot's subject more into line with the dramatic conventions of his age, and with the taste and understanding of the Golden Age audience. This is shown by:

i) The addition of numerous comic characters, who are the protagonists of the secondary plots created to amuse the audience;
ii) The change in class of some original characters: the eunuch servant, who showed a preference for Esther, and the two other eunuchs who conspired against the king, were converted into princes by virtue of Godinez' pen. Similarly, the king's harem receives no mention. Thus, the play's sexual politics are adapted to conventions appropriate to the author's culture;
iii) The names of the months correspond to the Gregorian rather than to the Hebrew calendar, used in the original;
iv) The characters who are created by Godínez bear names current in the theatre of his time, and some characters from the original are even rechristened in the play. Thus, for example, Hagai the eunuch becomes Egeo the prince, and Bagtán and Teres become Seleuco and Leoncio.

The second deviation is aimed at bringing the play into line with the New Testament, by creating genealogical links between Queen Esther and the Virgin Mary, which makes Christ a descendant of Esther. This reveals what I have termed the transcendental theological message of the work: the salvation of the Jewish people will ensure the coming of the Messiah.

It is possible, in my opinion, to see in this departure from the original the signs of an extraordinary effort to emphasize the Jewish roots of Christianity. The social and existential struggle of the New Christians finds in this dramatic work its own particular expression: Felipe Godínez is trying, in an indirect way, to present the condition of the *conversos* within Christianity as the chosen preference, the children of the people of the Messiah becoming the children of the faith of the Messiah. But Godínez' own fate shows us to what extent he failed in his aim and how pathetic this effort turned out to be. The Inquisition pursued him precisely because of his desperate efforts to lay a basis for his belonging to Christianity.

Mudéjares and *Repobladores* in the Kingdom of Granada (1485–1501)

MIGUEL-ANGEL LADERO QUESADA

The conquest of the Muslim Kingdom of Granada and its union to the Crown of Castile took ten years, from 1482 to 1491.[1] As the Christians gained control over the territory, the need to establish forms of *convivencia* (lit. living together) with the Muslim population which remained after the war and to organize new settlement and administration became apparent. For this purpose the pacts or *capitulaciones* (capitulations) agreed upon between conquered and conqueror were taken as the points of departure. Until the years 1500–1502, the Christian settlers, mostly of Andalusian origin, who were attracted to Granada by the allotment of lands and other real estate, lived together with the Muslim *mudéjares* of Granada.[2] From the beginning of the sixteenth century, baptized Muslims and their descendants, known as *moriscos*, remained a unique cultural community, distinct from that formed by the *cristianos viejos* (Old Christian) settlers and their descendants.

Finally, the rebellion of the *moriscos* from 1569 to 1571 provoked their expulsion from the territory and put an end to Islamic Granada, which had survived through them.[3] The aim of this article is to present a general outline of what we know about *convivencia* between *mudéjares* and *repobladores*, or settlers, during the initial period of the organization of the Kingdom of Granada, from its integration into Castile to the beginning of the sixteenth century.

This article was translated by Alan Paris

1. J. de M. Carriazo Arroquia, 'Historia de la guerra de Granada', in R. Menéndez Pidal (ed.), *Historia de España* (Madrid, 1968); M.A. Ladero Quesada, *Castilla y la conquista del reino de Granada* (Valladolid, 1967; Granada, 1987).
2. M.A. Ladero Quesada, *Granada después de la conquista. Repobladores y mudéjares* (Granada, 1988). This is a re-edition of works published in 1968 and 1969. It provides documentary information concerning most topics examined in this article.
3. M.A. Ladero Quesada, *Granada, Historia de un país islámico* (1232–1571) (Madrid, 1989), provides a general study and ample bibliography.

I. THE CAPITULATIONS AS A STARTING POINT

The capitulations regulated *convivencia* in Granada until the end of the fifteenth century. A distinction must be made between different types of capitulations applied following each conquest or military campaign. The first is unconditional surrender, which implies captivity, loss of all property, and, sometimes, exemplary punishments. There were few such cases, but one, rather notorious, occurred in Málaga where 11–15,000 persons were taken captive. Through the sale of most of these individuals and their property, the royal exchequer obtained 56,000,000 *maravedíes* (150,000 ducats).[4] The remaining types of capitulations have in common respect for the personal freedom of the Muslims, their social structure, legal and religious orders, in short, all aspects of Islamic culture. *Mudéjar* communities were only subject to a new political and military power. However, among these common features there was a spectrum ranging from lesser to greater generosity in other aspects, particularly economic ones, which enable us to speak of three types of capitulation: one, applied from 1484 to 1487 in the western part of the kingdom; another, in 1488 and 1489 in the eastern part, and a third, exclusively in Granada and the Alpujarra in 1491.[5]

The first type includes a general obligation to abandon real estate without any compensation for those among the defeated who put up armed resistance prior to capitulation. Only Muslims in villages and open settlements (*lugares abiertos*), occupied without resistance, retained their lands. In the other two types, this was not the case. The Muslims kept all of their real estate. However, the revolt in Baza, Gaudix, and Almería in the summer of 1490 deprived the residents in these areas of that right. This distinction is important because only in places where real estate had been abandoned was a large-scale repopulation possible. Elsewhere, it would be necessary to purchase the land.

The question of the captives was another issue where the criteria of the victors progressively became less strict. Prior to 1489, the return of Christian captives was demanded. From that year, they were ransomed; moreover, it was also promised that any Muslim captive who fled to Baza, Guadix, or Almería would regain his liberty. Thus, in 1491, inhabitants from Granada and the Alpujarra held captive in Castile were ransomed.

Other important concessions were various amnesties for violations of the codes of war, the inviolability of one's home, and respect for Muslims who had formerly been Christians and who could not be forced 'against the

4. M.A. Ladero Quesada, 'La esclavitud por guerra a finales del siglo XV: el caso de Málaga', *Hispania*, 105 (1967), 63–88.
5. Ladero Quesada, *Castilla y la conquista*, Pt. 2, 'Las capitulaciones, base de una nueva época', pp. 71–100, and *Granada después de la conquista*, where the texts of the main capitulations are presented.

law' to return to Christianity. The people of Granada to whom the second and third types of capitulations were applied kept their horses and side-arms, as did those subject to the first type, except when they capitulated after having resisted. Freedom to engage in commerce, the right not to engage in any type of work without receiving a fair salary, and the continuity of the fiscal or tributary administration of the previous period were recognized. Religious and judicial authorities kept their positions as did the *alguaciles* and other indispensable local administrators of *mudéjar* communities.

The capitulations served as the basis for a situation which did not last long, though it might have lasted. It was short-lived due to contemporary political tendencies, which highlighted factors of non-assimilation over those of *convivencia* between communities of different culture and religion and constantly created the temptation to use political power for resolving problems at the expense of the weaker group. It was short-lived due to pressure exerted by the conqueror, generally Andalusian, over the conquered Granadan, clearly manifested in the conqueror's eagerness to occupy lands, to revindicate preferential, though not unjust, treatment by the Crown. The capitulations showed, all told, the victor's position of dominance over an annoying enemy of many centuries. The Granadans, for their part, regarded the capitulations as an unavoidable evil, with an insincerity derived from their inability to limit themselves to acceptance of the new state of affairs. Although the conquest was restricted to the military sphere, the cultural core and social and religious organization of the dense Muslim population remained. There was the danger of an almost inevitable reaction, produced, in part, by the pressure of events, but also in part because, for the inhabitants of Granada, their new situation turned out to be profoundly unacceptable, as a similar situation had been over two centuries earlier to the Muslims who had risen against Alfonso X in 1264 throughout Andalusia and Murcia. None the less, in Granada at the end of the fifteenth century, there were several years during which the possibility of *convivencia* again existed.

II. THE GRANADA *MUDÉJARES*

Mudéjar, according to the correct etymology of the word, means the Muslim who lives under a non-Muslim power in a non-Islamic land. In medieval Christian Spain, the presence of such Muslims was common, in certain respects the counterpart of the Christian *mozárabes* who had lived in Al Andalus — the Muslim part of the Iberian Peninsula — between the eighth and twelfth centuries. The *mudéjares* were most numerous in the kingdoms of Aragon and Valencia, but were few under the Crown of Castile during the last centuries of the Middle Ages. The conquest of

Granada left a large number of Muslims in the former emirate, so many, in fact, that Granadans were rarely called *mudéjares*, though we shall do so in this article in order to differentiate between them and the approximately 25,000 Muslims who lived in the rest of the Crown of Castile, whose total population was just over 4 million.[6]

Two possibilities were open to the Granadan Muslim: emigration or remaining subject to the capitulations. Opportunities for emigration were many and were limited only in certain settlements awarded to nobles as seigneuries. There was generally a one to three year period during which ships provided free transportation to North Africa to those who wished to emigrate there. After that, a price was set at three Granada *doblas* of gold and a tenth of all movable possessions, except for emigrants included in the capitulation of the city of Granada who had to pay only one *dobla*. The possessions one might take were limited by the laws of Castile, which, following those in effect throughout Europe, prohibited removal from the kingdom to Muslim lands of gold, silver, iron and steel, arms, horses, and other materials considered precious or strategic, without special permission. Such permission was received by El Zagal, who emigrated in 1490, and by Muhammad XII, Boabdil, who moved to the Emirate of Tlemcen in October 1493. Among the items they took with them were 90,000 *doblas* of gold in indemnities paid by Castile. The transport of side-arms and horses by emigrants was another honorary concession sometimes authorized by the kings.

Many Muslims, particularly the notables of the land, were encouraged to emigrate by the Islamic precept against living in a land of infidels: the importance of this precept in explaining the decay of many of the cultural values of Granadan Islam and the gravity of events following the fall of Granada have not been properly assessed. We do not know how many people emigrated, for the data are very incomplete: 2,000 from Almería in 1491, who had already to pay the embarkation fee (which had not been the case in 1490); 2,000 Granadans in the autumn of 1492; another 1,370 prior to June 1493; more than 7,000 in the autumn, at the time of Boabdil's emigration ... these may represent but a few known numbers among many sailings which left no trace in surviving records.[7] There is other evidence: in the summer of 1493, according to the royal secretary Hernando de Zafra, only craftsmen and labourers remained in Granada. This is in accord with information from a few years later. If we add the losses in the war just past,

6. M.A. Ladero Quesada, 'Los mudéjares de Castilla en la baja Edad Media', in *Los mudéjares de Castilla y otros estudios de historia medieval andaluza* (Granada, 1989), pp. 11–132.
7. M Gaspar y Remiro, 'Emigración de moros granadinos allende' and 'Partida de Boabdil allende con su familia y principales servidores', *Revista del Centro de Estudios Históricos de Granada y su Reino* (1912), 1–13 and 57–111.

the contemporaneous emigrations and those just after the war, above all in the western part of the kingdom for which there is no information, and the emigrations during the tragic years of 1500 to 1502, it is evident that Granada lost more than half, perhaps 60 per cent of its former population and that the number of *moriscos* would not have exceeded 125,000 in the second decade of the sixteenth century.

Other considerations are the potential effect of such phenomena as the earthquakes of 1494 and 1504 or that of 1522, which devastated Almería, the grain shortages of 1503 to 1506 and 1521 to 1522, the plague of 1507 which affected all of Castile, and the measures taken against *morisco* crypto-Islamism in 1511. Bernard Vincent estimates that there were clandestine emigrations between 1502 and 1516 from at least 38 locations on the Alpujarra coast;[8] López de Coca and Gozalbes emphasize the role of the 'North African connection' and of the corsair activity organized from Tetuán and Xauen by leaders such as Al Manzarī and Ibn Rashīd which, although started in the *mudéjar* period, had greater effect during the first years of the *morisco* era, after 1502.[9] On the other hand, not all the emigrants withstood being uprooted. In 1489, for example, some emigrants from Vera who had gone to Orán returned to Spain via the port of Valencia.

The Muslim who did not leave for Africa could, if he so desired, settle in another part of Castile. Few wished to do so. For example, some Rondans of the lineage of the Banu al-Hakim went, in 1485, to Seville, from where their remote ancestors had originated, but shortly afterwards emigrated to Africa. The large majority of those who stayed behind remained in their places of origin, hoping to maintain the status quo gained from the capitulations. This cannot have been at all pleasant for them, though it must be borne in mind that the Crown and the noble owners of seigneuries and large estates favoured their remaining as they constituted an irreplaceable source of manpower for agriculture, particularly in the mountainous zones and in areas devoted to easily marketable crops and sericulture.

8. B. Vincent, *Andalucía en la Edad Moderna. Economía y sociedad* (Granada, 1985), and id., *Minorías y marginados en la España del siglo XVI* (Granada, 1987). These two books inccorporate many of the author's earlier studies concerning *morisco* Granada. See also 'Economía y sociedad en el reino de Granada en el siglo XVI', in A. Domínguez Ortiz (ed.), *Historia de Andalucía* (Barcelona, 1981), pp. 161–223; A. Domínguez Ortiz and B. Vincent, *Historia de los moriscos. Vida y tragedia de una minoría*, 2nd edn. (Madrid, 1984).
9. G. Gozalbes Busto, *Al-Mandari, el granadino, fundador de Tetuán* (Granada, 1988); J.E. López de Coca Castañer, 'Granada y el Magreb: la emigración andalusí (1485–1516)', in *Relaciones de la península ibérica con el Magreb (siglos XIII–XVI)* (Madrid, 1988), pp. 409–51, and his collected works in *El reino de Granada en la época de los Reyes Católicos. Repoblación, comercio, frontera* (Granada, 1989), 2 vols.

Among those remaining was a minority of *mudéjares* who collaborated with the Castilian authorities, either to facilitate *convivencia* or for their own interests. It is noteworthy, however, that practically none of these renounced Islam, though they permitted their descendents to do so. One example was Yahyā al-Najjar, defender of Baza, a member of the Granadan Royal family and ancestor of the marquis of Campotéjar,[10] another, the *alfaquí mayor* of Granada, Muhammad 'el Pequeñi', and the Granadan *alfaquíes* (though of Castilian origin) Yūsuf 'el Mudéjar' and Yūsuf de Mora, originally from Toledo. In Málaga, there was Alí Dordux, who negotiated the liberation or rescue of his fellow citizens and was accused by some of taking advantage of it,[11] and who attempted to disinherit his son, Fernando de Málaga, for having become a Christian. In Vélez Rubio and Vélez Blanco there were the brothers Muḥammad and 'Ali 'Abduladín, the first and *alguacil mayor* and the second a kadi. In various zones inhabited by *mudéjares*, local *alguaciles* such as Zulema el Baho in Almería, 'Ali Alhaje Farax in Baza, Hamete Uleylas in Guadix, and others, were appointed by the kings, certainly at the request of the local communities. Other positions, also local, were shared by Muslims and Christians. This was the case with some members of the first *'ayuntamiento'* of Granada, formed in 1492, or with the office of *alcalde de campo* of the city of Granada where one was Christian and the other Muslim. In 1495 the latter was Yaḥyā Fisteli.

The kings tried to attract *alguaciles*, *alfaquíes*, and local notables to their policies, appointing to these positions known personalities of wealth *bien abundantes* according to the term then in use, with the ability to lead the community. During the process of conversion in 1500 and 1501, many such individuals received new favours and positions after baptism. But their co-operation was not always sincere, nor did their co-religionists always accept it. In one case (Atajate, 1490), the *alguacil* was accused of being 'a Christian, having betrayed the Moors'. Others, such as 'Ali 'Abduladín', were killed during the revolts of 1500.

The capitulations were honoured, though always in the sense more

10. *Origen de la Casa de Granada, señores de Campotéjar* (Royal Academy of History), Madrid, 9/195. see also: M.A. Ladero Quesada, *Granada después de la conquista*, pp. 273–94, in *Los mudéjares de Castilla en tiempos de Isabel I* (Valladolid, 1969); A. Galáan Sánchez, 'Poder cristiano y colaboracionismo mudéjar en el reino de Granada (1485–1501)', in J.E. López de Coca (ed.), *Estudios sobre Málaga y el Reino de Granada en el V Centenario de la Conquista* (Málaga, 1988), pp. 271–89; id., *Los mudéjares del reino de Granada* (unpublished Ph.D. dissertation, University of Málaga); M. Espinar and J. Grima, 'Un personaje almeriense en las crónicas musulmanas y cristianas. El infante Cidi Yaḥyā Alnayar (1435?–1506): su papel en la guerra de Granada', *Boletín del Instituto de Estudios Almerienses*, 7 (1987), 57–83.

11. C. Pescador de Hoyo, 'Alí Dordux, un personaje controvertido', *Anuario de Estudios Medievales*, 17 (1987), 491–500.

favourable to the conquerors, for it was considered that that they sometimes left open dangerous possibilities, immediately taken advantage of by some Muslims, or that they limited the progress of Christian colonization. Permits for the possession of arms disappeared for various reasons. In the city of Granada, hunger ensured acceptance of this change in the capitulation in return for flour at the beginning of 1492. Real estate was kept, however, and at times this implied either obligatory residence or subsequent loss, as, for example, in Sierra Bermeja; alternatively, the *mudéjares* were permitted to sell though not to purchase, as was the case in the Granadan Vega. In the city of Granada in 1498, there was an agreement to the effect that the *mudéjares* would leave the *medina* to Christian neighbours and go to live in the suburbs of Albaicín and Antequeruela. The following year, Muslims from other parts of the kingdom, particularly the Alpujarra, were prohibited from residing in the city.

Towards 1500, the population movements and Christian colonization had produced a distribution of settlement in which the *mudéjares* were reduced to rural and mountain areas, in general following the scheme proposed by J.E. López de Coca.[12] In the west of the old emirate, the Muslims remained in the mountainous regions of Villaluenga, Havaral, Gaucín, Casares, and Marbella, while the Christian *repobladores* established themselves in the urban settlements and cultivated lands of Ronda, Marbella, Setenil, and El Burgo. The city of Málaga was completely repopulated as well as other nearby coastal sites (Mijas, Benalmádena, and Bezmiliana) and sites in the Valley of Guadalhorce (Alora, Cártama, Alhaurín, Coín, Alozania), while the Muslims remained in the villages and mountains of the Ajarquia and Garbia. A similar situation prevailed in Vélez Málaga, repopulated by Christians but surrounded by a mountainous area of *mudéjar* population up to the Zafarraya pass, which connected with the interior of the country. On the other hand, Loja and Alhama with their 'lands' were completely repopulated, as were the fortified *villas* of La Vega de Granada (Moclín, Illora, Montefrío, Piñar, Colomera, Iznalloz), where, moreover, Santa Fe was founded in 1491. On the other hand, the capitulation of November 1491 permitted many Muslims to remain in the city of Granada as well as its hamlets or nearby villages and farmsteads. It also permitted the Alpujarra to be converted into an authentic 'mountain refuge' populated exclusively by *mudéjares*, while the Christians only established themselves in coastal areas of strategic importance, such as Almuñécar, Salobreña, or Adra. In the eastern zone, the capitulations of 1488 and 1489 and the revolt of summer 1490 left the urban nuclei and their

12. J.E. López de Coca, 'El reino de Granada', in J.A. García de Cortázar (ed.), *Organización social del espacio en la España medieval. La Corona de Castilla en los siglos VIII al XV* (Barcelona, 1985), pp. 195–240.

districts in the hands of Christian 'repopulation', while Muslims remained in rural zones: Almería, Baza, Gaudix, Vera, Mojácar, and Fiñana in the Cenete. However, the rest of the territory continued to be inhabited by *mudéjares* in open areas while the fortifications had Christian garrisons, either of the king or of the nobleman to whom jurisdiction over that place had been granted.

The revolts of 1500–1501 were not preceded by a state of growing tension, nor by specific violations of the capitulations, for even the special levies on the *mudéjares* in 1495 and 1499, equivalent in each case to 16,000 *doblas* of gold, were familiar from previous practices of the exchequer of the Naṣrid emirs,[13] and found justification in the fact that the Granadan Muslims did not participate in the payment of the levies (*servicios*) and extraordinary direct contributions which were collected in the rest of the area under the Crown of Castile. None the less, the ordinary taxation was already very burdensome and one of the stimuli which led to capitulation and peace was, precisely, the hope that it would be alleviated. Thus, it is not surprising that the levies produced hardship and a certain amount of localized resistance.

The essential problem was the growing presence of the Christian *repobladores* and the cultural and ideological abyss that loomed between victors and vanquished. The lack of sympathy for the newly established order and fears concerning growing pressures were cause for sporadic and unrelated cases of unrest. Already in 1492 there were *mudéjar* bandits or *monfíes* in several mountainous regions.[14] The incursions of North African pirates on the coast were rather frequent. Such incursions cannot be said to have aroused much *mudéjar* opposition, although they themselves had to pay for guarding the coast and maintaining watch-towers with a special contribution or *farda*.[15] Among the cases of local violence may be cited, in chronological order, those of autumn 1485 and 1487 in Guacín and other points of the Serranía de Ronda due to the abuses of Christian tax collectors, Comares (1488), the Serranía de Ronda (1488 and 1495), and the revolt in Andarax and areas of Guadix and Baza during the summer of 1490, though the majority of these were still related to the war. In general, it can be said that the *mudéjar* period, prior to 1500, was overtly peaceful.

13. Ladero Quesada, 'Rentas de Granada', in *Granada después de la conquista*, pp. 361–71.
14. B. Vincent, 'Les bandits morisques en Andalousie au XVIe siècle', *Revue d'histoire moderne et contemporaine*, 21 (1974), 389–400 (and in id., *Minorías y marginados*, pp. 173–98).
15. A. Gámir Sandoval, *Organización de la defensa de la costa del Reino de Granada desde su reconquista hasta finales del siglo XVI* (Granada, 1988); Ladero Quesada, 'Defensa de Granada a raíz de la conquista (1492–1501)', in *Granada después de la conquista*, pp. 187–232; J.E. López de Coca, 'Financiación mudéjar del sistema de la vigilancia costera en el reino de Granada (1492–1501)', *Historia, Instituciones, Documentos*, 3 (1975), 397–416.

III. THE NEW ADMINISTRATIVE STRUCTURES OF THE KINGDOM OF GRANADA

The Christians who established themselves in the land brought with them a new military, administrative, juridical, and religious regime for their exclusive use and in order to maintain the conquest. Potentially, this regime could include the *mudéjares* or, at least, apply pressure upon *mudéjar* forms of organization. We now examine its principal characteristics prior to looking at the actual processes of colonization.

Military defence was guaranteed through the maintenance of garrisons and armaments at approximately 100 castles and towers; however their number and the costs of supporting them decreased considerably between 1492 and 1498 with the elimination of those considered unnecessary. There were also defensive armadas in the Sea of Alboran, despite their elevated cost, and a network of coastal defence paid for with the *farda* collected from the *mudéjares*. Territorial defence was completed by quartering some companies of mounted Royal Guards and cavalry of the *Ordenes Militares* at the Alhambra and other fortresses, and giving contracts of military service or *acostamiento* to some of the new settlers. In addition to the above, all settlers, depending on their wealth and the value of the lands which they had received, were obliged to bear arms, as *caballeros* or as *peones*, just as in the rest of Castile. It was even proposed to repopulate the Alhambra with about 100 families in order to lower the cost of maintaining the garrison, as had been done in previous centuries at other urban *alcázares*, but this was not implemented.[16]

This explains why Granada was able to provide considerable armament for the Neapolitan campaign of 1495–96. On the basis of this war, the Castilians also schemed their penetration of the Maghrib, which would assure better defence of the coast. Such campaigns were launched from the Atlantic Andalusian ports. Forays (*cabalgadas*) against the African coast proceeded from the ports of Santa María, Jerez, and Sanlúcar dc Barrameda. The Muslim response varied, depending upon the strength of their forces. Between 1490 and 1510 there was an increase in the danger posed by Tetuán, a 'hamlet' according to the Count of Tendilla, but a very aggressive one, re-established by 'Ali al-Manẓarī who was allied with the ruler of Xauen, 'Ali ben Rashīd. Both acted in complete independence of Fez and the actions of Castile on the Mediterranean coast of North Africa, which began with the taking of Melilla in 1479, nor are they directly related to any acts of piracy, which greatly increased after 1515 with the beginning of the presence of Turkish corsairs in Algiers.

16. Concerning the Castilian system of military organization, see Ladero Quesada, *Castilla y la conquista*, pp. 105–75.

Leading the military array was the *Capitán General* of the Kingdom of Granada and the governor of the Alhambra, Don Iñigo López de Mendoza, Count of Tendilla. His successors, the marquis of Mondéjar, kept the position of *Capitán General* until 1570.[17] He, together with Fray Hernando de Talavera, Hieronymite monk, confessor of Isabella I and first archbishop of Granada, and Hernando de Zafra, royal secretary, undertook the task of reorganization in the Kingdom of Granada. On the administrative and judicial levels, they sent *corregidores* and *pesquisidores* to the most important places: we recall the names of Andrés Calderón, first *corregidor* of Granada, and Juan Alfonso Serrano, *pesquisidor* in the Bishopric of Málaga, for Granada was the political capital while Málaga was the economic centre.

The ecclesiastic organization and restoration were the work of Cardinal Pedro González de Mendoza and of his nephew, the Archbishop of Seville Diego Hurtado de Mendoza, apostolic delegates sent to 'erect and institute all the dignities and benefices' in the new Granadan churches, thanks to the bulls which Pope Innocent VIII had issued in 1486, heeding requests made by the Catholic Kings in an extraordinary embassy led by the Count of Tendilla (Papal Bulls *Provisionis nostrae, Dum ad illam, Orthodoxae fidei*).[18] In addition to the two priests mentioned above, and continuing the task they had started: Hernando de Talavera, Archbishop of Granada, together with his suffragans of Málaga (Pedro de Toledo, also of *judeoconverso* origin), Almería (Juan de Ortega, an *absentista* who was royal chaplain), and Guadix Baza.[19] Rome had declared the Kingdom of Granada subject to *patronato regio*, which greatly increased Talavera's freedom of action, as well as the ecclesiastic subjection of the interests of the monarchic state. The kings actually paid two-thirds, later all, of the ecclesiastic tithe, which did not compensate them financially until the baptism of the *mudéjares* in 1500–1502, for they had to establish important economic endowments for the new churches, charged to the royal treasury: 4,000,000 *maravedíes* per year for the see of Granada, 2,192,000 for that of Málaga, 1,140,000 for that of Almería and an equal amount for that of

17. E. Meneses García, *Correspondencia del conde de Tendilla* (1508–1511) (Madrid, 1972–74), 2 vols; J. Szmolka Clares, *Los inicios de la Granada cristiana a través de la correspondencia del conde de Tendilla (1492–1516)* (Granada, 1975) (unpublished Ph.D. dissertation, University of Granada, 1975); E. Spivakovsky, *Son of the Alhambra. Don Diego Hurtado de Mendoza. 1504–1575* (London, 1970).
18. J. Suberbioloa Martínez, *Real Patronato de Granada. El Arzobispo Talavera, la Iglesia y el Estado Moderno (1486–1516)* (Granada, 1985); A. Garrido Arranda, *Organazición de la Iglesia en el Reino de Granada y su proyección en Indias. Siglo XVI* (Seville, 1979).
19. On Fr. Hernando de Talavera, see A. Fernández de Madrid, *Vida de Fray Fernando de Talavera, primer arzobispo de Granada* (1507) (in *Silva Palentina*, 2nd edn. [Palencia, 1932], Vol. 2, pp. 5–66); F. Márquez Villanueva, *Introducción a la 'Católica Impugnación' de Fray Hernando de Talavera* (Barcelona, 1961).

Guadix, in addition to the cost of maintaining over 400 benefices, or ecclesiastical offices, endowed in the kingdom, restoring churches, frequently through the transformation of old mosques, and endowing them with the objects necessary for worship. The establishment of convents of different religious orders (over 30 in 1504; 74 in 1591)[20] was a further expense. Prominent among these were the Franciscans, Dominicans, Carmelites, Augustinians, Hieronymites, Carthusians, and the Orders of La Merced, La Trinidad, and Santiago. It must not be forgotten that the conquest was regarded as a religious triumph and this, in turn, provided its ideological justification. The kings emphasized this, even in the appellations (*advocación*) given to larger churches in many villages, such as Santa María de la Encarnación, intended to manifest the divinity of Christ, born of Mary, which was incompatible with Islamic belief and with any syncretist concession.

Perhaps the ecclesiastical structuring of the new kingdom was one of the things best and most rapidly achieved, through conviction and because it suited the 'ideology of restoration' which guided the colonization of Granada. None the less, the pretensions of other episcopal sees over Granadan territories were rejected, not without difficulty, because they were incompatible with the new ecclesiastical policy of the monarchy, as was the case with the claims of Seville over Antequera or Toledo over Baza. At the local level, the main problems were, at times, the distribution among parishes of the *bienes habices* of the former mosques and non-compliance of nobles holding seigneuries of the regime of ecclesiastical *patronato* to which they were bound by royal decision.

IV. THE PROCESSES OF 'REPOPULATION'

Through the new channels which the military, administrative and ecclesiastical organization opened, the subjects of the Catholic Kings who arrived at Granada to obtain vacant real estate were able to conserve links with the world in which they had lived before while benefiting, at the time they acquired their rights of residency, from important tax exemptions. A minimum of 35–40,000 persons entered the territory between 1485 and 1498. The majority — perhaps more than half — were Andalusians. There were also many Castilians, people from León, the Extremadura, the Kingdom of Toledo, Murcia in the eastern sector and, as minority groups, immigrants from Valencia, Málaga and Vélez Málaga, Basques from the Cantabrian border, and some Portuguese from various coastal localities.

20. J. Sánchez Herrero, 'Monjes y frailes. Religiousos y religiosas en Andalucía durante la Baja Edad Media', in *III Coloquio de Historia Medieval Andeluza* (Jaén, 1984), pp. 405–56.

For the Andalusians it was much simpler: besides being nearer, a large part of the frontier area, no longer subject to the dangers of war, was repopulated at the same time. The settlers came from territories subject directly to the jurisdiction of the king (*realengo*) as well as from seigneuries. It was prohibited to create difficulties for inhabitants of the latter who desired to emigrate.[21]

The 'repopulation' aimed to exclude, though not always successfully, people without professions or family, individuals of bad reputation and marginal individuals ('onbre revoltoso ni cizañador'), *judeoconversos reconciliados* (cleared by the Inquisition), and, of course, Jews, who were expelled from Castile in 1492 and Muslims from other parts of that kingdom.[22] The projects of distribution were based upon fixed numbers of settlers which were not always met: 400 to 600 *pobladores* in Ronda, 506 in Loja, 600 in Vélez Málaga, 2,000 in Málaga, 800 in Guadix, 600 in Baza, 500 in Almería, 300 in Vera (where only 120 arrived), 80–150 in medium-sized fortified towns and 20–50 in many other minor ones. The frontier zones, previously almost empty, also saw great increases in population: Antequera, for example, which never had more than 200 inhabitants between its conquest in 1410 and 1480, witnessed the settlement of 620 and ended up with over 3,000, including its suburbs, in 1518. Something similar happened in Archidona, which grew from a population of 120 to over 500 in 1511. Likewise Olvera, which rose from 120 to 250 or Huelma, whose 'population register' lists settlements created from 1495 to 1509, and in the frontier localities populated by the city of Jaén from 1508, as part of a project for 800 inhabitants: Campillo de Arenas, Otiñar and Noalejo, among others. These are only a few examples to illustrte the human pressure which determined, whenever possible, the division of lands among numerous individuals in all parts of the former emirate.[23]

21. Ladero Quesada, 'La repoblación del reino de Granada anterior al año 1500', in *Granada después de la conquista*, pp. 3–88; R. Peinado Santaella, 'La repoblacion del reino de Granada. Estado de la cuestión y perspectivas de la investigación', *Actas del Coloquio de la V Asamblea General de la Sociedad Española de Estudios Medievales* (Saragossa, 1991), pp. 273–334.
22. J.E. López de Coca, 'Judíos, judeoconversos y reconciliados en el reino de Granada a raíz de su conquista', in *El Reino de Granada en la época de los Reyes Católicos*, Vol. 1, pp. 153–70; Ladero Quesada, 'Los judíos granadinos al tiempo de su expulsión', in *Granada después de la conquista*, pp. 245–60.
23. For an account of research concerning the distribution, see Ladero Quesada, *Granada. Historia de un país islámico*, pp. 373–80. Among the most important are: F. Bejarano Robles and J. Vallvé Bermejo, *Repartimiento de Comares (1487–1496)* (Barcelona, 1974); J.E. López de Coca Castañer, *La tierra de Málaga a fines del siglo XV* (Granada, 1977); id., 'El repartimiento de Vélez Málaga', *Cuadernos de Historia*, 7 (1977), 357–440; M. Acién Almansa, *Ronda y su Serranía en época de los Reyes Católicos* (Málaga, 1979), 3 vols; F. Alijo Hidalgo, *Antequera y su tierra, 1410–1510. Libro de repartimientos* (Málaga, 1983); A. Malpica Cuello, *El concejo de Loja en la baja Edad Media* (Granada, 1981); M. Barrios Aguilera, *Libro de los repartimientos de Loja* (Granada, 1988); E. Lapresa Molina, *Santa Fe. Historia de*

The distribution of lands (*repartimentos*) and other real estate was undertaken in virtue of the principle that the *bona vacantia*, as a result of the war and the capitulations, were property of the Crown which, therefore, controlled all forms of distribution, or all forms of acquisition, when property still owned was involved. There were three ways of obtaining houses and lands. First, purchase, which was the method utilized only when the Muslims abandoned their real estate voluntarily. In order to avoid monopolization, no one was permitted to purchase for a value of over 200,000 *maravedíes*. Secondly, royal favours, which were widely utilized in order to provide compensation for a multitude of services performed in war, as well as, on other occasions, division of booty. There is a report of around a thousand such favours prior to the end of the century.[24] Thirdly, there was the regulated distribution of lands in all areas where the Muslim population had left as a result of the stipulations of capitulation. The method of distribution differed little from that employed two and a half centuries earlier in the Guadalquivir valley. The Crown named distributors who, accompanied by a scrivener (*escribano*), surveyed and measured the lands being distributed and registered the lots in the names of the settlers, taking into consideration the number of inhabitants to be provided for and their different social classes. Málaga, Vélez Málaga, Ronda, Marbella, Loja, Alhama, the *villas* of Vega, Almería and Guadix, Baza and Vera, Almuñécar and Salobreña thus found new life, as did other *villas* and smaller settlements. The capital Granada, where there was no redistribution of lands, quickly gained an important nucleus of Christian inhabitants. Finally, the relatively great mobility of the settlers between different places in the kingdom is noteworthy. People moved in order to improve their position, although they lost their prior residency in the place they left for a fixed minimum period of time, generally five years. Logically, the Crown organized settlement in order to guarantee production and territorial defence and was not trying to promote vagrancy.

una ciudad en el siglo XV (Granada, 1979); A. Malpica Cuello, 'La villa de Motril y la repoblación de las costa de Granada (1489–1510)', *Cuadernos de Estudios Medievales*, 10–11 (1982–83), 169–206; M. Espinar Moreno and J. Martínez Ruiz, *Los Ogíjares. Estructura económica. Toponimia. Onomástica. Según el libro de Habices de 1547–1548* (Granada, 1983); C. Asenjo Sedano, *Guadix. La ciudad musulmana del siglo XV y su transformación en la ciudad neocristiana del siglo XVI* (Granada, 1983); J. Albarracín Navarro, *et al.*, *El marquesado del Cenete: historia, toponimia y onomástica, según documentos árabes inéditos* (Granada, 1986), 2 vols; C. Segura Graíño, *El Libro del Repartimiento de Almería* (Madrid, 1982); J.A. Grima Cervantes, *La tierra de Mojácar. Desde su conquista hasta la conversión de los mudéjares* (Mojácar, 1987); T. Quesada Quesada, *El Libro de Vecindades de Huelma* (Granada, 1989).

24. Ladero Quesada, 'Mercedes reales en Granada anteriores al año 1500', in *Granada después de la conquista*, pp. 89–185.

On the other hand, the settlement processes tended to reproduce social hierarchies extant in other parts under the Crown of Castile. In all cases of distribution, the distinction between *caballeros* and *peones*, who received, respectively, double lots and simple lots for houses, lands for cultivating grain, orchard, vinyard, olives, et cetera, was maintained. Triple lots were given to residents who were former members of the *Guardas Reales*. The endowment of the churches always assumed great importance in each distribution, as did the royal favours, which sometimes put the success of repopulation in jeopardy. The distortions caused by the distributions were frequently corrected in changes carried out in following years; however, in the end, in all fortified towns there was one or were several main landowners, for various reasons: Alvaro de Luna and Fernán Pérez del Pulgar in Loja, Gutierre de Cárdenas in Ronda and Almería, Enrique Enríquez in Baza, Diego López de Ayala in Guadix, Diego López de Haro in Mojácar, or the secretary Hernando de Zafra in various localities. But significantly, the jurisdictional seigneuries were reduced, as we shall see, to zones inhabited by *mudéjares*, for here the political interests of the Crown and popular exigencies coincided. The demographic growth, which not even a long and bloody war could stop, can be shown to have provoked a 'hunger for land'. The victory in the war created the means to satisfy this hunger. This was the impulse which dominated the conquest of Granada and the initial predominance of middle and small colonizing landholders in many territories, sometimes reinforced by redistribution at the beginning of the sixteenth century (Loja, redistribution of *rozas* in 1506–8), and with the presence of minor contingents of craftsmen and merchants who were given much smaller lots, due to the nature of their occupations.

The first colonizing impulse was accompanied, almost always, by tax exemption privileges, varying by area and which, at times, were renewed or modified in the last decade of the century.[25] These were intended to stimulate immigration and make social structures, whose basic traits, as already indicated, did not change in relation to other areas in Castile, more flexible. Moreover, the tax exemptions not only attracted people of modest economic means, but also richer individuals who saw increased possibilities of benefiting from establishing themselves on the new lands. In general, the foundations which the repopulation created led to a new rise of oligarchies and their concentration of land ownership, domination of economic activity and control of local power from approximately 1494 to 1495. This was strengthened during the period of difficulties of the first decades of the sixteenth century, during which many small and medium

25. J.E. López de Coca, 'Privilegios fiscales y repoblación en el reino de Granada (1485–1520)', *Baetica*, 2 (1979), 205–23; J. Suberbiola, 'Fisco, franquicias y problemas en la repoblación de Málaga (siglos XV–XVI)', *Cuadernos de Estudios Medievales*, 2–3 (1974–75), 149–96.

landowners were ruined and during which revolt was imminent: in Málaga in 1516, against the payment of admiral's dues; the anti-seigneurial uprisings of Huéscar against the duke of Alba between 1516 and 1521, encouraged by neighbouring nobles; or even the 'comunero' movement of Baza in 1520, which was such in name only.[26] Still, there was no failure of repopulation, just as the difficulties of the years 1275 to 1285 had not caused the failure of the repopulation of Guadalquivir Andalusia, but there was a decline by comparison to the expansionist phase of the previous years and a hardening of social structures. Many transitory advantages which the settlers had found at the time of their arrival disappeared. That the process of repopulation continued and culminated, above all, in urban nuclei and agricultural areas rich in subsistence crops is shown by the fact that, towards 1530, almost half the total population of the Kingdom of Granada, over 100,000, consisted of Old Christians.

Finally, the changes wnich the conquest and repopulation introduced to the landscape must be taken into consideration, as must the changes to the Granada economic system, although

> ... the repopulation and organization of space are dominated by the will to adapt to what already existed, without turning the material aspects of the past into a *tabula rasa* ... where no resistence is encountered, things are Castilianized; where there is resistance, adaptation is imposed and, at times, continuity of the prior situation.[27]

This helps to explain the permanence of landscapes and material elements of Islamic origin in a land whose population became Christian of Castilian origin. In the cities, suburbs began to grow, streets were improved, plazas planned, et cetera, all of which partially modified the previous landscape. In rural areas, *ejidos*, pasture-grounds and communal grazing areas were created, herding activity was organized and cereal cultivation promoted. Changes were slowly introduced, but systems of irrigation and orchard and fruit tree cultivation continued as in the previous period.[28]

Although there was an enormous transfer of property accompanied by the development of a certain mentality and productive techniques, in the fields of agricultural and craft in Granada, as indicated by sericulture,[29] for

26. F. Bejarano Robles, 'El Almirantazgo de Granada y la rebelión de Málaga en 1516', *Hispania*, 15 (1955), 73–116.
27. López de Coca, 'El Reino de Granada', pp. 195–240.
28. J.A. Luna Díaz, 'Apuntes para el estudio de la agricultura de Granada y su tierra en los primeros años de la conquista', *Chronica Nova*, 10 (1979), 232–48; J.R. Ramos Ibaseta, *Política ganadera de los Reyes Católicos en el obispado de Málaga* (Málaga, 1988); M. Espinar Moreno, 'Aproximación al conocimiento del regadío alpujarreño. Noticias de la taha de Jubiles', in *Sierra Nevada y su entorno* (Granada, 1988), pp. 121–67; M.T. de Diego Velasco, 'Las Ordenanzas de las Aguas de Granada', *En la España Medieval*, 4 (1984), 249–76.
29. F. Bejerano Robles, *La industria de la seda en Málaga durante el siglo XVI* (Madrid,

several decades the Muslim and Castilian lived together and an exchange of learning was possible.[30] The direction of foreign trade was also altered by the new circumstances, transforming the kingdom into a distinct unit. The port of Málaga, the best known example, continued to engage in the Barbary trade under conditions similar to those in Seville and Cádiz, from 1490 under a pontifical bull which excluded possible trade in the well-known prohibited items (*cosas vedadas*). Málaga continued playing the role of an intermediary and link with the North African Mediterranean markets. Up to 1508, trade was conducted with relative freedom, primarily with the ports of the Emirate of Tlemcen; however, the military conquest of *presidios* on that coast, beginning in 1505, forced concentration of Málaga's African commerce there, and imposed a monopolistic regime of contracts with the aim of making the Castilian presence in Mazalquivir, the Peñon de Vélez de la Gomera, Orán, and Bugia more profitable.

Málaga's mercantile role enabled it to continue as port of exit for traditional Granada products: silk, sugar, dried fruits, and anchovies, and as a port of entry for wheat and other goods imported to make up for deficits in local production. But the greatest novelty was the conversion of Málaga into a port for all of central Andalusia, the area of Córdoba and Jaén, through the Córdoba–Antequera–Málaga route. This was the route taken by imports of spices, metals, dyes, side-arms, paper, English woollens, and the exports of Andalusian wheat and wool. For all these reasons, the presence of Genoese merchants in Málaga and their intervention in trade continued to be of great importance, as it was in the Muslim period, though there were merchants of many other origins: Basque shippers, Portuguese interested in the grain and slave trades, natives of Nice, Catalans and Valencians who were engaged in extensive cabotage and had, from 1490, a consul at Málaga, though there was also a consul-general for all the merchants from other cities. Thus, the city maintained and renewed its mercantile status, despite the complete change of its population in 1487.[31]

1951); M. Garzon Pareja, *La industria sedera en España. El arte de la seda de Granada* (Granada, 1972).

30. A good example is the treatise by Gabriel Alonso de Herrera, *Obra de Agricultura*, first published at the beginning of the sixteenth century, Biblioteca de Autores Españoles, 235 (Madrid, 1970).

31. M.T. López Beltrán, *El puerto de Málaga en la transición a los tiempos modernos* (Málaga, 1986); J.E. López de Coca, 'Relaciones Mercantiles entre Granada y Berbería en época de los Reyes Católicos', *Baetica*, 1 (1978), 293–311; id., 'Los genoveses en Málaga durante el reinado de los Reyes Católicos', *Anuario de Estudios Medievales*, 10 (1980), 619–50.

V. COUNCILS OF *REALENGO* (ROYAL PATRIMONY) AND *SEÑORÍOS NOBILIARIOS* (HEREDITARY SEIGNEURIES)

An important aspect of the process of repopulation was the appearance of local administrative frameworks, except in the capital Granada, where this was not completely achieved until after the conversion of the Muslims although the constitutive charter of the city council dates to 20 September 1500. None the less, the kings had organized a Muslim 'council' and authorized statutes in 1492. In the following years an administration of the Christian community was organized, made up of notables who convened, starting in the year 1497 at the latest. This duality ended in 1500 with the formation of a town corporation of twenty-four *regidores* with two *alcaldes ordinarios*, one *alguacil* and twenty *jurados*. The term 'Granada', which until then had included only La Vega, was broadened to include the Alpujarrra and the coast, including Almuñécar, Salobreña, and Motril.[32] In the rest of the kingdom, already from the outset, the new councils were organized along the lines of Seville and Córdoba, maintaining, in general, the same rural terminology that was in use during the Muslim period, and the kings directly named *regidores*, *jurados*, scriveners and other officials from among the *abonados* (rich inhabitants). In 1494 and 1495, the Crown dictated several ordinances concerning the assignment of municipal posts — *regidores, jurados*, et cetera — which have been misnamed the New Statute (*Fuero Nuevo*), and which were in effect in the main municipalities such as Vélez Málaga and Málaga, Loja and Ronda, Baza and Guadix.[33] The number of *regidores* fluctuated between four and six, which was a reduction in their past numbers. These were assigned on the basis of a lottery or balloting for names from among a list of eligibles. The *alcaldes* and the *alguacil* would only act in the absence of the royal *corregidor*. The *jurados* disappeared, replaced by *personeros del común* elected by the taxed residents as a small compensation for the very rapid process of oligarchic domination emerging in those local administrations, but these probably also disappeared during the first decades of the sixteenth century, while many duties of the *regidor* were converted into hereditary positions, and the kings continued directly naming the council scriveners.

In other organizational respects, the Granada councils differed little from those already existing in Castile.[34] Its finances, for example, included

32. These figures in Ladero Quesada, 'La repoblación del reino de Granada', in *Granada después de la conquista*, pp. 47–51, based on documents in the Archivo Municipal of Granada.
33. A. Malpica Cuello, 'El fuero Nuevo en el reino de Granada y el Fuero de Gran Canaria (Notas para el estudio de la administración municipal)', in *III Coloquio de Historia Canario-Americana* (Las Palmas, 1980), Vol. 1, pp. 319–42; J.M. Pérez-Prendes, 'El derecho municipal en el reino de Granada', *Revista de Historia del Derecho* (Granada), Vol. 2/1 (1978), 371–460.
34. J.M. Ruiz Povedano, *El primer gobierno municipal de Málaga* (1489–1495)

private property; rents from urban and rural real estate; fees such as the *peso del concejo, almotacenía* (inspector's fee), *alhóndiga* (granary fee), soap monopoly, *tigual* on fish, and others on the production and commercialization of goods; and a share of royal rents — one quarter of the *hagüela* in Granada — up to the total amount which the Crown had estimated in each case as sufficient when it began the repopulation: 100,000 *maravedíes* for Ronda and Loja, 30,000 for Santa Fe, and 200,000 for Baza. Revenues were always insufficient to cover needs and projects, forcing imposition of extraordinary taxes or assizes, always with royal permission: Málaga collected 400,000 *maravedíes* in 1509 and Granada, which enjoyed an ample territory, collected the rents from numerous properties, public granaries, and urban structures used for commercial purposes, and assorted fees on sales. In addition, possessing the *villa* of Montejícar, the Granada councils went to great lengths to redeem the *villa* of Montefrío, in the hands of the Fernández of Córdoba, masters of Aguilar since 1491, as security for a loan to the kings of 7 million *maravedíes*.[35] The scarcity and weakness of financial resources was common to all the Granada councils.

The concession of seigneuries to members of the Castilian nobility was limited to a few of the many open areas of the country which continued to be inhabited by Muslims. It seems that the kings always reserved the right to resolve law suits at the level of appeals and considered the most important taxes and other attributions as prerogatives of the Crown. With the granting of seigneuries, a power was placed between the Crown and the *mudéjares* which was sufficient to prevent potentially harmful violations of the capitulation and was responsable for control of discontent. This enabled the kings to always adopt the role of mediators and not be seen as directly involved in possible disasters.

Around 1515 some 60 localities and farmsteads (*alquerías*) in the province of Granada, others in Almería, and 28 in Málaga had been ceded to seigneurial jurisdiction; however, only 50 of these were settlements of significant size.[36] Most noteworthy are those which were granted in the north-eastern sector of the kingdom (Castril, Huéscar, Orden, Galera,

(Granada, 1991); id., *Poder y sociedad en Málaga: la formación de la oligarquía ciudadana a fines del siglo XV* (Málaga, 1989); C. Segura Graíño, 'El concejo de Almería. Organización y bienes en su fundación (siglo XV)', *En la España Medieval*, 10 (1987), 445–57.

35. A. Collantes de Terán Sánchez, 'La formación de las haciendas locales en el reino de Granada', in *Estudios sobre Málaga y el Reino de Granada* (Málaga, 1987), pp. 185–97; R. Peinado Santaella, 'Financiación de la guera y señorialización del reino de Granada: Montefrío y la casa de Aguilar', *Baetica*, 4 (1981), 167–92.

36. M. Garzón Pareja, 'Señoríos del reino de Granada', *Boletín de la Real Academia de la Historia*, 74 (1977), 571–636; C. Segura Graíño, 'Realengo y señorío en tierra de Almería en el siglo XV', *En la España Medieval*, 3 (1982), 595–617; M.L. Villalobos Martínez, 'Señoríos de la provincia de Málaga (siglos XV–XVIII)', *En la España Medieval*, 9 (1986), 1299–1333.

Castilléjar, and both Vélezes); in the Almanzora river basin, Sierra de los Filabres and Cenete de Guadix, in the Alpujarra, in the Málaga region the mountains of Bentomiz, Comares, Tolox and Monda, and in the mountains of Ronda and Marbella (Villaluenga, Casares, Gaucín, Montejaque, Benahavís, and Daidín, et cetera). Despite what has been fantasized about them, it seems that none of those seigneuries was very important or profitable, and even less so in comparison to the seigneurial estates which then existed in Guadalquivir Andalusia and other parts of Castile. Their titularies, when they were important noblemen, regarded them as an element of prestige, but also as one more part, and never the main one, of their dominions. Some even sold them several years later to other noblemen with greater interest in increasing their Granada seigneuries. Examples include Rodrigo de Mendoza, marquis of Cenete and builder of the Palace of La Calahorra, and the Murcian Fajardos, marquis and lords of Vélez Blanco and Vélez Rubio.[37] In synthesis, the seigneuries were important in that they provided a framework for part of the rural Muslim population, though, after the revolts of 1500 and 1501, some also received Christian settlers, as happened in the mountains of Villaluenga.[38]

VI. THE END OF *MUDÉJAR* GRANADA (1500–1501)

The disappearance, legally, of Granada as an Islamic society occurred between 1500 and 1502. The events of those three years were unprecedented both in their all-encompassing character and the people who provoked them, as also in their end result, the mass baptism of the Muslims. Prior evangelistic activity had utilized other methods. The Archbishop of Granada, Hernando de Talavera, had brought ideas and solutions which made defence and expansion of the Catholic faith compatible with the privileges granted in support of respect for the religion of the *mudéjares*: prior to 1499, there were no complaints concerning violence against Granadan Islam. Fray Hernando had a clear standard for judging how conversion should be encouraged while respecting cultural habits and beliefs which were not incompatible with the Christian faith. Perhaps his religious sensitivity combined with his experience as an Hieronymite monk and as a person of *judeoconverso* blood made him more sensitive towards those extremes, and more favourably disposed to the

37. J.A. Tapia Garrido, *Vélez Blanco. La villa señorial de los Fajardo* (Almería, 1959); A. Franco Silva, 'La formación del señorío de los Vélez. Sus rentas y propiedades (1492–1540)', *I Coloquio de Historia medieval Andaluza* (Córdoba, 1982), pp. 197–206.
38. M. Acién Almansa, 'Un ejemplo de repoblación señorial: la Serranía de Villaluenga', *I Congreso de Historia de Andalucía, Andalucía Medieval* (Córdoba, 1978), Vol. 2, pp. 357–73.

procedures of faith, *festina lente*, than to those of compulsion, *compelle intrare*, which his colleague the Archbishop of Toledo, Francisco Jiménez de Cisneros would apply towards the end of 1499; this despite the many similarities in the ecclesiastical preparation and political careers of the two men.

Talavera founded the College and Seminary of San Cecilio, serving as a model among the pre-Trentists.[39] Before and after the events of 1499–1501, he attempted evangelical proceedings which utilized the Arabic language and Granadan customs, stimulated translations of biblical and evangelical passages, even utilized *zambras* and other *morisco* dances in the Corpus Christi procession, substituted Arabic for Latin at a certain point in the liturgy of the Mass (*baraka fīcūm* instead of *Dominus vobiscum*, for example) and affirmed 'that they had to adopt our faith and our customs', referring to the Granadan Muslims. Fray Pedro de Alcalá, protégé and confessor of the archbishop, composed the brief *Arte para ligeramente saber la lengua araviga*, published in Granada in 1505, for the use of priests who exercised their ministry among the *moriscos*, in which he made a strenuous attempt at explaining the commandments and deadly sins as an exercise prior to penitence. In 1501 the Archbishop had the *Improbatio Alcorani* of the Dominican Ricoldo de Montecroce translated and edited so that his priests would be armed with convincing theoretical arguments. It is, however, somewhat doubtful that these tracts of doctrinal polemics could have promoted a spirit of conciliation.[40]

It is evident that such a spirit suffered a major setback between 1500 and 1502, when the elements of compulsion and cultural confrontation returned to prominence. In 1499 the Catholic Kings returned to Granada and remained there several months, until the end of November. Their stay ended when the Archbishop of Toledo, Jiménez de Cisneros, arrived in the city with the authority of the Inquisition to investigate the *helches*, or former Christians converted to Islam, who resided at Albaicín and other parts of the kingdom and who continued practising Islam because the city's capitulation did not oblige them to do otherwise: 'It is written and agreed that none dare defame or insult in any way any Christian man or woman who had become a Moor at some time in the past. In such a case, the person defaming or insulting them should be punished by Their Highnesses.'

Their situation was not identical to that of Muslims by birth, but it seems that another of the articles of capitulation could apply to all: 'It is written and agreed that no male or female Moor should be made to become a

39. F. Martín Hernández, *Un seminario español pretridentino. El Real colegio eclesiástico de San Cecilio de Granada (1482–1842)* (Valladolid, 1960).
40. See points of comparison with other medieval cases in B. Z. Kedar, *Crusade and Mission: European Approaches toward the Muslims* (Princeton, 1984).

Christian by force', unless this referred only to *moros* by birth and not those *tornadizos* who were of Christian nationality, according to the concepts of the time.

Thus, it was felt that the capitulation, a royal prerogative from the juridical point of view, could not override the canons of ecclesiastical law concerning apostates, the responsibility of the Inquisition. Although this should have been realized from the beginning, nothing was done about it until the end of 1499 and it was decided to respect the situation and to try the path of evangelical faith which Archbishop Talavera favoured.

After the kings went to Granada, Cisneros began his intervention with several weeks of voluntary *pesquisa y atracción*, of course in his own personal style; he was, to quote the description of the chronicler Pedro Mártir de Anglería, 'hombre de condición armígero y aun desasosegado'. This understandably provoked such unrest and fear among the Muslims that it finally produced a revolt at Albaicín on 18 December 1499. The revolt lasted three days, while the count of Tendilla and Archbishop Talavera tried to calm the protestors and called reinforcement troops which did not arrive to assist because, by the twentieth day, the *mudéjares* surrendered in return for the promise that only those directly responsible would be punished and that amnesty would be given to all those who would accept baptism.[41]

When the monarchs learned of what had happened, they suppported the measures adopted by Tendilla and Talavera. These measures were carried out without delay and at times using methods which were not 'those prescribed by law', the only ones acceptable to the kings. For the *mudéjar* of Albaicín, a decision not to accept baptism under these circumstances was a very difficult one. The pressures mounted, and in the following months the need to separate *moriscos* from Muslims was discussed. In the face of the combined absence of guarantees and threats of punishment, there was a mass conversion of Muslims in the city of Granada and its nearby suburbs. We know the names of over 9,000 of those baptized, but this record is incomplete and there were certainly many more. The clergymen recorded the Muslim name, the new Christian name, as well as the names of the godfathers, other details concerning family and, at times, information concerning the profession of the neophyte. Meanwhile, the last remaining mosques were converted into churches. The Great Mosque of Albaicín was transformed into the Church of El Salvador prior to 11 January 1500, and immediately afterwards into a collegiate church

41. Described in Ladero Quesada, 'Las rebeliones de 1500–1501 y el fin de la Granada mudéjar', in *Granada después de la conquista*, pp. 295–306; J. Meseguer Fernández, 'Fernando de Talavera, Cisneros y la Inquisición en Granada', in *La Inquisición española. Nueva visión, nuevos horizontes* (Madrid, 1980), pp. 371–400.

receiving 202,000 *maravedíes* in annual rent, half of which came from the *bienes habices* of former mosques.[42]

It is not true, therefore, that baptism was forced indiscriminately, but, of course, all those in political control opted to promote widespread baptism under the circumstances. At least two factors influenced those who might have opposed the promotion of baptism: first, their religious mentality forced them to believe, or at least to say, that baptism was the greatest good, though they may have been aware of the insincerity of the converts (the king is quoted as having stated: 'my wish and that of the queen is that these *moros* be baptized; if they do not become Christians, their sons or grandsons will'). Secondly, it was felt that baptism would destroy barriers which prevented acculturation and fusion, barriers maintained with religious arguments.

With time it became clear that both hypotheses were erroneous, but it would be wrong to think today that the kings or even Cisneros did anything other than apply the dominant religious standards of the time to political practice in order to legitimize their actions to others and to themselves. In medieval theological thought since the thirteenth century, as B. Kedar has written, the crusade either preceded the mission, or the mission justified the crusade, though conversion through preaching was generally preferred and forced conversion rejected: 'per prediçació millor que per forsa', as Ramón Lull wrote. Thomas Aquinas was of the opinion that if the infidels hindered him with blasphemies, *persuasiones demoniacas* or open persecution, then war against them was licit. Duns Scotus went a step further: the Christian prince could take children from infidel parents if these did not convert to the Christian faith, because 'their descendants, if correctly educated, would be true believers by the third or fourth generation'. The above phrase (attributed to King Ferdinand) is taken almost literally from this theological position. We even know that in 1501 there was a draft of *pragmatica regia* intended to establish this separation of families, at least among *judeoconversos*, but that there was sufficient political wisdom not to give it legal force, contrary to what had happened shortly beforehand in Portugal.

Let us return now to Granada following this brief theoretical interlude.

The events in Albaicín changed the spirit of the remaining Granadans who, deprived of the most of their former leadership, believed less in the promises of the kings of Castile than in what they thought the latter might eventually do. The letters sent by the queen to the *mudéjares* of the Serranía of Ronda giving assurances that their capitulation would be respected, maintained the peace in that zone for a few months. In the Alpujarra, on

42. M.A. Ladero Quesada, 'Nóminas de conversos granadinos (1499–1500)', in *Los mudéjares de Castilla y otros estudios* (Granada, 1989), pp. 133–68.

the other hand, the people rose in arms in January 1500, despite the fact that the events of Albaicín did not directly affect them. It took three months to subject them and the kings only granted a general pardon in exchange for baptism and a change to the juridical and administrative systems in use in the rest of Castile. Six months after the end of the war of Alpujarra, in October 1500, a new uprising in the far eastern area of the country began: Níjar and Velefique were besieged and taken by force. The end of these uprisings coincided with the start of another uprising in the Ronda and Villaluenga mountains in January 1501. Resistance was concentrated in Sierra Bermeja and was not decisively overcome until May.

The Granadan uprisings may be regarded as having ended then. The kingdom was pacified and the *mudéjars* were either baptized or on the verge of baptism. Many localities established new capitulations with the Crown in order to regulate their transition to this new situation. In July 1501, with the baptisms completed, the kings prohibited any Muslim entry to the Kingdom of Granada, so that they would not be able to disturb the indoctrination of the new *morisco* Christians, and ordered the destruction of all Islamic religious books. In February 1502 the *mudéjares* of the rest of Castile had to chose between baptism and exile, like the Jews in 1492. Nearly all of them were baptized.

One point to consider is the question of the forced nature of the conversions. The case of Albaicín, complex in itself, has resulted in other facts being long hidden. It seems that no one compelled the *mudéjares* of the rest of the country, certainly not to an extreme which would justify their violent attitude. In fact, in many places, particularly in the eastern region, the *mudéjares* agreed to new capitulations which incorporated them into the Castilian legal system and accepted baptism without being required to do so or being subjected to violence. There are reports of 14 capitulations of this type, including the list of baptisms of Vélez Rubio between 19 and 21 March 1501 (329 persons).[43] Why, then, the uprisings in other parts of the former emirate? We are led to imagine a primitive and anarchic reaction of the Muslim community, ill-informed, lacking qualified leaders and counsellors, uncomfortable and fearful in the wake of the situation established on the basis of Castilian military conquest. This collective movement did not require leaders and surely the kings did not want any leadership to form. However, in each uprising, the kings had to act on the basis of criteria dictated by their political understanding. The result was captivity of rebels in Níjar and Velefique, mass conversion for the people of the Alpujarra, emigration to Africa in the case of Sierra

43. J.A. Tapia, 'Primera Iglesia y primeros cristianos de Vélez Rubio', *Revista Velezana*, 2 (1983), 35–54.

Bermeja. These are criteria rooted in the mentality of a European prince of that period and do not imply previous violence against the Muslims who, with their uprisings, had automatically broken the earlier capitulation agreements. These points of view have already been explained by Pedro Mártir of Anglería, sent as ambassador to Cairo shortly afterwards.[44]

Conversion, however, also brought enormous problems. How could threats to the faith be avoided in the *convivencia* between Muslims and *conversos*? The situation of the *mudéjares* and of the Crown would have become progressively more difficult had preservation of the earlier capitulations been attempted. The recent example of the expulsion of the Jews in 1492 to avoid their *convivencia* with the *judeoconversos* carried great weight. However, that situation was not repeated because the *mudéjares* accepted baptism *en masse*. Many accepted it on false pretences, but, was this false acceptance entirely due to the pressures to which they were subjected? It would seem that in many places this was not the case, though this cannot be proved or disproved with certainty. Occidental criteria of genuine acceptance and Islamic criteria of concealment or *taqiyya* were here juxtaposed.

The kings encouraged the conversions, but with the hope that they would be genuine. If the contrary was true, they would have proved to be both a religious and political error. In the first years of the sixteenth century, the wealth of missionaries and the novelty of conversion made it impossible to predict future problems. Baptism was considered the first stage in Christianization, an important stage because it was accompanied by the change in legal status, and law was the framework of society. With baptism and new laws, it could be expected that the Granadans would eventually become Christians and loyal vassals like the rest of the Castilians. This was the reasoning of the political leaders of Castile. However, the history of the Granadan *moriscos* up to 1571, the continuity of their cultural identity and crypto-Islamism, and their rejection by the Old Christians showed that this was not the case.[45]

44. L. García y García de Castro, *Una embajada a los Reyes Católicos a Egipto según la Legatio babilónica y el Opus epistolarum de Pedro Mártir de Angleria* (Valladolid, 1947); J. García Oro, Fray Mauro Hispano O.F.M. (1504–1506): un portavoz del Soldán de Babilonia en Europa', in *Homenaje ... Cabanelas* (Granada, 1987), Vol. 1, pp. 345–54.

45. In addition to the works cited in n. 8, other fundamental books are: P. Longas, *Vida religiosa de los moriscos* (Madrid, 1915); J. Caro Baroja, *Los moriscos del reino de Granada. Ensayo de historia social* (Madrid, 1957); N. Cabrillana Ciézar, *Almería morisca* (Granada, 1982). More complete bibliographies may be found in Ladero Quesada, *Granada, Historia de un país islámico*, pp. 389–97; M.A. de Bunes, *Los moriscos en el pensamiento histórico. Historiografía de un grupo marginado* (Madrid, 1983).

CONCLUSION

The baptism of the *mudéjares*, their conversion and becoming *moriscos* closed the initial period in the relations between the victors and vanquished of the war which ended in 1492: the importance of juridical and institutional changes in religious status resulting from the events of 1500 and 1501 cannot be minimized. However, since there was continuity in what may be called structural aspects of relations between both communities and in the processes of transformation which the territory of Granada experienced, it should be added that there were also many links uniting the period preceding to that following the turn of the century. Although the major processes of settlement were already complete, there was continued settlement activity, in frontier areas and previously abandoned interior areas, for example the hills of Granada north-east of the city.[46] Progress continued on the path towards the formation of a social oligarchy. In the political sphere, movement in the direction of ever greater integration with Andalusia and the rest of Castile proceeded: the installation of a second *Audiencia* or *Chancilleria Real* (Royal Court of Appeals) in Granada in 1505 was a decisive event. The problems posed by the *moriscos* were a continuation of those already experienced with the *mudéjares*, with the difference that assimilation had become possible and was persistently sought: sometimes via legal measures forcing compliance, as occurred in 1511–13 and again in 1526; at other times, practising a kind of moderate *laissez-faire*, the efforts of missionaries and the passage of time were relied upon. Finally, the dangers of the maritime frontier with North African and Mediterranean Islam, of privateering and assaults on the coast, increased considerably as did, at certain times, the banditry of certain *moriscos* in the interior of the territory. Nothing was formally resolved when the baptism of Muslims was completed in mid-1501. From that time on, however, problems no longer belong to the context of the period which lasted until the end of the fifteenth century which we have examined here, a period still dominated by the capitulations which ended the war and by the great settlement or repopulation which followed.

46. R. Peinado Santaella, *La repoblación de la 'tierra' de Granada: Los Montes Orientales (1485–1525)* (Granada, 1989).

Negotiations between the Georgian and Spanish Kings at the End of the Fifteenth Century

ELGUJA KHINTIBIDZE

In 1823 a Russian translation of a letter from Konstantine II, king of Kartli (Georgia), sent in 1495 to Queen Isabella of Spain, was discovered in the Warsaw archives and published.[1] Subsequently, the letter in question became the subject of lively discussion, although it was not known whether it had ever reached its destination and, if so, what the response of the Spanish Court had been.

In 1965 Antonio de la Torre published the documents of the Royal Archives of Aragon, which include the text of the reply of Ferdinand and Isabella to Konstantine.[2] Georgian historians were first informed about the existence of the letter by the late Hungarian scholar Layosz Tardy.[3]

Following research undertaken in Spain in 1984, I translated the reply of the Spanish sovereigns into Georgian and studied it *vis-à-vis* the above-mentioned letter of the Georgian king. Prior to this, the document compiled at the Spanish Royal Court had not been studied either by Spanish, Georgian, or other scholars.

The correspondence between the Georgian and Spanish monarchs, following the expulsion from Spain of the Moors and Jews and the rise of the authority of the Spanish kingdom in the Christian world, reflects the bid of the Georgian king to form a Mediterranean military coalition against the Muslim world with the active participation of Spain, in particular by launching military operations against Turkey.

The second half of the fifteenth century was a momentous period for Georgia. The capture of Constantinople by the Turks in 1453 sealed the fate of the Byzantine Empire. The loss of coreligionist Byzantium proved a mortal blow to Christian Georgia. From that day, Georgia spared no

The final English text of this article was prepared with the kind assistance of A.A. Tchanturia, lecturer in English at Tbilisi University.

1. *Trudy vysochaishe utverzhdennogo volnogo obshchestva rossiskoi literatury* [Works of the ... Society for Russian Literature], Vol. 2 (1823), pp. 213–19.
2. A. de la Torre, *Documentos sobre relaciones internacionales de los Reyes Catolicos*, Vol. V (Barcelona, 1965), pp. 114–15.
3. L. Tardy, *Ungret-Sakartvelos urtiertoba XVI saukuneshi* [Hungarian-Georgian relations in the sixteenth century] (Tbilisi, 1980), pp. 30–31.

effort in her attempt to ally herself with Christian Europe against Turkey. In 1494, Konstantine sent his envoy for talks with the Egyptian Sultan Al-Ashraf Saif al-Dīn Qā'ītbāy. Conceivably, the object of these talks was the creation of an anti-Turkish coalition. It should be noted that the Georgians had traditionally good relations with Mamlūk commanders and rulers, doubtless partly due to the fact that many of the Mamlūks were of Georgian extraction. On his way back, the Georgian envoy met the ambassadors of the Spanish king in Jerusalem, and from them he learned about the expulsion of the Moors from Spain. Konstantine sent his envoy, who had returned from Egypt, by the northern route to Spain, together with the ambassadors of the Spanish king. He sent along a valuable present and a letter to Queen Isabella in which he lauded the queen of Spain, conveying his admiration for her conclusive victory over the Muslims. He expressed his grief at the oppression of Christians in the East, stated his readiness to fight the Muslims, and informed the queen about the dispatch of his envoy to Cairo as a step in this direction, asking her to continue the great battles started in Spain against the Muslims and to arrange an expedition against Constantinople. The Georgian king declared his readiness to march against the Turks the moment he received news from the envoys about a campaign started from Spain. Konstantine stated in his letter that his envoy would personally convey the more confidential details to the queen.

The full text of King Konstantine's letter to Queen Isabella reads:

> The autocratic King Konstantine, the All Powerful, faithful to Christ Our Lord, king of the land of all Iberians, the Jikis and Megrelians, the Abkhazians, Imeretians and Armenians, Kartvelians and Turbanians, to the renowned and highest, most resplendent and serene, autocratic Royal Queen Isabella, appointed by God over the Spanish citadel and all the Christians. Let the same all-gracious God protect and guard the realm of Your Sacred Majesty from grief and malady, keep you in good health of soul and body for many seasons for the fortification of all Christians. For we have heard of the wholesome state of your kingdom through your effort and God's might, and that wherever you marched you have vanquished your enemy, which has caused us boundless joy, and we glorify our almighty God ... Let His grace strengthen you in your dominion and patronage of the entire world, so that with God's power you destroy all other infidel rulers. And you have put to shame that very impious Saracen with his counsellors and have captured his great cities; none of your victories have been won through the aid of others; God granted you the crown and victory over the godless Hagarians for being wise, reigning ably and adorned with wisdom and splendour, doing credit to your position. We have heard of many kings

having united under your position as ruler, which has given us great joy, and we thanked God for granting love and unity to all Christians and orthodox kings. Now we are conscious of and see another sun shining from the East all over the world and a second apostle, God-crowned and great King Constantine who destroyed Greek godlessness and the snare of idols. And your God-fearing Majesty has vanquished and taken captive the godless Hagarian nation, thereby liberating our nation that was first taken captive by them. Where are the glory and honour the Christians enjoyed? Where are the praise and hope? Where is the splendid and glorious city of Constantine? Where are the pious princes that resided there? Where is its joy and gaiety? Where is the 'holy of holies' — God's wisdom? ... Oh, how much grace our nation is deprived of! Where is the kingdom of Trebizond? or the land of the Serbs, where is it? or what has become of other lands? And now we have learnt and are firmly hopeful that our God Jesus Christ, the true God, has granted us victory over the Hagarians. To this end, we have sent the saintly monk and confessor Kir-Nilo [*sic*] to the sultan of Egypt, in Cairo, who received him with great joy. In other respects we have nothing in common with the infidels. On his return to Jerusalem from Egypt, he found there, in Jerusalem, the envoys of Your Majesty, and he arrived together with these ambassadors from whom we learnt the news of Your Majesty's victory, which caused our great joy. We immediately dispatched our confessor and confidant, the same Hieromonk Kir-Nilo to you; you will hear my confidential verbal message from him, for there are numerous enemies on the road. As Your Majesty is aware, we are continuously harassed by the godless Persians who have brought many evils upon us. They demand three things from us: either to pay them an enormous tribute, or to fight together with them, or to become converted to their faith. They are numerous. Of Constantinople, Trebizond, and other Christian countries, we remain alone, being very few in number. Now the entire multitude of the godless nation of the Hagarians, as well as the Turks and the Arabs, having heard your name, are dismayed and afraid. Our time has come now given by God, and we entreat you to set out against Constantinople; do not tarry, but take the Cross of Our Lord Jesus Christ as a shield, and let the Lord grant your desire. And I, with my son, and troops as many as I can muster, shall be ready to follow suit without delay, as soon as Your Majesty sends an ambassador to us. In token of our love for Your Majesty we have sent you small gifts: 19 lengths of Jazd brocade, 35 *zolotniks*[4] of large pearls, and 9 lynx furs. And we, as equal brothers in love for Christ, are writing to Your

4. Old Russian measure or weight, about 4.25 gr.

> Majesty in great hope and pray to God for this to come true by God's will, and for you to reign over all nations; may God never part us for all time. Amen. On the 10th day of March, 1495 CE.[5]

On 30 June of the same year, Ferdinand and Isabella drew up a reply at Burgos. Conforming to the rules of the Spanish Royal Court of the period, the letter is signed by the king's secretary Luys Gonçales. The Catholic Monarchs acknowledge the receipt of King Konstantine's letter and call on him to continue his struggle in defence of the Christian faith. They recount their own exploits in Spain. In Bethica, the most beautiful province of their dominions, with God's help, they have recovered the kingdom of Granada, possessed, as they declared, for 800 years by Arab kings who had come from Libya, and forced the Arabs to cross the Libyan Sea. In addition, they have completely expelled all the Jews from their kingdoms and dominions. The monarchs inform Konstantine about their desire — if God should grant aid — to continue their struggle against the Arabs in Africa

The reply of King Ferdinand and Queen Isabella to the Georgian king is preserved in the Royal Archives of Aragon in Barcelona.[6] A translation of the text published by Antonio de la Torre follows (see n. 2):

> Ferdinand and Isabella, by God's grace, King and Queen of Castile ... cordially greet the most serene Konstantine Pancratios, the best and most esteemed sovereign of the Georgians. We have received your letters sent to us by your serenity through your venerable, religious ambassadors, the Brothers Nilus and Zacharias; through them we at the same time learnt something from their verbal communication on your behalf. In the first place, we are immensely glad to hear about your faith, study and observance of the Catholic religion; and contrary to this, we felt grieved in no small degree at learning from them about the annoyance, anxiety and hazards your serenity is suffering daily. Notwithstanding this, under all circumstances, we shall give great praise to God, pray much to Him that He deliver not only you but the entire Christian population from the infidels, guard you, and grant you victory. We have no doubt that gracious God will effect this if your serenity will in the future too — as today — make every effort to defend yourself from the enemies and will strive in ardent spirit to defend and augment the Catholic faith. And in God's name we ever beg and urge your serenity to do this and not to desist from this undertaking. As for us, up to now we have acted precisely in this way and, God willing, shall

5. Translated by the author from the above-mentioned Russian text of the letter. See I. Tsintsadze, *Maselebi polonetisa da sakartvelos urtiertobis istoriisatvis* [Sources illustrating the history of Polish-Georgian relations] (Tbilisi, 1966), pp. 73–80.
6. Official Files, reg. 3607, fols. 186–7.

continue in the future too. Indeed, in Bethica — one of our most illustrious and beautiful provinces — with God's help, we have retrieved the Kingdom of Granada which was ruled for 800 years by the Arabs who had come from Libya, and we forced the Hagarians across the Libyan Sea. Further, we have completely expelled from our kingdoms and dominions all the Jews who wandered about in great and pernicious multitudes. But if, as we wish and hope for, the great God will aid our aspirations, we shall by no means desist from our intention, for our King Ferdinand has no stronger desire than to cross into Africa with his troops and wage constant warfare for the Catholic faith. Concerning your gifts, which your ambassadors delivered to us on your order, we convey our great thanks for your humanity. Joyous in spirit and contented, we place our power, means, and abilities, our kingdom and dominions at the disposal of your honour and for your benefit, in exaltation and augmentation of the Catholic faith. As to the other matters about which your ambassadors told us, we answered them comprehensively; let your serenity believe kindly what they will tell you. Let the immortal God guard and favour your royal person, your state and your kingdom ...[7]

The value of this correspondence can hardly be overestimated. Its importance goes beyond shedding light on the highly interesting details of Georgian history of the period and the evidence of the fact that in 1495 the aggrandized Spanish kingdom carried on correspondence with remote Georgia.

The facts recorded in the above-quoted letters point to some important conclusions regarding the political situation at the end of the fifteenth century. In the first place, note should be taken of the change in the Mediterranean world's perception of geographical boundaries following the Ottoman conquest of Constantinople and the downfall of the Byzantine Empire: the correspondence points to the search for new routes linking the Christian countries of the East, particularly Georgia, with the West.

The Georgian king sent his ambassadors to distant Spain via the northern route. From western Georgia the envoys had to journey northwards along the Black Sea coast; then, by the Dnieper and its eastern tributary, they reached the Lithuanian-Polish royal court. The choice of this route was not accidental. The Georgians seem to have been well aware that in the Kingdom of Lithuania and Poland they would encounter strong anti-Turkish feeling. This Christian kingdom, which had grown in power from the beginning of the fifteenth century, had, via the Dnieper route,

7. Translated by the author with the kind assistance of the Spanish scholars, the late Antonio Tovar and Antonio Bravo.

acquired access to the Black Sea, from where it traded with the western states via the Genoese colonies in the Crimean peninsula. On this route, Turkey was the principal adversary of the Kingdom of Lithuania and Poland. It was at this royal court, with its anti-Turkish sentiments, that the Georgian embassy stopped over *en route* to Spain. King Konstantine's letter to the Spanish queen seems to have been discussed at the court. Here it was translated into Russian and the translation was entered — along with other documents of business correspondence — in the so-called Metric Book.

The great changes that took place in the political geography of the fifteenth-century Mediterranean world induced Georgian Christian kings to look not only for new routes to the West but also for a new language of communication with the Christian countries of the West. Until then the Georgians had communicated with the world in Greek. King Konstantine's letter to the Spanish queen appears to have been compiled in Georgia in Latin. This conclusion is based on the only extant Russian translation of this letter. In this translation the Spanish queen is referred to as 'established by God over the citadel of Spain' (*bokhom utverzhdaemogo khrada ispaneiskogo*). There can be no doubt that this is a rendering of the Spanish royal title (as used in the reply: *rex et regina Castelle*). The name of the Spanish province Castile came from the Latin *castellum* (Spanish *castilio*), denoting a fortress or castle. This warrants the assumption that the Russian translation of Konstantine's letter, preserved in the Polish archives, was of the Latin original version. It is conceivable that the Russian translator of the letter into Latin would have rendered the Queen's title as 'established by God over the citadel of Spain'.

The momentous developments in the Mediterranean world of the latter half of the fifteenth century (the fall of Constantinople, the ascendancy of the Catholic religion in Spain) raised the question of the eastern Christian countries showing a more tolerant attitude to the symbols of faith. This conclusion also follows logically from the correspondence studied here. When asking for aid from the Catholic Monarchs of Spain, King Konstantine of the Orthodox Christian kingdom of Georgia must have felt the need for a loyal and conciliatory attitude to the Catholic faith. There is an intimation of this in the reply of the Spanish monarchs: 'We are immensely glad to hear', they wrote to the Georgian king, 'about your faith, study and observance of the Catholic religion'. This sentiment is reiterated elsewhere in the letter: '... gracious God will effect this if your serenity will ... strive in ardent spirit to defend and augment the Catholic faith'. The question arises as to whether 'Catholic religion' and 'Catholic faith' in this letter refer generally to the Christian religion, or to Catholic Christianity in particular. It would seem wrong to deny the emphasis of the Catholic faith proper in the letter of the Spanish sovereigns. True, King

Konstantine's stand in this regard is not clear from the extant Russian translation of his letter, but there are other facts indicative of his loyal and tolerant attitude to the Catholic religion. As is known, Konstantine also sent a letter with the embassy to Pope Alexander VI. From Spain Konstantine's ambassador Nilus travelled to Rome where he delivered the Georgian king's letter and gifts to the Pope, congratulating the latter on the Spanish queen's victory over the Moors and asking for aid in his struggle against Turkey. On 10 July 1496, the Pope sent Nilus on his return journey to Georgia with a message for the king. It is obvious from the Pope's letter that in his letter to Alexander, King Konstantine — and also his envoy Nilus in his audience with the Pope — evinced a definite orientation or loyalty to the Catholic religion. In his letter, King Konstantine asked the Pope to send him the Act of the Florentine Council, which the representatives of the Georgian Church had refused to sign (1437–39), and pledged its observance. 'Through your envoy Nilus, you beg us', the Pope writes to Konstantine, 'to send Your Majesty the decision regarding the faith, taken at the Council of Florence at the time of my predecessor of blessed memory the Pope Eugenius IV, and with which the oecumenical Church — whether Latin or Greek — agrees, so that you and your kingdom might become worthy of eternal life by placing trust in it ...'[8]

The correspondence between the Georgian and Spanish monarchs points clearly to the emergence of religious coalitions in the Mediterranean world, facilitated by the developments of the end of the fifteenth century. It should be noted, however, that the coalition that King Konstantine sought to form was not solely and strictly religious, viz. anti-Muslim, in character. Political interest predominated in Konstantine's project, for his aim was not to form an anti-Muslim coalition in general, but an anti-Turkish one in particular. While urging the kings of European Christian countries against the Turkish state, he did not regret having an African Muslim sultan, namely the Mamlūk sultan of Egypt, as a member of the conspiracy. He does not conceal this intention from the Christian kings of Europe: 'To this end, we have sent ... our confessor, the Hieromonk Kir-Nilo, to the sultan of Egypt in Cairo, who received him with great joy. In other respects we have nothing in common with the infidels'. From this point of view, King Konstantine's policy did not constitute an exception; neither did other Georgian kings subject their political interests directly and solely to religious needs. Within a century of Konstantine's attempt to form an anti-Turkish coalition, the question of organizing a crusade against Turkey was again placed more urgently on the agenda of the European states. At the end of the sixteenth century, the initiative for such a coalition

8. M. Tamarashvili, *Istoria katolikobisa kartvelta shoris* [The history of Catholicism among the Georgians] (Tbilisi, 1902), pp. 64–7.

came from Pope Clement VIII and the German emperor Rudolph II. The proposed coalition was hailed with great enthusiasm by Simon I, the king of Georgia, who — like his predecessor — looked hopefully to Spain. King Simon's letters to King Philip II have come down to us. In them Simon informs the Spanish king about the readiness of King Alexander of the second Georgian Christian kingdom, and — what is most important — of Shah 'Abbas of Muslim Persia to fight against Turkey. Thus, in this case, too, the Georgian king was not forming a strictly religious — anti-Muslim — coalition, but primarily a political anti-Turkish alliance.

The correspondence with the Spanish royal court failed to yield any substantial improvement in Georgia's complex political situation. Ferdinand and Isabella sympathized with the Georgian king: 'Joyous in spirit and contented, we place our power, means, and abilities, our kingdom and dominions at the disposal of your honour and for your benefit'. It is hard to say from the letter what lay behind these words, for in those days views on important state affairs were, as a rule, exchanged verbally. King Konstantine wrote to Queen Isabella: 'In addition to the letters, you will hear my confidential message from him [Hieromonk Nilus — E.Kh.]'. The king and queen stated in their reply: 'As for the other matters about which your ambassadors told us, we answered them comprehensively'. What did the Spanish monarchs reply to the envoys? What was the connection between the grave political situation in Georgia and the plans of the Spanish royal court, about which the king and queen informed Konstantine: '... our King Ferdinand has no stronger desire than to cross into Africa with his troops and wage constant warfare for the Catholic faith'? This intention is probably related in some way to the negotiations of King Konstantine with the Mamlūk sultan of Egypt, about which the Georgian king so readily informed Isabella. It seems impracticable at present to try to guess the intentions and assumptions of the Spanish and Georgian monarchs of the period under discussion.

Historical developments proved these projects to have been mere dreams that could not be put into practice. Europe could not busy herself with Georgia; Constantinople was too far from Spain for the latter to embark on such a campaign; and at that time Spain was concentrating her military efforts in North Africa and would not expand her political initiatives to the eastern Mediterranean.

The *Almogataces*: A Historical Perspective

FELIPE MAÍLLO SALGADO

This paper discusses the historical role of the *mogataces* or *almogataces*. First, however, it is necessary to define the term briefly yet precisely, in order to understand its exact meaning before approaching its referent, that is the individual to whom it applied.

This will not be difficult, as I shall begin with an article that I wrote some years ago,[1] fairly precisely clarifying the meaning in Spanish of this term (derived from Arabic), and based on how it is used in various Spanish texts of the modern period.

ETYMOLOGY, MEANING, USE, AND REFERENT OF THE WORD *(AL)MOGATAZ*

The term *almogataz* or *mogataz*, as Dozy has indicated,[2] comes from the verb *gaṭṭasa*, 'to baptize' (second conjugation), the passive participle of which, *mugaṭṭas*, 'baptized', passed into Spanish in two variants, one with and one without the Arabic article being incorporated; thus we have the forms *almogataz* and *mogataz* which refer, as we shall see, to a special category of Moors [North African Muslims — but 'Moors' is the closest term to the Spanish *moros* — *Tr.*] that did not exist in the Iberian Peninsula during the medieval period.[3]

The term *(al)mogataz* alludes to a reality that existed neither in the Iberian Peninsula, nor in medieval times, but did exist in North Africa and in more modern times; the ability to state this fact is due not only to the

This article was translated by Judith Roumani

1. F. Maíllo Salgado, 'Breves notas sobre la historia y el significado de la palabra almogataz', *Studia Zamorensia*, 5 (1985), 471–80.
2. W.H. Engelmann and R. Dozy, *Glossaire des mots espagnols et portugais derivés de l'arabe* (Leiden, 1869), p. 171.
3. Concerning medieval ethnic-religious categories in the Iberian Peninsula, see F. Maíllo Salgado, 'Esbozo tipológico étnico-religioso de los grupos humanos peninsulares en la Edad Media', *Studia Philologica Salmanticensia*, 7–8 (1984), 259–67.

texts in which the Arabism is recorded, but also to other clear indications that will be explained below.

In all the many medieval narrative sources that I have examined, I have never come across the word *(al)mogataz*. It is only in texts written in post-medieval times that this Arabism can be documented and is used; moreover, such works almost invariably issue from or relate to the region of Orán. Just as when I encountered the word for the first time, the work that provided me with further documentation also came from that region. In the latter case, however, there was a difference: the word was explained in the text.

Diego Suárez, a soldier for over 30 years in the city of Orán, and an undoubted authority on the customs and life of the region, used the term in his book written during the last decade of the sixteenth century, explaining its meaning and indicating that it applied to the Moors of the region of Orán who requested protection from the *capitánes generales* (commanders) of the Spanish fort. Such Moors were thus obliged to contribute to the maintenance of the fort with appropriate victuals at agreed-upon prices and 'the Moors who thus desire and request his friendship, which they themselves call *temín*,[4] generally receive all sorts of favours and assistance against their enemies the Turks and other hostile Moors, who thus dislike them; and they often come to blows because of the rumour that they are bad Moors who transgress the precepts of Muḥammad, since they make friends with the Christians, and therefore they are called *almogataces*, which means traitors (*tornadizos*) who desert and convert to another religious law.'[5]

This revealing context makes us aware of a reality different from that of the medieval Iberian Peninsula, not only because it belongs to another place and time, but also because such Moors, in view of the characteristics defining them, cannot be compared or associated with the *mudéjares* [Muslim subjects of a Christian sovereign — *Tr.*] and thus could not be included under the latter term; this was so especially since they were not subject, as were the *mudéjares* of the peninsula, to the tax burdens and compulsory tribute imposed by Christian conquerors occupying Muslim lands in the Iberian Peninsula.[6] In the case of the *almogataces*, Moors and

4. *Tamān› temīn* means in Arabic, among other things, 'protection'. '*Romía* and *temín* are taxes that they pay to HM for protection they receive, in order to remain in the areas near the forts and sow their crops there.' L. Galindo y Vera, *Memorias de la Real Academia de la Historia, Historia, vicisitudes y política tradicional de España respecto a sus posesiones en las costas de Africa* (Madrid, 1884), p. 410.
5. Diego Suárez, *Historia del Maestre último que fué de Montesa*, ed. F. Guillén Robles (Madrid, 1889), p. 66.
6. One must also bear in mind that from the sixteenth century onwards and definitely in the seventeenth century the term *mudéjar* became more specialized in Castilian Spanish and came to denote the Andalusian Arabs who had settled in North Africa. See F. Maíllo Salgado, 'Acerca del uso significado y referente del término mudéjar

Christians would conclude arrangements that were mutually beneficial and that, in principle, and in general, were agreed upon by both parties.

Moreover, the Christians lived in walled towns on the coast and did not occupy the land of peaceful Moors; nevertheless, in their incursions, they would occasionally succeed in controlling some fairly extensive territory, including hostile Moors whom they had subjected by force (since there were sometimes expeditions to more distant territories).

From the above, especially the last point, it emerges that, despite being labelled *tornadizos*, the *almogataces* were not so in the strict sense, as they continued being Muslims. These two traits that distinguish them (co-operation with the Spanish and loyalty to their own religion) enable us to understand the logic behind the term that was used to name them, since there was an unavoidable need for a term that would take into account specificities that had not existed earlier in the framework of the socio-professional and ethnic-religious categories of medieval Spain.[7]

Nevertheless, considering the term's etymology, it must have retained some connection with the meaning of the Arabic word which, as we know, means 'baptized'. We should thus assume that some of the *almogataces* at least were baptized, that is, that they became *tornadizos*.

This is Dozy's opinion, since he assures us that in Arabic *mugaṭṭas* already referred to the Moor who received baptism, the apostate.[8] Eguílaz, for his part, states that *almogataz* means the 'baptized Moor who serves Spain in the garrisons of Africa.'[9]

Leaving aside for the moment the accuracy of these assertions, we realize that each definition is based solely on etymology, rather than on the meaning the Arabism acquires in a particular Spanish context. Both Arabists in this case unwittingly fall into the etymological fallacy, by believing that they can arrive at a word's meaning by uncovering its origin.

Dozy, the Dutch scholar, takes a fourteenth-century text and ascribes the meaning of the word *tagtīs* (baptism) in its language of origin to another isolated word, one which is more or less neutral and not dated (stemming from the same root), and that has been incorporated into Spanish and about which we now know that it was not used in those days, in fact not until two centuries later; while the Spanish philologist,

in *Actas del IV Congreso Internacional Encuentro de las tres Culturas (Toledo 1985)*, ed. C. Carrete Parrondo (Toledo, 1988), pp. 109–11.

7. Studied in Maíllo Salgado, 'Esbozo tipológico'.
8. R. Dozy states, in fact, that *mugaṭṭas* in Arabic denoted 'a Moor who has been baptized a Christian, an apostate', on the basis of a single document found in a non-clarificatory text by al Ḥaṭib in which, moreover, this passive participle form does not appear. See *Supplément aux dictionnaires arabes*, 2nd edn. (Paris–Leiden, 1927), pp. 216–17.
9. See L. Eguílaz y Yanguas, *Glosario etimológico de las palabras españolas de origen oriental* (repr. Madrid, 1974), pp. 232–3.

following in the Dutchman's footsteps, is guided by intuition as much as by etymology, and, perhaps, by some other piece of information that he knew.

Obviously, due to chronological impossibility, we cannot hold against these scholars the fact that they were not aware of the distinction between the diachronic and synchronic aspects of language, established subsequently and today a methodological necessity; but, for that very reason, when these eminent scholars take, as they often do, texts from periods far apart in time, and treat them as examples of the same language, we cannot fully subscribe to the considerations and definitions that they deduce from this, however eminent they were and however great was their knowledge of Arabisms imported into Spanish. If we do not bear this in mind, we will never be able to understand the true nature of the loan words.

And so, leaving aside the earlier definitions, let us look at what the actual context provides. Fortunately, Diego Suárez himself includes additional clarifications, indicating that when the troops of Orán made a seizure, 'and the Moorish spy who sold it to them is aware that the others know about it, he does not dare to remain among them and comes to Orán with the Christians, where they free his wife and children and the rest of his family when the lots are drawn as they sell off the booty; and to the dragoman involved they likewise give his due, a certain number of doubloons for each prisoner. This Moorish spy stays in Orán, where he receives a simple wage; of such Moors there are usually half a dozen, called *almogatazes* which, in their language, means traitors. The hostile Moors, if they take them, will cut such people in pieces, because they always act as spies for the kingdom and many seizures are made with their help.'[10]

We thus see that the *almogataz* was also a kind of spy who received wages and a percentage of the prisoners who had been captured through his efforts. We may thus deduce that the *almogataces* were indispensable and most essential elements in Spanish domination of the territory around Orán. This way of behaving, however, despite what one might think, did not necessarily lead to the abandoning of Islam nor to the conversion to Christianity of these subjects. That this was the case can easily be seen when one reads the texts, which say nothing about their supposed baptism.

It is obvious that these were not baptized Moors, but rather friendly Moors or Moorish spies who for various reasons collaborated or made agreements with the Spanish in Orán; it is nevertheless possible that occasionally someone did get baptized but, interestingly, in the cities of North Africa the convert is never called an *almogataz*, this term being used initially only in the area around Orán.

This confirms that the diachronic analysis of a term or, to put it in another way, the study of its semantic trajectory and of its eventual

10. Suárez, *Historia del Maestre*, p. 8.

changes between two temporal points in a particular language cannot be studied independently of social, economic, and cultural history (nor of the dimensions of time and geography) of the community in which it arose or has been used.

In summary, the word *almogataz*, originating from an Arabic passive participle meaning 'baptized', probably reflects the fact that in principle the Moor who entered the service of the Spaniards and betrayed his own people became Christian, in the sense that he was a renegade. The consequences of this were two-fold: given the individual's characteristics, that is, those of the referent, the word *(al)mogaṭṭas* in Arabic acquired pejorative connotations as a process of semantic depreciation unfolded; in the eyes of the Muslims, such an individual could only be despicable. From the Christian point of view (it is already significant that the terms *tornadizo* or *renegado*, generic terms laden with pejorative shades of meaning are not applied), it is equally plausible that the Arabic term *almogataz* should become more ennobled, semantically speaking, due merely to the fact that it refers to a converted Moor (a rather infrequent occurrence in those parts),[11] and one, moreover, who provided the Christians with opportunities for booty.

Thinking along these lines, it is easy to realize that the Arabism *almogataz*, in order to adapt to its new system, underwent a process of semantic restriction, since from the generic meaning of 'baptized', the meaning that the word had in its Arabic original, it acquired the more specific meaning of 'baptized Moor' in the new language, into which it incorporated itself as a new, specialized appellative, thereby filling a lexical void due to the absence of a word to denote a specific reality that did not exist previously for Spanish-language speakers.

Thus *almogataz* in its earlier meaning refers to the 'Moor who, baptized or not, served as a spy, guide or auxiliary in the forces of some Spanish garrisons in Africa'. In its other meaning it refers by extension to the 'friendly Moor in the vicinity of some Spanish garrisons who, in exchange for his security, would make agreements to coexist with the Christian troops of the garrisons'.[12]

11. This was infrequent but not rare. José Vallejo, the governor of Orán between 1733 and 1738, in his *Memoria sobre el estado y valor de las plazas de Orán y de Mazalquivir*, written in 1734, relates that he succeeded in drawing up statistics on christenings of Muslims, and that they amounted to an average of 30 a year, although he was told that only those who received the sacrament before the age of seven continued to live as good Catholics. See J. Cazenave, 'Contribution a l'histoire du vieil Orán', *Revue africaine* (1925), 368.
12. Maíllo Salgado, 'Breves notas sobre la historia', 480.

THE IMPORTANCE OF THE *ALMOGATACES* IN MAINTAINING THE SPANISH STRONGHOLDS IN AFRICA

In the late fifteenth century the North African strongholds were conquered mainly by the Portuguese or by special expeditions subsidized by Andalusian magnates (Azamor and Melilla were taken in this way).

In the early sixteenth century Ferdinand and Isabella — the Catholic Monarchs — decided to send a fleet organized by the count of Tendilla, governor of Granada, to North Africa, with the immediate aim of cutting off the communications that the *moriscos* of Granada — who then were in rebellion — maintained with their fellow-Muslims in the Maghrib. The expedition had other purposes as well, such as subjecting the Muslim kingdoms to Christian rule and, through a military presence, controlling them in an economic sense by monopolizing their foreign trade.

These plans were not carried out, however, because the wars of Naples and the illness and death of Queen Isabella led to their postponement. Finally the enterprise was launched again by Cardinal Cisneros, and an expedition which set out in 1505 conquered Mazalquivir. Another fleet under the command of Pedro Navarro in 1508 took the Peñon de Vélez, the fortress which had served as a base for piracy and for attacks on the Spanish coastline. Orán was conquered in 1509 and Bejaïa in 1510.[13]

In the face of this series of Spanish triumphs, the sovereigns of Tunis, Tlemcen, and other cities signed capitulations of vassalage to King Ferdinand. When Tripoli was taken in the same year, it marked the crowning achievement of Spanish expansion in North Africa, preceding the defeat at Gerba, which dampened Spanish spirits. This disappointment and other more important and urgent matters gradually pushed the African campaigns into the background, so that after Gerba and the failure of Charles V's expedition against Algiers, the Spanish strongholds that remained could only maintain themselves through a system that carefully balanced a defensive attitude with military intervention, a system that enabled Spanish rule over those cities to last for centuries.

From the time that they first occupied the cities of Orán and Mazalquivir, the Spanish realized that life would only be possible for them in the recently taken strongholds if they could achieve alliances and collaboration with the friendly Moors[14] and the service of the *almogataces*.

13. See A.C. Hess, *The Forgotten Frontier* (Chicago, 1978), pp. 35–44.
14. Diego Suárez writes: 'When the Moors of the Kingdom of Tlemcen are at peace with the forts of Orán, they supply Orán generously with many provisions, especially wheat and barley for the army, some for money, at moderate and low prices, and others for the protection that His Majesty gives them; in addition to the great quantity of wheat and barley which they mainly supply to the region there is often too much, and some of it is exported to Spain and elsewhere, even though Orán

We know that the friendly Moors, or rather those subjected to Spanish rule, paid the authorities of Orán an annual tribute called *rūmiyya*, normally a quantity of grain amounting to up to two *doblas* (a measure) per *aduar* (village or tent settlement); by paying this tribute, the *aduar* acquired freedom from war for a period of a year (renewable in the month of August).

The volume of this *dobla* varied according to the size of the *aduar*: 'In 1708, the year in which Algiers recovered the city of Orán, the *dobla* was equivalent to 112 Arabic *almudes* (about 4.6 litres), that is 28 Spanish *barchillas* (*barsal*), or 43 Castilian bushels, or the equivalent of 2,000 kilos.'[15] In accordance with this, in order to be exact, bearing in mind that the Castilian bushel is a measurement of volume equivalent to 55.5 litres, each *aduar* paid a tribute each year of at least 4,773 litres of wheat.

The *aduares* or tribes that requested protection from the Spanish in Orán had to send their sheikh's *(jeques')* sons to the city as hostages, to be kept there until the tax payment had been made in full (if it was not, after several extension periods and if the *aduar* was moving away to escape reprisals, the sons would be sold into slavery and the proceeds from the sale would go into the royal coffers). In exchange for the payment, the Spanish authorities in Orán committed themselves to protecting the *aduar* from the attacks of hostile Moors, and later on from those of the Turks, and allowed its members to conduct trade with Orán. The Spanish authorities would also practise a sort of arbitrating role in conflicts between tribes that had surrendered (these tribes were obliged to provide troops if the Spanish were attacked by hostile Moors or Turks).

The treaties of protection were granted only for a year, and had to be renewed each August. *Aduares* wishing to continue to be allies of the Spanish had to request a new treaty at sowing time; the *capitan-general* of Orán would then specify any increase or decrease in tax, depending on whether the *aduar* had more or fewer tents than before.[16]

The amount of the *romía* (*rūmiyya*) and the price — generally a very low one — for the grain that the Moors were obliged to bring to Orán was established by the sheikhs of *aduares* or tribes, at a meeting with the Spanish commander in the city; at this meeting, following a ceremonial meal and the giving of gifts to the sheikhs, the commander would lay down his requirements.[17]

requires 40,000 bushels of wheat each year and 12,000 bushels of barley to satisfy its needs.' *Historia del Maestre*, p. 50.

15. A. Tawfīq al-Madanī, *Harb al-Thalathimi'a sana bayna al Djazā'r wa-Isabāniyā 1492–1792* (Algiers, 1969), p. 448.
16. These issues are examined in detail in Vallejo's *Memoria*, translated and annotated by azenave, 'Contribution à l'histoire du vieil Orán', 358–61.
17. According to an original report dated 6 Feb. 1634, upon the invitation of the commander of Orán, 'the sheikhs came on the 20 July last to fix the price of the

The friendly Moors of the *zafinas*[18] of Chafai, Hamian, and Habra, a group of tribes, now more or less sedentary, that lived near the city of Orán, were under a different regime: they were obliged neither to provide hostages nor to pay the tax itself. In the sowing season, the Spanish authorities in Orán would send a few *almogataces* to the *aduares* of these tribes to observe how many fields the villagers were planting. Afterwards the commander called in the sheikhs and the names of the contributors were provided in a register. Each tent had to supply one *dobla* of grain and 30 *almudes* of straw for each field sown. The sheikhs — like those of other friendly tribes — received in recompense one silver *real* for each *dobla* of wheat that reached Orán.

The Moors of the neighbouring villages of Ifre and Canastel, serving as they did as auxiliaries for the Spanish of Orán, seem to have had a special status. The people of Ifre received part of the booty won in the expeditions[19] or forays (*cabalgadas*) that were conducted against the hostile Moors.

Whenever regular tax levies did not cover the expenses made by the aforementioned Spanish towns in North Africa, the Spanish had to turn to forays or expeditions of plunder. On these occasions, in addition to acquiring booty and provisions for the fort (livestock and cereals) they achieved other purposes, such as spreading terror among the indigenous people so that they would submit to Spanish rule,[20] while the winnings were topped off with recruits for a flourishing slave trade. Orán was the

romía. Those who came were of the tribes of Uled [<awlād]-Muza; Ben Abd-Allah, Uled-Arvía, Uled-Jubara, Uled-Brahen, the inhabitants of the *zafinas* of Xafá and Gamayán with many mounted soldiers; they came up to the Alcaçaba and the marquis received them handsomely, pointing out to them the advantages and benefits that they gained with peace and how bad war would be for them, and exhorting them to set a moderate price for the *romía*; they all answered him that they would comply with HE's wish and agreed to provide the three bushels of wheat and six of barley at 8 *reales*. HE gave them gifts and ordered that the money that is customarily given on such occasions be distributed among the sheikhs and nobles'. Galindo y Vera, *Memorias de la Real Academia de Historia*, p. 406.

18. 'When the lineages of Moorish nobles ... assemble with their vassals at a particular place designated for the purpose and possessing water, wood and pasture for their flocks ... which congregation and assembly they commonly call a *zafina*'. Suárez, *Historia del Maestre*, p. 47
19. When the booty was distributed we find '97 Moors on foot from the town and from Ifre, people who do not capture Moors and serve only to rob and create disorder; half of them are to be freed'. Galindo y Vera, *Memorias de la Real Academia de Historia*, p. 419.
20. During the sixteenth and seventeenth centuries the expeditions of punishment or plunder were very frequent; in the eighteenth century, after the cities of Orán and Mazalquivir had been reconquered in 1732, the Crown moderated and even at times prohibited this type of expedition.

main supplier, if not the only one, during the whole of the modern period for the east of the peninsula.[21]

Such a situation was encouraged by the very nature of the cities, for once the conquests had become established, the coastal fortified cities of Orán and Mazalquivir became the setting for urban communities revolving around military life, since many of their inhabitants were soldiers or their families, obtaining a large proportion of their income from the pillaging expeditions that the cities' governors or commanders ordered or led. This income at the same time encouraged the development of crafts, commercial transactions, and other activities.[22]

In a society with characteristics such as these, the role that the *almogataces* played was necessarily an extremely important one, since their very existence was one of the original differentiating traits of these Spanish-African communities. Such communities would have found it difficult, or well-nigh impossible, to come into being without friendly Moors in the immediate neighbourhood. It seems that, by extension, these were consciously confused with the *almogataces*, being given this name whenever they collaborated in any way at all with the Christians living in Orán and Mazalquivir.

As mentioned previously, the name was originally derogatory: to summarize, it meant 'renegade', 'turncoat', 'traitor'. Such individuals

21. R. Torres Sánchez, 'La esclavitud en Cartagena en los siglos XVII y XVIII', *Contrastes: Revista de Historia Moderna*, 2 (1986), 93–4.
22. Due to the tribute paid by peaceful Moors and the booty collected thanks to the *almogataces* the governors of Orán enjoyed pomp and splendour. When John of Austria arrived in Orán, according to an account by D. Suárez, he said to the city's commander, 'Maestre, I warrant that you put on a far better show than my brother the King' (*Historia del Maestre*, p. 215). Thus the city became, and was even called, the 'Little Court' (as opposed to the Great Court in Madrid). The process was assisted by the fact that, from the time of Philip II, Orán also became the 'headquarters' for nobles and important figures who had fallen into disgrace. As punishment for their crimes, they had to contribute there to the public good through their own efforts or through those of their servants. Many of them would go out and fight the Moors and Turks with the city troops and would even be active participants in expeditions and forays (*cabalgadas*).
This élite of exiles living in close proximity to Muslims and Jews — a unique experience at the time in the Hispanic world — recreated an environment similar to that of Spain in the late Middle Ages. Thus situations recurred and scenes were relived similar to those that in other ages had inspired the poetry of the *Romancero* (folk ballads), an echo of which reaches us still in the song of one of our most famous writers of the Golden Age of Spain, in a beautiful *romance* (ballad with refrain) beginning like this: 'Servía en Orán al Rey/ un español con dos lanzas,/y con el alma y la vida,/a una gallarda africana ...' (In Orán there served the King/A Spaniard with two lances/With his soul and with his life/He served a lovely African lady [the phrase *dos lanzas* implies that he provided two servants for expeditions and also means 'service' or a 'tax paid in defence of the city'], Luis de Gongora y Argote, *Obras Completas* [Madrid, 1961], p. 72).

(whether subjected by the Spanish, as were the so-called friendly or peaceful Moors , or active collaborators with them, as were the *almogataces* and the Moors of Ifre) turn up from the same date of the occupation of Orán, in 1509. They existed by the logic of alliances. Thus many *aduares* or villages began to pay taxes to the Spanish because they were subject to raids from more warlike tribes or bands. They thus sought support and assistance from the Spanish forces of Orán, and willingly joined their side in order to repel attacks or get their revenge over their enemies. On other occasions they took refuge in or near the Spanish stronghold when their livestock had been stolen, their tents or villages destroyed, and some of their family members killed. It also happened that some of them, who wished to encourage razzias against enemy villages, came to inform the Spanish, offering to serve as guides in the hope of winning some good booty.

This diffuse situation, with its complex psychology, its loyalties and changes of heart, gave rise to the many-faceted and particular group of people known as the peaceful Moors, and within it to the socio-professional category of the *almogataces*, a category which would acquire a distinct personality in the sixteenth, seventeenth, and eighteenth centuries, especially in the garrisons of Orán and Mazalquivir.

Indeed, they would constitute one of the main factors enabling the Spanish presence to survive in North Africa. They contributed in great measure towards maintaining the Spanish cities, whose immediate hinterland consisted of the space inhabited by the friendly Moors, and also by hostile Moors who refused to pay the *romía*; this hinterland extended about 100 kilometres[23] around both cities (at times it reached as far as Tlemcen) and could only be controlled by means of continual demonstrations of force and military expeditions. That was where the important function of *almogataz* would come into play for, due to his knowledge of the country, its customs and language, the *almogataz* was able to spy on behalf of the Spanish; without this the military forces of Orán would have found it difficult to be effective in defending either their allies or their own interests, which were at risk whenever a portion of the Spanish economic territory was blockaded or captured.[24] The *almogataz* made it possible to carry out accurate and effective surprise attacks, striking the enemy in places that the latter considered too far away for Oranese reprisals or capture.

23. Some areas had a large number of villages and thousands of inhabitants. See al-Madanī, *Ḥarb al-Thalathimi'a sana*, p. 447.

24. 'The peaceful [Moors] who are friends of Orán and the hostile Moors seldom agree, and the hostile Moors frequently carry out raids and assault them, preventing them from bringing supplies and other merchandise to Orán, sometimes bringing the Turks to their assistance to wage full war and do as much damage as they can.' Suárez, *Historia del Maestre*, p. 66.

These forays or expeditions against hostile Moors who, obviously, did not pay tribute to Orán, began almost always with a denouncement by a Moor or spy who suggested to the Oranese authorities that they capture one or two *aduares*. After the governor or commander had received information on the matter and weighed the pros and cons of a particular expedition, he would send *adalides* (scouts) and *almogataces* to reconnoitre the terrain and guide the raiders later on; once he possessed reliable, verified information, the cost of the operation would be calculated and, after all these factors had been taken into account, if it seemed feasible and profitable, the expedition would be launched; if it succeeded, commanders of *escuadrillas* would be selected to distribute the booty.[25]

José Vallejo, in his *Memoria* (Report) written in 1734, states that avarice and the desire for booty led the Spaniards of Orán to organize raiding expeditions without reasonable motives on *aduares* that had requested assistance and protection. 'That was the original and principle cause for losing it [Orán] .'[26]

Don Antonio de Zũñiga, the marquis of Flores de Ávila and governor of Orán, sent an expedition 24 leagues from the city (equivalent to 114 kilometres) to a place on the edge of the desert, where the inhabitants 'had never seen any Christians nor did they wish to be granted protection by His Majesty, as they considered themselves hardly in need of it, both because they were far away at the edge of the desert and because they possessed more than 25 armed villages'.[27] The operation succeeded due to the precise

25. Until 1565 the commanders would keep the booty, but after that year, following the reform by Hernando Tello de Guzmán, the booty would be shared between all the participants in the operation and others who, though they had not participated, had a right to some of it for various reasons. See Ch. 7 in Suárez, *Historia del Maestre*, pp. 124–31; see also Appendix 21, 'Manera de repartir las presas', Galindo y Vera, *Memorias de la Real Academia de la Historia*, pp. 415–23, where articles 53 and 54 read:
-'Some Jews (*hebreos*) who receive a salary here and Moors who do not and other people are given, without participating and without recourse to the booty, the following portions:
-Jacob Çaportas, Jew, 3 parts and 12 fourteenths, as a salary.
-Yaho Çaportas, a Jew, 5 parts and 8 fourteenths, due to his salary.
-Yuna Cansino, a Jew, 2 parts and 4 fourteenths, as a salary.
-Abrahén Canssino, a Jew, 4 parts and 5 fourteenths, in proportion to his salary.
-Çaid Sheikh of Yfre, one part by custom.
-Mulao, a Moor of Yfre, one part.
-The *alferez* of Yfre, one part.
-Another Moor whom they call the sergeant of Yfre, one part.
-Jacob Canssino, son of Ayon Canssino, one part.
26. Vallejo, *Memoria*, p. 367.
27. *Relación verdadera de la gran victoria que el Sr. D. Antonio Zũñiga tuvo con los moros venarajes*, in *Guerras de los españoles en África*, F[uensanta] del V[alle] edn. (Madrid, 1881), pp. 388, 391, 395.

information that the spies or *almogataces* had acquired for the expedition's leader, and had several purposes. The primary and most important one was to make Spain's presence felt, so that no one would doubt the effectiveness of the Spanish in Orán, however distant they might be, or dare to interfere or create difficulties for the trade of its economic territory or hinterland. Secondly, as a punitive expedition, the purpose was to secure sufficient booty to finance the cost of the operation (concretely, during the 1632 expedition about 300 prisoners were taken as part of the booty, and, when sold by public auction brought 42,000 ducats, in addition to livestock 'which went to *almogataces* and the Moors of Ifre',[28] and other items). Lastly, actions of this type served to raise the prestige of Spanish arms in the region in relation to Turkish power, and thus marked off areas of influence.

The same report records the immediate effects of the operation, stating: 'The Blad-Adax Moors, horseriders dwelling at Tlemcen, fearful due to the little security that they had on their lands, came to request protection. And Corte, a fortified place with a tower and Turkish garrison 15 leagues to the south, they have depopulated, through fear of Your Excellency.'[29]

Besides being spies and auxiliaries, the *almogataces* fulfilled the function of guides, like the *adalides*, as is clearly shown in the seventeenth-century texts. One reoort among many that exist concerning the deeds of the marquis of Flores de Avila states: 'His Excellency sent the *almogataces* and *adalides* to try them (certain *aduares*) and to look and reconnoitre from a distance.... Once that was done, they waited until the *almogataces* and *adalides* came back, and until day broke, before the attack was launched' ('para darles con el Santiago').[30]

In this case the *almogataces* also appear to have been expert guides or scouts, sent ahead before the bulk of the troops into unknown or uncertain

28. As Félix Nieto de Silva, the marquis of Tenebrón, and commander of Orán in 1691, explains in his memoirs, Yfle or Yfre was 'a small village of 70 souls about a harquebus' shot away from Orán'. *Memorias de D. Félix Nieto de Silva*, ed. A. Cánovas del Castillo (Madrid, 1888), p. 188.
29. *Relación verdadera de la gran victoria*, p. 396.
30. *Relación muy verdadera de dos presas y aerrotas que ... tuvo el Señor Marqués de Floresdavila*, in *Papeles de mi archivo: Relaciones de Africa, III (Argel-Tunez-Trípoli)*, ed. Ignacio Bauer Landauer (Madrid, 1922), p. 36. The expression 'darles con el Santiago' refers to 'shouting the name of St. James' in order to further disconcert the surprised Moors, who would become easy prey for the Spaniards who had prepared in advanced for the attack. The goal was to capture the Moors rather than kill them, which happened only to those who resisted; livestock and valuable property would also be stolen. In this case, those being attacked with the Santiago were the Benaraje Arabs, hostile Moors who were always feared in Orán. Their most important town was Carte, which was defended by a Turkish and Algerian garrison in a fortified tower. This place was ravaged several times by the Spaniards.

terrain to examine the area. Their fulfilling this task, in view of the difficulties that it entailed and the responsibilities it required, leads us to suppose that the *almogataces* were known subjects of proven trustworthiness, since it was they who would indicate to the expedition's leader the goal he was seeking and the path he should follow. Contexts such as the following prove this: 'Then came the *almogataces* and said that the *aduar* that His Excellency was seeking was at the place that they call Hamu of the Benarahes.'[31]

The *almogataces* were, obviously, warriors possessing tactical and strategic skills. Probably in some cases, together with the *adalides*, they formed an advisory council to the leader of the expedition, largely made possible due to the guidance and skills of the *almogataces*, and the information they provided.

To summarize: of considerable value and importance, the *almogataces* were much appreciated and one might say almost cultivated by the authorities of the Spanish strongholds in North Africa.[32] The reason for this favour and high estimation may be seen in the functions inherent in the position, which were both varied and fundamental: a) They provided information on possible booty; b) They carried out spying and reconnoitring of possible targets; c) They were well practised in laying and avoiding ambushes; d) They served as experienced guides and scouts; e) They would serve as advisers to the expedition leader; f) As fighting men, they were usually expert in attacking and withdrawing; g) For the purpose of determining how much tribute was owed, they could be sent to asess the number of fields sown by peaceful Moors.

As we see, their services were extremely valuable, and thus the *almogataces* comprised a socio-professional category — typical of these walled towns[33] — that flourished during the sixteenth, seventeenth, and

31. *Relación muy verdadera de dos presas*, ed. Landauer, pp. 39–40.
32. So much so that one of the few arguments given as to why the Jews should be expelled from Orán (which happened in 1669), is the following: 'They cause the *almogataces* to absent themselves from the city, with great risk to it, by persuading men who are conversant and well-informed to go and live with the enemy, where they are persuaded to work for our enemies whenever we are slightly less vigilant'. Galindo y Vera, *Memorias de la Real Academia de la Historia*, p. 430.
33. We know that there were *almogataces* not only in Oran and Mazalquivir, but also in Melilla, that is the other garrison town which was purely Spanish (Ceuta was a Portuguese stronghold and opted to become Spanish in the time of Philip IV). The *Relación de lo que don Juan Rejón de Silva Sotomayor, Comendador de Villarruvia, de la Orden de Calatrava, obró de la ciudad y fuerca de Melilla en el tiempo que estuvo gouernando y visitando por orden de su Magestad...*, dated Melilla, 31 March 1645, states *inter alia:* 'As for the *almogataces* who come to serve His Majesty after being forced to flee from their land, he ordered that they be seized, so that they could not leave the fort, because they could do much damage to it, being aware of what went on there and having had experience of it.' See I. Bauer Landauer, *Papeles de mi Archivo: Relaciónes de Africa I (Ceuta y Melilla)* (Madrid, n.d.), p. 320.

eighteenth centuries, though during the last of these centuries its social importance diminished as its tasks and functions changed.

THE *ALMOGATACES* IN THE EIGHTEENTH CENTURY: THEIR HISTORICAL TRAJECTORY UP TO THE MODERN PERIOD

Manuscript sources in the library of El Escorial refer to the position and employment of the *mogataces*, and inform us that whenever one of these *moros mogataces* distingished himself by his loyalty and bravery, 'he received appropriate orders to distribute to others of his kind, and even if he were a simple soldier with no higher grade or wages than the others, they would name him their captain'. They were usually incorporated into the cavalry and and would often show outstanding conduct and bravery.

It seems that, without intending to do so, the *almogataces* as a group also consolidated themselves in the form of a special military unit, since from the late seventeenth century various grades appear among their members.

This evolution into a military body was cut short by the events of the War of Succession, when the garrisons found themselves abandoned to their own devices. The commander of Orán, Melchor de Avellaneda Sandoval y Rojas, saw that resisting would be impossible, and in 1708 decided to retreat, ordering the evacuation of civilians and troops. Despite this order, a portion of the troops and some of the *almogataces* refused to leave, and buttressed themselves in various strongholds within the town, the last defenders being flushed out three months later after a heavy siege. Those who remained alive were taken captive. There must have been many of them: they appear repeatedly in the documents concerning redemptions carried out by the various Christian orders in North Africa.[34]

After both garrisons, Orán and Mazalquivir, had been recaptured on the orders of Philip V, they both flourished for a while under the competent administration of José Vallejo, which lasted for five years (1733–38). Through his managerial dynamism, he succeeded in totally overhauling the way the city of Orán was organized, among other things rebuilding the road network, paving and lighting, streets and squares, improving the infrastructure, particularly the water distribution, putting up buildings, repairing the city walls, reinforcing defences. He also

34. The detailed report on the three redemptions carried out in 1723, 1724, and 1725 by the Order of Mercy, the first two in Algiers and the last in Tunis, shows that in 1723, the first year, 64 captives taken mostly at the castles of Almarza, San Felipe and San Andres in Orán were freed; 60 were freed in 1724, and 69 were freed in 1725, all captured when Orán was taken in 1708, in the previously-named castles, or others nearby. See the work written between 1724 and 1729 by Fr. M. Garcia Navarro, *Redenciones de cautivos en África*, ed. Fr. Manuel Vazquez Pájaro (Madrid, 1946), pp. 485–577.

> ... tried to improve relations between the city and the inhabitants of the neighbouring areas, rejecting policies based on terrorism and systematic destruction of the farmers' resources. It seems that he achieved some results in the small villages close to Orán, but did not manage to put an end to the large tribal federations armed by the governor of Algiers and determined not to tolerate a foreign presence on their land.[35]

Governor José Vallejo was not only outstanding as a manager and administrator, but also excelled in his reorganization of the city's army: in 1734, he decided to set up and organize the *Compañía de Moros Mogataces de Orán* (Company of *Mogataz* Moors of Orán). The Ministry of War approved its founding by means of a royal decree in May 1734, granting this auxiliary force an official statute.[36]

The Company of *Mogataz* Moors was actually constituted in July of 1734, consisting of a captain, a lieutenant and two sergeants, four corporals and 92 *mogataces*. These troops were equipped and rules were drawn up for the care and maintenance of the horses, bonuses for soldiers, and other matters. However, due perhaps to the fact that the name *mogataz* was then considered an insult or, most probably, to the *almogataces*' dislike of fighting in a unit subject to military discipline, by 1735 only 46 *almogataces* could be recruited, and among these only a few cavalry.[37]

It was not until some time had passed that it became a title of honour for Moors loyal to Spain to serve in forces carrying out such brilliant and remunerative enterprises, when finally the effectiveness of the *Compañía de Mogataces* and the benefits that its incursions and surprise attacks brought became apparent.

35. M. de Epalza and J.B. Vilar, *Planos y mapas hispánicos de Argelia* (Madrid, 1988), pp. 69–70
36. See Cazenave, 'Contribution à l'histoire du vieil Orán', p. 327.
37 José Vallejo gave a complete account of this in his *Relación de todas las obras de fortificacion correspondientes a ella, que se han executado en las Plazas de Orán, Mazalquivir y sus Castillos ... del número y estado actual de su Guarnición y de las demás desposiciones que se observan para su Gobierno Militar y Político*, MS, report preserved in the El Escorial Library. It states:

> Although this Company was formed in July of last year, 1734, and was supposed to consist of 100 mounted Moors ... up to now it has not been possible to recruit such a number, only 24, due to the aversion that these inhabitants of the Barbary Coast have to serving under this name and in this class; nor has it been possible to buy more than this same number of horses for them, despite diligent efforts to find them, because of how difficult it is to buy not even good horses but passable ones; because the Arabs are constantly on horseback, and are continually at war between themselves, they want to have good horses, and will pay exhorbitant prices to acquire them. *Relación*, Sec. 'Legajos de Orán', pp. 39 *bis*–40, El Escorial Library.

It is important to indicate that José Vallejo's initiative was the first attempt, in modern times, to create a colonial army in Africa. Thus, through various denominations and permutations, the *almogataces* formed the nucleus of the famous regular indigenous troops, the loyalty and heroism of which can be traced down from the eighteenth to the nineteenth and twentieth centuries. Testimonies to the warlike and valiant enterprises of the *almogataces*, either alone or in conjunction with Spanish troops, occur from 1748 to 1775.[38]

The history of the *almogataces* did not come to an end when the strongholds of Orán and Mazalquivir were finally abandoned in February 1792. Although some of them decided to stay in the country, the majority chose exile, first in Cartagena and later in Ceuta. There they experienced many difficulties due to the well-known lack of understanding on the part of the Spanish authorities.[39] In both Ceuta and Melilla, where there had already been *almogataces* for some time, they would serve as a basis and model for the creation, initially in Melilla and later in Ceuta, of the company of *Moros Tiradores del Rif* (Moorish Riflemen of the Rif). This unit served from the time of O'Donnell (1860)[40] up to the mid-twentieth century in the protectorate of Morocco and in the cities of Ceuta and Melilla under the name *Regimiento de Regulares* (Regiment of Regulars).[41]

Thus, the historical and existential trajectory of the *almogataces* stretches over five centuries[42] and constitutes yet another proof of the necessity for living together (*convivencia*), through both misery and grandeur, on the part of two peoples which have had to share a single geographical space for half a millenium.

38. T. García Figueras, *Presencia de España en Berbería Central y Oriental, Tremecén-Argel-Túnez-Trípoli* (Madrid, 1943), p. 254.
39. Ibid., p. 278.
40. This was the time when Tetuan was occupied, though it had to be abandoned later due to British pressure. See M. Tuñón de Lara, *La España del siglo XIX*, 8th edn. (Barcelona, 1976), Vol. 1, p. 219.
41. Military units with the following names still exist in the two cities: *Regimento de Infantería Motorizada de Regulares de Melilla No. 52* (Fifty-Second Motorized Infantry Regulars of Melilla), and the *Regimento de Infantería Motorizada de Regulares de Ceuta No. 54* (Fifty-Fourth Motorized Infantry Regulars of Ceuta). They are no longer composed of indigenous or volunteer soldiers, as they were until the mid-twentieth century, but rather of draftees doing their national service.
42. Three in Orán and Mazalquivir and five in Melilla.

Autobiographical Elements in the Writings of Kabbalists from the Generation of the Expulsion

MICHAL ORON

In his book *Zakhor: Jewish History and Jewish Memory*,[1] Yosef Hayim Yerushalmi notes that the revival of Jewish historical writing during the sixteenth century 'constituted an innovation that was unprecedented throughout the Middle Ages'. This observation leads us to another, namely that, after the Expulsion from Spain, we find a vigorous resurgence of autobiographical writing among the Kabbalists.

So far, the lives and history of Spanish Kabbalists up to the end of the fifteenth century have been shrouded in mystery. Except for R. Abraham Abulafia[2] who did record a certain amount about his life at the end of his books *Otzar 'Eden Ganuz* and *Sefer ha-'Ot*, we know nothing at all about the Kabbalists who were active during the flowering of Kabbala in Spain.

On the other hand, we find in Kabbalist writings after the Expulsion more and more biographical details, both in the compositions of the Kabbalists and in their personal diaries: the latter stand as autobiographical works in their own right.

For researchers of autobiography,[3] a central issue is the question why and on whose behalf a person tells his life story. In general, they note the existence of two types of autobiography. The first is 'personal', primarily introspective, in which the author conducts a kind of dialogue with himself

1. Y.H. Yerushalmi, *Zakhor: Jewish History and Jewish Memory* (Tel Aviv, 1988), p. 78.
2. On R. Abraham Abulafia see G. Scholem, *Major Trends in Jewish Mysticism* (New York, 1962), pp. 119–55; M. Idel, 'Abraham Abulafia's Works and Doctrine', Ph.D. thesis (Hebrew University, Jerusalem, 1976); id., *The Mystical Experience in Abraham Abulafia* (Jerusalem, 1988); *Otzar 'Eden Ganuz*, ed. A. Jellinek, Beth Ha-Midrash, Part 3, (repr. Jerusalem, 1967), pp. xl–xlii; *Sefer ha-Ot*, ed. A. Jellinek, Ateret Zvi, in honour of Rabbi Heinrich Graetz (Breslau, 1888), pp. 66–85 (Hebrew).
3. A. Momigliano, 'Marcel Mouss and the Quest for the Person in Greek Biography and Autobiography', in M. Carrithers, (ed.), *The Category of the Person: Autobiography, Philosophy, History* (Cambridge, 1985), pp. 83–92; H. Lloyd Jones *et al.* (eds.), *A Medieval Jewish Autobiography, History and Imagination: Essays in Honour of H.R. Trevor Roper* (London, 1982), pp. 30–36; P. Roy, *Design and Truth in Autobiography*, (Cambridge, MA, 1960); L.W. Schwarz, *Memoirs of My People through a Thousand Years* (Philadelphia, 1960).

for purposes of soul-searching, self-examination, and clarification of his self-identity. The second is the 'public' autobiography, which is addressed to the external reader. Here the writer tells the story of his life, sometimes in order to enhance his reputation and celebrate his deeds, and sometimes in order to clear his name and defend himself against various accusations.

In this article, I will focus upon the writings of two Kabbalists of the generation of the Expulsion, R. Judah b. Jacob Ḥayatt[4] and R. Abraham Saba.[5] I shall attempt to answer the question of why and for whom these Kabbalists wrote their personal histories.

Jewish historiographic works, such as those of Ibn Verga[6] and Elijah Kapsali,[7] and the Jewish chronicles which have been found tell us of the lives and trials of Jews in Spain on the eve of the Expulsion. But it should be noted emphatically that this information consists of second-hand accounts based upon testimony and hearsay gathered and documented by the authors. In contrast, among Kabbalists of the generation of the Expulsion who experienced these events at first hand, there is no mention at all of their lives while still in Spain. Their story begins, as it were, with the exodus from Spain, starting at their first station in Portugal.

In the introduction to his book *Minḥat Yehuda*,[8] R. Judah Hayatt tells the story of his life. This book is a Kabbalist interpretation of one of the

4. On R. Judah Hayatt and his Kabbalistic doctrine, see G. Scholem, 'The Kabbalistic Responsa of Rabbi Joseph Alcastel to Rabbi Judah Ḥayatt', *Tarbiz*, 24 (1956), 167–206 (Hebrew); s.v. Judah b. Jacob Ḥayatt, in *Hebrew Encyclopedia*, Vol. 19 (Jerusalem–Tel Aviv, 1968), p. 212; E. Gottlieb, *Studies in the Kabbala Literature* (Tel Aviv, 1976), pp. 399–402, 422–6, 439–75, 493–505 (Hebrew); E. Rachel, 'The Messianic Re-evaluation and the Alteration of Religious Perspective in the Light of the Spanish Expulsion', in S.N. Eisenstadt (ed.), *Proceedings of the International Conference for Jewish Society in Transformation in the Sixteenth Century* (Jerusalem, 1985, Hebrew).
5. On R. Abraham Saba and his writings, see H. Graetz, *Geschichte der Juden*, Pt. VI (Warsaw, 1898), p. 393 (Hebrew); I.M. Toledano, *Ner ha-Ma'arav* (Jerusalem, 1911), pp. 80–83 (Hebrew); S.A. Rosanes, *History of the Jews in Turkey*, Pt. I (Tel Aviv, 1930), p. 107 (Hebrew); N.S. Libowitz, *Abraham Saba* (Brooklyn, 1936); H.H. Ben Sasson, 'The Generation of the Spanish Exiles on its Fate', *Zion*, 26 (1961), 23–64 (Hebrew); G. Nagel, 'A Chapter in the Philosophy of the Spanish Expulsion Generation,' *Sinai*, 74 (1974), 77–80 (Hebrew); D. Manor, 'Abraham Saba: His Life and Work', in *Jerusalem Studies in Jewish Thought*, Volume II (2), 1982/83, pp. 208–31 (Hebrew); id., 'The Writings of Rabbi Abraham Saba,' *Sefunot*, 3/18 (1985), pp. 317–38 (Hebrew). After this article was written, Abraham Gross called my attention to two of his articles with which I was familiar at the time, and for which I thank him. See A. Gross, 'Rabbi Abraham Saba: Ha-Megurash Beshnei Ha-Gerushim', in *Studies in Memory of the Rishon Le-Zion Rabbi Yitzhak Nissim*, Vol. 4 (Jerusalem, 1985), pp. 205–24 (Hebrew); id., 'The Iberian Expulsions as Reflected in the Commentary on Esther', in *Proceedings of the Ninth World Congress of Jewish Studies* (Jerusalem, 1988), pp. 153–8 (Hebrew).
6. Solomon Ibn Verga, *Shevet Yehudah*, ed. A. Shoḥat (Jerusalem, 1957, Hebrew).
7. Elijah Kapsali, *Seder Eliyahu Zuta*, trans. and ed. A. Shmuelevits *et al.* (Jerusalem, 1976, Hebrew).
8. *Minḥat Yehuda* by Rabbi Judah b. Jacob Hayatt in *Ma'arechet* (reprint edn.,

most important Kabbala works written in Spain (probably at the beginning of the fourteenth century), titled *Ma'arechet ha-Elohut*.

The period of his life in Spain is mentioned by R. Judah casually when, in his introduction, he writes:

> While I was in Spain, I tasted a bit of honey, and my eyes were opened [a reference to the study of the Kabbala]. I allowed my heart to seek and search out the wisdom [of the Kabbala]. I went from strength to strength, gathering all that was to be found in the book I have mentioned [the Book of the Zohar], and I gathered a little bit here and a little bit there until I possessed most of what it had to offer. And I am completely convinced that it was this privilege which stood me in good stead and helped me overcome all the trials and tribulations that befell me at the time of the Expulsion from Spain.[9]

The story of R. Judah's life thus commences after he was expelled from Spain to Portugal, and in fact, he begins it in the winter of the year 1493, when he was expelled from Portugal by order of the king. He and his household left on a boat carrying 250 persons. Upon their departure, an epedemic broke out among the passengers, and attempts to disembark anywhere were met with unyielding refusal. He reports: 'And it was that nowhere would they receive us.'

For four months they sailed back and forth, existing on only scanty food and the 'water of affliction'. While at sea, Spanish pirates attacked them, carrying off their possessions and towing the boat back to Spain — to Málaga. In Málaga the authorities would neither allow the boat to enter the port and its passengers to disembark, nor would they allow them to continue on their way.

For about three months the boat remained outside the port. Every day priests came on board and tried to convince the exiles to convert to Christianity, but they remained adamant in their refusal, rejecting all attempts at persuasion. In response to this, 'the cardinal declared a ban, forbidding anyone to give us bread and water or any other supplies'.[10]

For five days those on the boat were prevented from receiving food and water, and during that time 'the judges and important men of the city, along with its priests and sages came on board in order to persuade them, so that in one day close to 100 persons were converted'.[11]

During those terrible days, 50 people died, including the wife of R. Judah. After the conversion of some of the exiles, the ban on supplying the

Mantua, 1558; Ferrara Press, 1558); *Ha-Elohut*, Jerusalem Manuscript, HEB, 8⁰ 2020.

9. *Minḥat Yehuda*, Mantua, p. 2/2.
10. Ibid., Ferrara Press, p. 2/4.
11. Ibid., p.3/1.

boat was lifted, and two months later the authorities finally permitted the exiles to resume their journey. The boat continued on its way and headed for Morocco. When it reached Berberia, the exiles, including R. Judah, disembarked. But matters did not improve for R. Judah Ḥayatt when he finally disembarked. Immediately upon his arrival in the city, he met an Arab who had previously been his neighbour in Spain. This Arab testified in court that, when he had been in Spain 'while the holy community [that is to say, the Jews] were celebrating the conquest of the large city of Granada [namely the conquest of the last Muslim stronghold by the Christians], I [R. Judah] had the Jews there, to whom I had taught Torah, make an effigy of the Ishmaelite prophet [Muḥammad] and drag it about the marketplaces and through the streets — and they believed him [the Arab] since he had three witnesses.'[12] From this account, in addition to the story of R. Judah who was accused before an Arab court in Morocco, we also learn the important fact that in 1492 (when the Expulsion edict had already been signed), the Jewish communities in Spain celebrated the Christian victories, and that in one place (according to this testimony), in an outpouring of joy, they even ridiculed Muḥammad.

Because of this libel, which was believed in court, R. Judah was beaten and thrown into prison. He sat in prison for 40 days awaiting his sentence, but the local Jews came to his aid and paid a ransom for his release. R. Judah continued his wanderings and reached Fez. In Fez there was a great famine, and he was forced to hire himself out as a miller. 'Every day in an Ishmaelite [Arab] household I would work very hard in order to receive a small, thin slice of bread, not fit even for dogs.'[13] In the evenings, he felt the hunger and cold, 'and against the bitter autumn cold, and because there was no shelter from the cold and we also had no houses in which to dwell, we would dig into the rubbish heaps in the city and lay down our bodies there.'[14]

From this description, R. Judah Ḥayatt jumps ahead to recount his life in Italy. He reached Naples (how and by what means he does not say) 'after vicissitudes and countless misfortunes. There too, I did not find peace and quiet, and I incurred the wrath of the king of France.' R. Judah was taken prisoner. Beaten, wounded, and penniless, he reached Venice. He was aided by the Spanish Jewish community there which welcomed him, gave him clothing, and provided for all his needs.

From Venice he continued to Mantua where he found a large centre of Spanish Jews, some of whom had arrived even before the Expulsion. He

12. Ibid., Mantua, pp. 2/2, 3/1; Ferrara, p. 3/1. Note the text here: 'And an Ishmaelite, who came from Spain, lived in my place and happened to pass by, *since they too were banished*, testified against me...' In this connection see also Ben Sasson, 'The Generation of the Spanish Exiles' and the note on p. 27.
13. *Minḥat Yehuda*, Mantua Press, p. 3/1.
14. Ibid.

met R. Joseph Jabetz,[15] one of those expelled from Spain and one of the city's wealthy and honoured citizens. At the request of the latter, a group of prominent Mantuan citizens, R. Judah wrote his book *Minḥat Yehuda*.

At this point, the story of the life of R. Judah ends. Along with the fact that we have here an autobiographical account which includes testimony concerning the experiences of one of the exiles, this story also has a kind of psychological explanation for the conservative thinking of R. Judah as expressed in his composition. *Minḥat Yehuda* is written in the theosophic, symbolic, Kabbalistic spirit of the author of *Ha-Zohar*. R. Judah Ḥayatt tries to preserve the conservative Spanish Kabbalist teachings and, in his composition, he attacks any manifestation of positions which differ from his own. R. Judah fought spiritedly against that trend of philosophic Kabbalists who tried to remove the myth from the talmudic *aggadot*. The problem of the nature of the spheres acquired important significance for the exiled generation, especially as related to the problem of reward and punishment, and it is this problem which occupied R. Judah in his composition.

Another example of his conservatism appears in his introduction to the book. After telling his story, he finds it necessary to divide the Kabbalistic literature into books which are to be read and books which are to be avoided.[16] Forbidden books are mostly those in which there is an attempt

15. R. Joseph Jabetz was the father-in-law of R. Joseph Ibn Shraga: on Jabetz see Scholem, 'The Kabbalistic Responsa', 168–9: Gottlieb, *Studies*, p. 400.
16. On this he writes:

> Because I saw that misleading books on Kabbala have been distributed in this region, I felt it my duty to warn you of those books which you should avoid and to advise you of those which should be studied. The words of the heavenly sage R. Isaac Ibn Latif of blessed memory, in his books *Ha-Shamayyim*, *Ha-'Olam*, and *Tzror-ha-Mor*, and *Ginzei Melech* are dearer than pearls. But as for all that pertains to the wisdom of Kabbala, he has one foot in and one foot out, so that you may read the end of the book but should not read the whole book through. And with the Lord's help, I shall explain some of Ibn Latif's opinions and thereby separate the wheat from the chaff. I have also seen books written by a sage called R. Abraham Abulafia who in his stupidity and his pride has referred to himself as 'Abraham Ha-Moreh' [the teacher]. And just as only fools prophesy since the temple was destroyed, in this case, the appellation 'Ha-Moreh' actually means fool, for in Greek the word *moreh* means 'fool'. Beware of him; do not listen to what he says, because his books are filled with his own inventions and fabrications. He has written a book called *'Or ha-Sekhel* [The Light of Wisdom], whereas he himself wanders in darkness, and another book *Moreh ha-Moreh* [Guide to the Guide] which is an interpretation of the Guide [*Guide to the Perplexed* of Maimonides — M.O.] about the ways of the Kabbala. See how great is his stupidity that he would interpret the words of Maimonides using imaginary Kabbalistic precepts which never entered the head nor the heart of Maimonides of blessed memory ... and the sage Ha-Motot [R. Samuel Ibn Motot] — keep away from him, and don't go near his door; for in his hand he holds a wicked staff to lead you away from the straight path to the crooked one ... *Minḥat Yehuda*, p. 3/2.

at compromise between philosophy and Kabbala, and especially the books by Abraham Abulafia.

It seems to me that R. Judah's objection to philosophy and his great conservatism stem from his biography. The generation of the Expulsion tended to accuse the philosophers and their theories of being one of the factors which gave support to the converts. Furthermore, as one who had undergone trials and tribulations and had lost his family, his property, and his books, R. Judah tried to preserve the Kabbalistic heritage which he had acquired in Spain and to fight against those who tried to change it.

Along with the personal story of R. Judah Ḥayatt, mention should be made of another autobiographic account, that of R. Abraham Saba,[17] who was, apparently, a preacher in Zamora in northern Castile. In one of his writings he tells how he preached in the synagogue in Zamora and chastised the people for their negligence in keeping the Torah commandments,[18] how he was praised by the sages for his sermons, and how he began to write his compositions there.[19]

A document published by Yitzhak Baer[20] confirms that R. Abraham Saba did indeed live with his family in Zamòra. In this document we find that his brother Saul Saba was tried there for giving false testimony and sentenced to death. R. Abraham Saba appealed against this judgment, claiming that the trial was held without the presence of a defence lawyer.

Except for those details which I have mentioned, we have no information about the life of Abraham Saba in Spain. He begins his autobiographical story with the Expulsion order: 'And King Ferdinand expelled all the Jews from his kingdom, and some went to Navarre, and some went by boat to the kingdom of Portugal, and some died from plagues and hunger ... and I remained in the diaspora in Portugal in the city of Guimaraes ...'[21]

From here R. Abraham Saba goes on to describe his literary work. We learn that between 1492 and 1497 he completed most of his writing — in Portugal.[22] In a continuation of his story which appears in his book *Tzror ha-Mor*, Oxford manuscript 16723, 254, [copied and published by Dan

17. On R. Abraham Saba and his writings, see n. 5 above.
18. *Tzror ha-Mor* (Venice, 1546); 'Pikudei', 89/3.
19. Ibid, 'Vayishlaḥ', 47/3.
20. F. Baer, *Die Juden im christlichen Spanien*, Vol. II (Berlin, 1936), pp. 371–6; id., *A History of the Jews in Christian Spain* (Tel Aviv, 1959), p. 418. See also A. Gross, *R. Abraham Saba*, p. 26.
21. *Tzror ha-Mor*, 'Vayetzei', Oxford MS, 16723, 254; Parma MS, p. 8/5; Microfilm 3075 according to the copies of Dan Manor. For R. Saba's biography see n. 5, and see Gross, 'Rabbi Abraham Saba ... ', and id., 'The Iberian Expulsions'.
22. Ibid: 'And there I settled and finished the exegesis of the Torah which I had begun in Castile and constructed it beautifully and truthfully for the perusal of any questioner, and I decorated it with shining pearls [of the Zohar — '*Margaliyot ha-Zohar*'] in each chapter ... and I set with it the interpretation of five Megillot included and decorated above and below....' (*Tzror ha-Mor*, see n. 21).

Manor] as part of his commentary on the Book of Genesis, he tells of what befell him after 1497. In that year King Manuel decreed the expulsion of all the Jews in his kingdom. This was preceded by a decree concerning infants, according to which infants and youths were to be removed from their homes by force in order to change their faith.[23] Thus, it is possible that some of R. Abraham's children were taken from him as a result of this order. Upon the publication of this decree, R. Abraham went to Lisbon, as ordered by the king. He writes: 'Then I left my home and my abode and all that was dear to me and my books as well in Portugal'.[24]

When he, his wife, and his two married sons who remained with him left for Lisbon, he endangered himself by taking his writings with him. Upon reaching Lisbon, he heard the crier's voice calling for all the Jews to bring all their books to the Lisbon synagogue:

> And the crier went all about the country declaring that all the books and phylacteries must be brought to the great synagogue in Lisbon on threat of death. And they already took one Jew who loved his books and beat him severely with straps.... As I listened, I stood trembling, walking on with trepidation and fright, and I dug into the middle of a large olive tree which had extensive roots in the ground, and there I hid these three books which I had written.[25]

Afterwards, relates R. Abraham Saba, all the Jews were gathered 'into one courtyard, about 10,000 Jews, and they were all pressured and enticed to change their faith, so that after four days, not 40 men and women remained.'[26] Thus we learn that, of 10,000 Jews, the large majority converted, including the wife and sons of R. Abraham Saba. The rabbi and the 40 Jews who remained steadfast and did not convert were thrown into jail.

Elsewhere in his commentary on the Torah (the exegesis of the Book of Leviticus),[27] R. Abraham relates that when this group sat in jail it was led by R. Simon Mimi. The latter, who was a community leader and rabbi in

23. About the decree ordering the taking away of infants and the trauma it caused, see J.R. Hacker, 'Pride and Depression: The Polarity of the Spiritual and Social Experience of the Iberian Exiles in the Ottoman Empire', in *Culture and Society in Medieval Jewry: Studies Dedicated to the Memory of Haim Hillel Ben Sasson* (Jerusalem, 1989), pp. 541–86 (Hebrew). See id., 'On the Intellectual Character and Self-Perception of Spanish Jewry in the Late Fifteenth Century', *Sefunot*, 2/17 (1983), 30 (Hebrew); I. Tishby, *Messianism in the Time of the Expulsion from Spain and Portugal* (Jerusalem, 1985), pp. 24–8, 38, 40 (Hebrew). Also see Kapsali, *Seder Eliyahu Zuta*, Vol. 1, p. 238.
24. *Tzror ha-Mor*; Manor, p. 227; Gross, p. 221.
25. *Tzror ha-Mor*; according to Gross, 'and in fear and trepidation I went with a Jew...', p. 222.
26. Ibid.
27. Ibid. *Be-ḥuqqotay*, p. 105/4.

Portugal, was severely tortured and died a martyr's death. The story of R. Abraham is corroborated by other chronicles that relate to the death of R. Simon Mimi (in the *Sefer ha-Kabbala* by R. Abraham of Torrutiel,[28] in *Seder Eliyahu Zuta*,[29] and in selections from R. Joseph Sambari).[30] After the death of R. Simon Mimi, the authorities released R. Abraham Saba and permitted him to leave Portugal: 'And after six months, the king ordered that we be given one damaged ship and taken to Arzilia, and from there we came to the land of Ishmael.'[31]

R. Abraham's first station was at 'el Ksar el Kebir,[32] on the north-west coast of Morocco. Here, mourning the loss of his family and of his books, he began to rewrite the books.[33] This project was completed in the city of Fez in the year 1500. Bitterly he observes that old age and forgetfulness have come upon him, but he is still able to recall and reconstruct his writings.[34]

His commentaries on the Torah and the Megillot, while they seemingly do not diverge from the Kabbalistic exegetic framework, none the less contain new elements, different from anything found previously in that exegetic literature. Into them, R. Abraham Saba wove his life story and the events of his time. He searched the Bible for answers to his personal questions and tried to find allusions to his personal fate. The time gap between the Bible and his own era was easily bridged, when he found a parallel between the fate of the people then and his own fate. Connecting the text to his reality and the latter to the text created points of contact between his concrete reality and that of the distant past.[35] Thus, for example, he found similarities between the story of the Book of Esther and

28. *Sefer ha-Kabbala* by Abraham of Torrutiel (Jerusalem, 1967), p. 113.
29. R. Elivahu, *Seder Eliyahu Zuta*, Ch. 82, pp. 236, 337.
30. *Excerpts from the Writings of R. Joseph Sambari* (Jerusalem, 1967), p. 143 (Hebrew). On this incident, see M. Benyahu, 'A New Source Concerning the Spanish Refugees in Portugal: Their Move to Saloniki after the Edict of 1506', *Sefunot*, 11 (1971–77), 233–65 (Hebrew); Toledano, *Ner ha-Ma'arav*, pp. 54, 226; Tishby, *Messianism*, p. 33, and also Libowitz, *Abraham Saba*, pp. 13–14 (where he presents the different sources which describe the event).
31. *Tzror ha-Mor*; Manor, *Abraham Saba: His Life and Work*, p. 228.
32. 'And from there we came to the land of Ismael, and I remained here in the city of 'el Ksar el Kebir, and this holy public clothed my nakedness and supplied all my needs, and I tarried long here, weeping when I recalled my sons and my books which I had left behind, and in particular those books which I had myself written'. (*Tzror ha-Mor.*)
33. See Manor, 'Abraham Saba: His Life and Work', pp. 219–20.
34. *Eshkol Ha-Kofer* (Commentary on the Book of Esther), p. 22.
35. This is especially noticeable in the exegesis of R. Abraham Saba on the Book of Esther, but also in his interpretation of the Torah and of the Book of Ruth we find the analogies which the author draws between what is written there and contemporary reality.

the story of his own generation.[36] But the denouement of the story of his generation differs from that in the Book of Esther.[37]

In his Torah exegesis (the Book of Deuteronomy), he observes that, in fact, the forced conversion edicts were already hinted at in the Torah.

> And he said: 'These words which I command thee this day shall be in thy heart' [a hint at the attributes mentioned of the oneness of God and also the readiness to martyr himself for *Kiddush Ha-Shem*]. And this matter is as I have written: that the Torah foresaw the troubles and edicts which would be decreed with regard to the Jewish people so that they would change their faith and forsake the Torah, as was the case during the Expulsion from Portugal, when it was forbidden to teach in public or instruct the children. And they took all the books and the synagogues to keep us from praying and from studying the Torah, until the Jews all but forgot the Torah.[38]

It was for his generation that *Parashat Shem'a* was written, for in it is found the creed of monotheistic faith:

> And all that is left for them is that they be taught 'Hear O Israel!' and that they know how God is one, and that He is worthy of our love and martyrdom for His name's sake. Therefore, God gave the Jewish people of that day the short chapter of *Shem'a Yisra'el* which contains the entire Torah, so that if they were not able to know the entire chapter, they would at least know the *Shem'a Yisra'el* verse and would teach this verse to their sons.[39]

As in the case of R. Judah Ḥayatt, R. Abraham Saba did not belong to those Kabbalists who produced innovations in the sphere of Jewish thought. His uniqueness lies in the synthesis he created in his writings between his own personal life experience and the biblical text.

The question I posed at the beginning of this article remains, and must finally be answered: why, then, did these Kabbalists weave their life stories into their exegetic compositions? Two answers may be suggested:

(a) It seems to me that the question of the continued existence of the Jewish people, its fate, and its future after the Expulsion was the central reason for describing the life history of those who were

36. During the period of the exile, it was very common to find in the Book of Esther an analogy for the situation of the Jews at the time.
37. R. Abraham Saba describes at length the bitter end of the wicked, namely, Haman and his sons, in contrast to the rise and glory of Mordechai (*Eshkol ha-Kofer*, p. 97): 'And perhaps this may be of some small consolation for those who have undergone great hardship and who now wait and hope for some other resolution, such as that found in the Book of Esther'.
38. *Tzror ha-Mor*, 'Va-Etḥanen', 139/1.
39. See ibid.

banished from their homes, from their positions, from their culture, from the graves of their fathers, and from their families. In writing their life stories, they erected a memorial to themselves and to their history, presenting their readers with the genealogical chart of their past.

(b) We have at hand testimony as to the negative attitude of rabbis from various communities throughout the Ottoman Empire and from Palestine towards the Spanish exiles.[40] These rabbis regarded all those who came from Spain and Portugal first and foremost as people who had changed their religion and then returned to the fold. They treated the exiles with contempt and blamed them for converting and betraying. (Evidence of this is found, for example, in the response of R. Elijah Mizraḥi to the writings of R. Joseph Algazi from the island of Naxos, and in the words of R. Gershon Mizraḥi in opposition to R. Levi Ibn Ḥabib.)

R. Judah Ḥayatt and R. Abraham Saba tell us their life histories and emphasize their suffering and their fortitude in the face of religious persecution. It may be that these words contain the roots of an apologetic for their generation. Their story, interwoven in their Kabbalistic compositions, is thus intended for the public and is meant to commemorate the events and the suffering of their lives and of all the other Spanish exiles, and to solicit encouragement and vindication for their actions from both contemporary and future readers.

40. See Hacker, 'Spanish Jewry in the Late Fifteenth Century', 35 (n. 77), 36; Hacker presents the response of R. Levi b. Ḥabib as opposed to R. Jacob Berab, and states: 'This phenomenon of Jews who had converted under duress and then returned to the fold was widespread and well-known in the Ottoman Empire at the beginning of the sixteenth century'. Hacker also notes the comments of R. Elijah Mizraḥi written in opposition to R. Joseph Algazi from the island of Naxos and emphasizes: 'In any case it is clear from his words [those of R. Elijah Mizraḥi — M.O.] that the greatest Romaniote sages at the beginning of the sixteenth century considered all the Iberian Jews as having converted and then returned to the fold'.

Sephardic Settlements in Sixteenth-Century Italy: A Historical and Geographical Survey

RENATA SEGRE

The sixteenth century witnessed one of the greatest upheavals in the history of the Jews in Italy. The community which existed on the eve of Charles VIII's descent into the peninsula is barely recognizable a hundred years later. The original and most relevant (though not the only) cause was the policy of the Catholic Monarchs. The events of 1492 were immediately followed by a mass emigration of Spanish Jews fleeing conversion to preserve their national and religious identity. The same year saw the influx to the mainland of a considerable part of the large Jewish community of Sicily, with its own characteristics and rituals, which was reinforced by a growing number of Spanish Jews dislodged by the massacres (*matanzas*) of the fourteenth and fifteenth centuries. To these two transfers of population we can add a third, which almost completely emptied the Kingdom of Naples of its ancient Jewry, harassed by the French army in 1494 and chased by the new Spanish regime a decade later.[1]

This first phase was followed, broadly speaking, by a second with the arrival of Jews from Portugal in the early years of the pontificate of Clement VII (1523–34), their numbers rapidly increasing after the Inquisition started operating in Portugal in the late 1530s. The third phase consisted not so much of a further inflow as in a sudden reshuffling of the settlements of the Spanish Jews, due to the tragic events of 1556 at Ancona. The expulsion in 1569 of all the Jews from the Papal States — not only those who had recently immigrated from the Iberian Peninsula and thus were less deeply rooted, but also those long established there — and their

Source abbreviations: ASA (Archivio di Stato, Ancona); ASM (Archivio di Stato, Modena, Cancelleria estense); ASP (Archivio di Stato, Pesaro); ASR (Archivio di Stato, Rome); ASV (Archivio di Stato, Venice); ASVat (Archivio Segreto Vaticano); BAV (Biblioteca Apostolica Vaticana); BNP (Bibliothèque Nationale, Paris); BUG (Biblioteca Universitaria, Genoa); b. (busta); reg. (registro).

1. For a general picture of Jewry in fifteenth- and sixteenth-century Italy, see R. Segre, 'Gli ebrei nell'area padana fra Quattrocento e Cinquecento', in G. Tamani (ed.), *I tipografi ebrei a Soncino 1483–1490. Atti del Convegno*, pp. 13–25; V. Bonazzoli, 'Gli ebrei del Regno di Napoli all'epoca della loro espulsione', *Archivio storico italiano*, 137 (1979), 495–559; 139 (1981), 179–287.

resettlement in the two great communities of Rome and Ancona — brought about analogous ethnic and demographic changes.

To the four stages briefly outlined here we must add a continuous influx of refugees, not always directly from Portugal, of varying dimensions which lasted till the end of the century, though it became more pronounced during the last tragic years of the kingdom's independence and the early years under the Crown of Spain.

Significantly, on the basis of the Christian sources, it appears that at least six different terms existed to identify those sectors of the Jewish community recently arrived from the Iberian Peninsula. The term 'Marrano' occurs chiefly in denunciations, legal documents, expulsion orders, in short in cases where there was to be no quibbling about the fact that the subject was one who, after undergoing baptism, had turned again to judaizing. The case of 1550, in which a Roman notary did not hesitate to state that his client Salvatore the cap-maker, son of Solomon, 'alias the Marrano' took his oath *tacto calamo*, namely *more hebreorum*,[2] was probably an exception due to the climate of unusual indulgence that the refugees enjoyed during the pontificates of Paul III (1534–49) and Julius III (1550–55). The sources, however, are not usually so explicit. Certainly no one mistook a Sephardi for one of those 'theutonic' (Ashkenazi) or 'old' (*antiqui*, *vecchi*, viz. settled in Italy for many generations) Jews whose features had been familiar to Italian eyes for centuries — and yet a need was felt to distinguish between them. The term 'New Christian' was used in contrast to the old and more familiar 'neophyte': both were of Jewish birth, both baptized, but the former was of Iberian origin while the latter had grown up in an Italian community. 'New Christian' was a broad term covering all immigrants irrespective of whether they practised the Jewish or the Catholic faith. None the less, this more prudent generic term was almost always intended to designate those who had turned again to Judaism. To swing between the qualifiers 'Spanish' and 'Portuguese', or indeed use them both linked by an *et* or an *aut*, has more than a geographic-linguistic relevance. A notary registering his client or witness with one or the other, would not have made the distinction on the grounds of whether he spoke Castilian or Portuguese.

A man who called himself or was known as a 'Spanish' Jew had not necessarily been baptized and was certainly a practising Jew. On the other hand, to be a 'Portuguese' definitely implied baptism and meant being a Marrano without any of the injurious connotations of the word. The two other more commonly used terms, 'Levantine' and 'Oriental' (any distinction between them is questionable) had another meaning. They were

2. ASR, Archivio dei 30 notai capitolini. Uff. 31, Sano Perelli, reg. 8, 17 March 1550.

intended to designate a Jew according to his place of origin, be it Alexandria, Jerusalem, Safed, Aleppo, Constantinople, Bursa, Salonica, Skopje, or Bitola (Monastir) — in short anywhere in the Mediterranean under Ottoman rule[3] — and probably implied possession of a Turkish safe-conduct (even in the case of that Abram of Narbonne who was held by the Inquisition at Rome in 1555).[4] However, to be a Levantine or Oriental Jew did not only indicate place of origin but also inferred, even more strongly than the term Spanish Jew, the claim to a totally Jewish background uninterrupted by baptism.

To these six denominations (all of which, it must be said, may at times be used to describe one and the same person) a seventh is to be added: *retagliato* which had broader connotations and is best taken on its own. The literal meaning is 'circumcised' and it can therefore in theory also be used to indicate a Muslim. In the Venetian sources of our period (where it occurs most frequently), it was used exclusively for the New Christians who had been circumcised and were therefore, legally speaking, to be persecuted as Marranos.

To take the analysis much further would be unwise, for one runs the risk of dispelling that ambiguity and lack of clarity which was a real fact of contemporary life. In the summer of 1550 the Christian merchants of Rialto were upset by a decree of the Senate introducing draconian punishments for anyone caught trading with a Marrano. The problem was how to decide 'who is a Marrano and who not': this question applied to different groups of individuals such as those 'who are called New Christians and live in Apulia', those who bring in Spanish wool through Florence and Livorno, the corn-merchants of Sicily, the money-changers from Lyons and from Flanders, or the Portuguese who trade in sugar, pepper, and pearls. How could one tell? 'We know neither the background nor the way of life of these men, nor what they believe or what they do not', cried the merchants in dismay, and the Senate reassured them, explaining

3. The Maghrib Jews were usually defined only by the land or the place of origin: for instance, 'Israel Giagion of Morocco and Chigia of Abraam of Fez Jews, subjects of the Sherif' in Venice; Angelo Sed of Morocco in Rome. ASV, Collegio. Notatorio, fol. 76r, 5 Dec. 1572; ASR, Archivio dei 30 notai capitolini. Uff. 16, Bernardino Pascasio, reg. 6, 22 Sept. 1585.

4. ASR, Archivio dei 30 notai capitolini, Uff. 30, Pietro Bardetti, reg. 14, 19 Nov. 1555. An inventory of the possessions found in the Roman home of the 'Levantine Jew' Abram of Narbonne gives us a remarkable insight into his world: 'a water-jar and a leather pouch containing pieces of material such as Jews wear on their heads, also a pair of phylacteries, a little pig-skin brush, a leather flask, a writing-quill ... Turkish style black overcoat and gown' ('una vettina d'acqua et una saccoccia di corame con certe tele dentro da poner in capo a modo de hebrei con un paro de tephilini, una scopetta di pelo di porco, un fiascheto di corame, un calamaro da scrivere ... gabbano e sottana nera alla turchesca').

that its purpose was not to impede commerce with traders of uncertain faith, but simply to prevent Marranos from settling in the republic.[5] However, those same Venetian merchants would have had no less difficulty in ascertaining the faith of their partners had they met them in the flesh and not just done business with them: it was no easy matter to produce evidence of a return to Judaism, and those governments who wished to welcome Marranos without seeming to infringe canon law took advantage of this fact.

Veiled in bitter sarcasm, just such an accusation was made in an audience Gregory XIII gave the Venetian ambassador at the Curia, Giovanni Correr, in 1579. The Inquisition had informed the Holy See that the Serenissima refused to lend 'the secular arm to arrest certain Portuguese who lived here [in Rome] as Christians, and now live as Jews', and this was especially deplorable 'since it was a question of the service of the Lord God and of the universal good'. The ambassador presented a weak and uneasy defence, basing his position on the lack of legal proof. The Pope held the trump card and answered with a smile, 'well, what can we do more than tell you, as we have done, that these men lived here as good Christians, while in your city they have declared they are of Jewish faith'.[6] If not even the word of the Head of all Christendom counted, it would be quite impossible to open legal proceedings against Marranos.

But were the rites of forced baptism enough to cancel the 'national' and religious identity of a Jew? Even in the early years of the seventeenth century the cautious duke of Urbino was not completely sure. In response to an application made by 'certain Portuguese Jews' to come and operate international trade (apparently at Pesaro), he immediately expressed his

5. ASV, Senato, terra, filza 11; document published by D. Kaufmann, 'Die Vertreibung der Marranen aus Venedig im Jahre 1550', *Jewish Quarterly Review*, 13 (1900–1901), 527–30. In reporting the banishment order to the Pope, the nuncio emphasized the Venetian government's determination to combat heresy, and justified the measure by the fact that the number of Marranos had grown to 'around ten thousand souls ... a great scandal and a wrong, both for religion and for the evil they do in their business': F. Gaeta (ed.), *Nunziature di Venezia*, Vol. 5 (Rome, 1967), p. 88. The Este ambassador, however, explained to his duke that 'the causes are various; firstly ... these Marranos are worse than the Jews since they are not Christians nor are they Jews, who live all together in the ghetto separate from the Christians'; ASM, Ambasciatore Venezia, b. 36, Venice, 20 July 1550. For the most recent research on the New Christians in Apulia, see M. Sirago, 'L'inserimento di una famiglia ebraica portoghese nella feudalità meridionale: i Vaaz a Mola di Bari (1580–1800)', *Archivio storico pugliese* (1987), 83–158.
6. ASV, Capi del Consiglio dei Dieci, Lettere ambasciatori, b. 26, lett. 207, Rome, 6 June 1579. For an epitome of the life of Correr, an able politician and respected ambassador who gained experience of Jews from various nations during his term as *bailo* at Constantinople (1575–78), see A. Baiocchi, s.v., in *Dizionario biografico degli italiani*, Vol. 29 (Rome, 1983), pp. 495–6.

fear 'that they may be Marranos'.[7] The Jews' answer left him perplexed, for it struck him as not unreasonable: 'They say they have always been Jews for, though in the Kingdom of Portugal ... they made an outward show of being Christians, yet in the secrecy of their homes they lived as Jews'. Only when they are haunted by the wish to show 'their greater zeal for their own law' do they decide to emigrate and, abandoning all their possessions, move to 'Ferrara, Livorno and other places under other princes'. Francesco Maria II recalled that his father had allowed Marranos to settle in the dukedom, and that a brief by Paul III 'quite clearly' authorizes him 'to grant admission to these people'. The issue was not, however, completely straightforward and to overcome 'petty scruples' the duke asked his envoy to put the problem to some cardinals in the Curia. Their answer has not come down to us, but it is not hard to imagine the discouragement it expressed.

'I was born a Jew, brought up a Jew, have lived as a Jew, neither must I nor will I abjure, nor can I for any reason whatever be asked to do so' declared Abram 'Righetto', otherwise Henrico Nuñes, detained for being a Marrano by the Holy Office in Venice. The accusation, he maintained, apart from being groundless in itself, was therefore unwarranted in his case, since forced baptism never destroyed the true identity of the Portuguese Jews who, even before 'leaving that kingdom, as gradually they did, lived as Jews in every way as ever they had before, and there was not a man among them in those times who baptized his children'.[8]

It was not without good reason that, in 1613, the Servite theologian Paolo Sarpi came to state that on principle the problem of the Marranos was baseless and without any canonical foundation whatever: since baptism is by its very nature an act of faith, to administer it by force must inevitably render it null and void.[9] It was therefore of no use to debate, as certain princes' jurists still did, whether the bulls of Paul III and Julius III had been annulled or were still in force despite Paul IV's 1555 proclamation, for all such argument was based on the groundless presumption that the

7. The letter containing the instructions of the duke of Urbino to his agent in Rome was published by C. Roth, 'Les marranes à Venise', *Revue des études juives*, 89 (1930), 177–8, 210, but not dated. Elements inherent in the text lead me to place it definitely between 1604 and 1607, when Malatesta was Duke Francesco Maria II's agent. Paul III had awarded the *tolleranza* to Duke Guidobaldo II on 1 Jan. 1539, ASVat, Arm. XLI, Vol. 12, fol. 4r.
8. ASV, Sant'Uffizio. Processi, b. 36. P.C. Ioly Zorattini (ed.), *Processi del S. Uffizio di Venezia contro ebrei e giudaizzanti, 1570–72* (Florence, 1984), pp. 142–3; B. Pullan, 'A Ship with Two Rudders: Righetto Marrano and the Inquisition in Venice', *Historical Journal*, 20, 1 (1977), 25–58, *passim*, especially 45. See also n. 42 below.
9. Gaetano Cozzi insists on the significance of Sarpi's theses concerning the validity of baptism, especially in the case of minors, even after the issue of the Marranos had cooled down. See 'Società veneziana, società ebraica', in G. Cozzi (ed.), *Gli Ebrei e Venezia (secoli XIV–XVIII)* (Milan, 1987), pp. 354–6.

Jews of Ancona had for a time been Christian. Hence the inability to decide clearly, the vacillation between severity and permissiveness, and the deliberate ambiguity characterizing persecution, tolerance, and protection alike. To find an answer, we now turn to the policies of the princes towards the New Christians, above all to that pursued by the Papal States and the way it affected the Republic of Venice, the main Italian secular power confronted with our issue.

The Jewish policy of the popes from Alexander VI to Julius III was continually shifting, and yet it was not without a certain line of its own: baptism, whatever way it is received, is binding, and all the same Christian charity invites us to show magnanimity towards the doubts and hesitations of converts. The Catalan Pope Alexander VI (1492–1503), who undoubtedly was acquainted with the Iberian Jews more intimately than his successors, considered Ferdinand and Isabella's assiduousness in casting perfidious Jewry from their realm *in celestem utilitatem*, a good reason for solemnly proclaiming them Catholic Monarchs.[10] Yet the papal court knew that winning converts must be a long-term project. For instance, on 10 January 1506 we find a married couple from Seville, then living in the Curia at Rome and who had been detained because 'humani generis inimico suggerente, nonnullos ritus judaicos observaverant', receiving a pardon.[11] On a spiritual plane the same policy that the Crown of Portugal followed in the case of New Christians was employed, allowing them a period of 20 years to become true Catholics. Perhaps that Jew Leo of Toledo had been baptized; a brief of 1507 grants him the licence to practise medicine in Benevento (which he had in effect already done for the 14 years following his banishment from the Iberian Peninsula); however, his colleague Alexander of Lisbon, who practised in the same town and to whom in 1522 Adrian VI confirmed the franchises already awarded him by Leo X, was almost certainly a Marrano.[12] After all, the *Università* of Catalan Jews in Rome itself had been made self-governing under this same Pope,[13] without any attempt to distinguish between refugees and converts,

10. S. Simonsohn, *The Apostolic See and the Jews: Documents: 1464–1521* (Toronto, 1990), p. 1447. Among the other merits of the Spanish sovereigns were numbered the war of Granada and the opposition to the French army in Italy and Africa. On the Marranos in Rome in 1493 and the case of a general pardon for 180 of them in 1498, see L. Thuasne (ed.), *Johannis Burchardi...Diarium sive Rerum Urbanarum Commentarii, 1483–1506* (Paris, 1884), Vol. 2, pp. 82, 491–2.
11. Simonsohn, *The Apostolic See: 1464–1521*, pp. 1487–8. Among other privileges which as 'Christi fideles' they again had the right to enjoy ('alias eorum statui convenientes') was the right to wear precious silks and jewellery, and ride on horseback.
12. Ibid., p. 1503; id., *The Apostolic See and the Jews: Documents: 1522–1538*, pp. 1628–9.
13. A. Esposito, 'Le "comunità" ebraiche di Roma prima del Sacco (1527): problemi di identificazione', *Henoch*, 12, 2 (Sept. 1990), 169–70, 176–85: the author emphasizes the process of a gradual breaking up of the 'communitas hebreorum hispanorum in

and in the 1520s we find banks run by Iberian Jews scattered throughout the Papal States, especially in Romagna, at Imola, Cesena, and Ravenna.[14] A soft-line policy was followed, confirmed by papal reluctance to give way to the Portuguese Inquisition,[15] one that only the firm hand of Charles V was able to win from a wavering Paul III.

In this period, namely from the beginning of the pontificate of Paul III, when the Spanish Jews were mentioned in curial circles it was with immediate reference to the extraordinary economic changes they were causing in Ancona. The bull of 23 December 1534 lumped 'turce, judei vel alii infideles' under one charter,[16] yet one month later (25 January 1535) a consul for the Levantine merchants was appointed[17] — the third to receive recognition after those of Venice and Florence — and the Levantine community emerged to full social status on 26 April 1536 when the Jews 'partium Orientalium' were given permission to have their own synagogue.[18]

Urbe commorantium' (already existing in 1496) and the special role played by the wealthy Catalans right from the beginning of the sixteenth century. For the disputes among the Spanish communities, see S. Schwarzfuchs, 'Controversie nella comunità di Roma agli inizi del secolo XVI', in *Scritti in memoria di Enzo Sereni. Saggi sull'ebraismo romano* (Milan–Jerusalem, 1970), pp. 97–8; A. Toaff, 'The Jewish Communities of Catalonia, Aragon and Castille in Sixteenth Century Rome', in A. Toaff and S. Schwarzfuchs (eds.), *The Mediterranean and the Jews: Banking, Finance and International Trade (XVI–XVIII Centuries)* (Ramat Gan, 1988), pp. 253–7.

14. Just a few instances: in Ravenna Leone Samorano, 'Portuguese', in 1520 (and his charter is again renewed for five years in 1531); and Leone Abulafia in 1531 under a five-year charter. At Imola, Master David ben Joseph and his son Joseph, 'Spanish' (alias the famous Rabbi David Ibn Yahya) in 1529 are awarded a sixteen-year papal *tolleranza*. At Cesena, Davit *hispano* has been operating since the 1520s. In Bologna, in the years 1519–1533, Rabbi Master Abram Cohen and his son Salvador, both 'Spanish' physicians and bankers (and relatives of Joseph Ha-Cohen, he too a banker in Genoa and Genoese, and the author of *Emeq ha-Bakha*); at Viterbo, Salomon of Moise Alfandari, a Sephardic Jew, in 1533 under a five-year charter: Simonsohn, *The Apostolic See: 1464–1521*, pp. 1593–5, 1606–7; *The Apostolic See: 1522–*1538, pp. 1753, 1759, 1775–6, 1827, 1835, 1843, 1866, 1870, 1929.
15. For an accurate and detailed essay on the Curia's efforts to control the working of the Portuguese Inquisition, see C.-M. de Witte, *La correspondance des premiers nonces permanents au Portugal: 1532–1553* (Lisbon, 1980–86), 2 vols.
16. Simonsohn, *The Apostolic See: 1522–1538*, p. 1924. The charter was confirmed and extended on 12 April 1535, and its validity was prorogated at the Pope's will, subject to four months' notice; it covered all merchants and brokers, thus authorizing import and commerce via third parties; and it revoked the clause requiring families of traders travelling 'ad partes infidelium' to remain in Ancona as guarantors. See ibid., p. 1960.
17. The Siennese merchant Gerolamo Maretto, appointed consul of the Oriental traders, called 'levantini', spoke Turkish ('multasque linguas, etiam turcicam callentem'); his successor, the Florentine Baldino Alessandri, appointed on 22 Jan. 1536, was a polyglot ('insignis tua in mercatura peritia, ad fidem, probitatem tuam et singularem Sclavoniae, Thurciae ac Grecae linguarum scientiam tuam adiuncta'), ibid., pp. 1931–2, 2016–17.
18. Ibid., p. 2027.

At this point the Holy See does not seem to have attached any importance to deciding whether or not they were Marranos. Any such distinction was ruled out by Julius III's brief of 5 December 1553, which granted the Ponentines (*Universitas hebreorum lusitanorum seu portugallensium*) those same privileges and immunities already awarded to the Levantines (*hebreis et Christi fidelibus orientalibus levantinis nuncupatis*):[19] well aware of the ambiguity inherent in these denominations, the papal chancery took care to bring them all together.

A glance through the notarial deeds of the period offers a very good picture of how this highly tolerant trend in papal policy was reflected in everyday life. The Jews never failed to state their origin, describing themselves as *portugallensis* or *lusitanus*, *levantinus*, *orientalis* or *gallus*, *de Salonicchio* or *de Safet*. For them a distinction still existed: the extraordinary indulgence of this brief did not yet remove it. As a representative example from among the many documents that have come down to us, we can mention an act dated 26 July 1554 certifying an agreement between two partners both with Spanish names. The one takes his oath *ad sancta Dei evangelia*; the other, who states he is from Constantinople and is one of the four heads of the Portuguese community, *super legem Moisis manibus tacto calamo more hebraico*. A year later, in the thick of the persecutions, a fiscal notary made an inventory of the possessions that had been confiscated from him as a Marrano.[20]

And yet not even at this time, when the papacy so clearly expressed its desire to be tolerant, did the upper echelons of the Curia mean to depart from the old tradition of hostility. In a letter to Julius III, Bishop Mignanelli insisted that, should Turkish sails be sighted off Ancona, in order to prevent 'any intelligence, either secret or open' of a landing, measures be taken to 'cast all Jewish Marranos, New Christians and Levantine scum out of the city'.[21] This is an attitude that will again come to

19. ASVat, Arm. I–XVIII, n. 1583. It was the Portuguese, merchants and others alike, 'de genere hebreorum, christianorum novorum nuncupatorum, aut alias qualificati essent, seu ab hebraica natione quomodolibet originem trahent', who were entitled to the safeconduct. R. Segre, 'Nuovi documenti sui marrani d'Ancona (1555–1559)', *Michael*, 9 (1985), 131–2. The 'Università delli hebrei portughesi in Ancona' had entrusted the negotiations with the Apostolic Chamber to a 'certain' Salomon Sarfati, actually the agent of Isaac and Jacob Abrabanel, who was present in Rome with letters of introduction from the duke of Ferrara: ASM, Ambasciatore Roma, b. 59, 22 Jan. 1551; 23 Feb., 21 May, 3 Dec. 1552.
20. Joseph Brudo 'de Constantinopoli' and Antonio Gomes 'de Ispania' (alias 'Lusitanian') took an oath to accept arbitration on the dispute over the will of the late Fabiano Rodrighes, Gomes's son-in-law and agent in Antwerp, who, on his deathbed in Cremona, had named Brudo sole heir. It is worth noting that Gomes and his guarantor were regarded as Portuguese Jews. See ASA, Notarile, 754, 26 July 1554; 189, 243, 19 June 1554; Segre, 'Nuovi documenti', pp. 188–9.
21. ASVat, Lettere principi, reg. 19, fols. 21r-v, 25r, Macerata, 28 June 1551. The other proposal made by the bishop was to bar any ship sailing from the Levant from entry

light at Ancona within a short time. After all, the greedy nephews of Pope Carafa were not alone at the Curia in urging Paul IV's hand: Cardinal Alessandrino (the head of the Holy Office and future Pius V) wrote on 20 June 1556 to his subordinate in Genoa that the Inquisition was reluctant to condemn men to the galleys as it was 'a sentence fit for demons and desperate men' ('penetentia da disperati o da dimonii') and better reserved for 'those Marranos since by many, nay, by almost all of them, have we been taken in' ('quasi da tutti siamo aggabbati').[22]

The change of policy introduced by Paul IV (1555–59) was to be long-lasting and, even when the violence of the persecutions had died down, it was never formally withdrawn. A quarter of a century later in fact it was evoked by the Inquisition, under the leadership of Cardinal Savelli, when he called for harsher measures against the Marranos, that resulted in three being put to the stake in Campo dei Fiori at Rome in 1583.[23] Even though no further evidence of formal executions appears in the records, the doctrinal position was thenceforth clear. For the post-Tridentine Church a principle was established which could no longer be questioned: the New Christian to all effects and purposes is a Catholic, the Marrano an apostate to be persecuted and condemned to the stake or to the galleys.

Until the mid-sixteenth century the Republic of Venice found it necessary to bring its attitude towards Spanish Jewry into line with its foreign policy. While Alexander VI was welcoming men of uncertain faith at his court, the Venetian Senate set forth the decree of banishment of 1497 in no uncertain terms: 'omnes marani tam qui venerunt ex Hyspania quam aliunde'.[24] Here is no generic reference to New Christians, but a clear declaration that the Marranos are forbidden to enter and ordered to quit the dominions of

into the harbour of Ancona on the pretext of anti-plague precautions. He further suggested (19 July) that all the costs 'be extracted from the Marranos and Jews of Ancona, thus sparing the treasurer general any pain for the time being'. Only the week before, Fabio Mignanelli had been appointed papal commissioner for the defence of the Adriatic coast, and was to become cardinal before the year was out. See C. Eubel, *Hierarchia Catholica medii et recentioris aevi* (repr. Padua, 1968), Vol. 3, pp. 32, 229.

22. BUG, E VII 15, fol. 54r. In a letter, dated 6 July 1571, Pope Pius V refused the doge and governors of Genoa permission to charter the 'Levantine Contardo' as physician for the city, since he had been sentenced to the galleys: ASVat, Arm. XLIV, Vol. 16, fols. 180v–181r.
23. In 1581 a stiffer papal policy towards the Marranos ('be they from Portugal or elsewhere') was echoed in a brief by Gregory XIII and by the stepping-up of repressive action by the Inquisition, also at the State level and especially in Tuscany and Piedmont, on the authority of the bull *Antiqua judaeorum improbitas*. See BAV, Barb. Lat. 1369, fol. 184r; R. Segre, *The Jews in Piedmont*, Vol. 1 (Jerusalem, 1986), pp. lxii, 618–9; also n. 44 below.
24. ASV, Senato. terra, reg. 13, fol. 22r-v; published by Kaufmann in 'Die Vertreibung der Marranen', 525–6.

the Serenissima. This clarity and resolution of intent is not inspired by papal, but rather by Spanish policy. The alignment with the Catholic Monarchs arose from Venice's need to consolidate the precious bridgeheads she was acquiring on the coast of Apulia. The great Levantine merchants had probably not yet arrived in sufficient numbers to cause a scandal in the lagoon city. It is a line of action that smacks entirely of politics. But when over half a century later, in July 1550, the Senate confirmed the ban, it was perhaps less due to the example Charles V had set at Antwerp the year before than to the strong wave of Catholic orthodoxy that was pervading Italy at the time. On this occasion, quite apart from the odd physician of Spanish-Jewish origin found here and there in the Venetian territories for some decades, there was a clear target: 'wandering Levantine merchant-Jews' have put down solid roots: all the commerce from Romania — this the Senate had been bemoaning somewhat insistently since 1541[25] — is in their hands. Whether these Levantine go to church or synagogue is a question that will begin to be asked only after the predicament of 1550, and then without much insistence, giving the benefit of the doubt. When in July 1555 the nuncio Archinto called on the republic to follow the example set by the Pope at Ancona,[26] the *Collegio* declined his threatening invitation in terms veiled with legalistic formalities: Julius III's brief was still in force and the Holy See had not repealed it. With this argument the republic seemed to wish to put forward the thesis that the papal document had exonerated her from applying the severe decree ordained by her Senate five years before. But this was not the true reason. The Venetian government's interests lay in encouraging the 'wanderers'' — namely, the great travelling merchants' — fruitful visits to the city. Of other Marranos in her territories she had no need. This was the reason why in 1559 the heads of the Council of Ten reproved the rectors of Padua University for letting Marrano

25. Ibid., mar, reg. 20, fols. 27v–28r; B. Pullan, *Rich and Poor in Renaissance Venice* (Oxford, 1971), p. 512; D. Jacoby, 'Les Juifs à Venise du XIVe au milieu du XVIe siècle', in *Venezia centro di mediazione tra oriente e occidente (secoli XV–XVI). Aspetti e problemi* (Florence, 1977), p. 214. In Nov. 1550 Venice took strong measures to crush Lutheranism (especially in Padua and Brescia) and heresy in general, and gave orders to reinforce the Inquisition in all her dominions with erudite laymen, ASV, Consiglio dei Dieci, Parti secrete, reg. 6, fols. 56r–57r.

26. The nuncio reported to Cardinal Carafa that when he asked the *Collegio* to take judicial action against one of Brianda Mendes' agents who was 'far-gone in perfidy, one of the *Savi grandi* leaped to his feet, upbraiding me for my punctiliousness, adding that the Apostolic See would have it otherwise, and he cited that misunderstood brief of the late Pope Julius III, which lets the New Christians of Ancona judaize as they like'. Archinto's embarassment grew when he was asked to prove that the brief had been revoked ('when I told the *Collegio* today that His Holiness had annulled it as being passed by mistake, some asked me to give evidence of the annulment, which could not be, though the summary held good'): ASVat, Lettere principi, reg. 22, fols. 42v–43r, 45v, 3 and 17 Aug. 1555.

students circulate wearing the black caps and not the yellow caps of the Jews. The rectors' defence was significant: they claimed that within the precincts of the university Jewish students had never been bound to wear the badge and thus eluded the necessity of deciding who was a Marrano and who a Jew.[27]

And yet it was not this home policy, this business, as it were, of policing the state, that best manifested the republic's position. What was of real concern was the creation of a credible image in foreign affairs. This is why, in May 1561, the *bailo* received orders to refuse a request by the son of the Grand Signior (the future Selim II) for a safeconduct for two Levantines, on the grounds that when they were living in Venice they posed as Christians, and had lately reverted to Judaism:[28] in other words, they were Marranos. In this period the repercussions on Venice's foreign policy as regards Levantines and Marranos came not from Rome but from Constantinople. In the decade leading up to the war of Cyprus (1570–73), fear and mistrust of the Porte daily conditioned Venice's policy. In 1565 the Este's ambassador noticed a reluctance among the Venetian patriciate to renew the Jews' charter and a propensity to seek their expulsion, be they 'old', Levantine, or 'those few Portuguese'. This attitude was due (he felt) to a desire on the part of 'many of the nobles' — the more enterprising, that is — 'to encroach stealthily on their trade'.[29] The envoy of Ferrara distinguished between the Levantines and the 'few Portuguese' who were definitely Marranos. In all likelihood many of the former were too, but the fact was ignored. The Venetians' greedy resentment of foreign merchants (ever present and shrewdly held in check by the republic) was one motivating force. Another was undoubtedly the symbiosis with the Turkish world which was at once commercially precious and politically more than suspect. It was 'for most high matters and policy of our State' that, on 9 April 1567, the Council of Ten revoked the sentence of banishment passed against João Miguez (now everywhere respectfully known as the duke of Naxos) for the abduction of his cousin Beatrice Mendes: an affair which we will touch on again later.[30] The powerful

27. ASV, Capi del Consiglio dei Dieci, Lettere, filza 61, 18 April, 19 May, 8 and 22 June 1559, to the rectors of Padua; Lettere di rettori e altre cariche, b. 83, 10 June 1559. The rule that the badge was not obligatory for Jews only when travelling was confirmed.

28. Ibid., Senato, secreta, Dispacci ambasciatori, Constantinopoli, reg. 7, fols. 45v–46r, 19 May 1561.

29. ASM, Ambasciatore Venezia, b. 52, 8 and 17 Aug. 1565; B. Ravid, 'The Socioeconomic Background of the Expulsion and Readmission of the Venetian Jews, 1571–1573', in F. Malino and P. Cohen Albert (eds.), *Essays in Modern Jewish History: A Tribute to Ben Halpern* (London—Toronto, 1982), pp. 36–40.

30. ASV, Consiglio dei Dieci, Parti secrete, reg. 8, fol. 82v. The vote for a pardon was taken unanimously. B. Ravid, 'Money, Love and Power Politics in Sixteenth Century Venice: The Perpetual Banishment and Subsequent Pardon of Joseph Nasi', in *Italia Judaica*, Vol. 1 (Rome, 1983), pp. 179–80.

Marrano councillor of the Grand Signior was pardoned, while at the same time the Pope was secretly warned that he was scheming to land a Turkish force at Giulianova near the borders of the Papal States. In communicating this information to their ambassador at Rome, the Ten showed that they had forestalled the consequences of this threat: 'We have sent orders to our fortresses in the Levant to seize the said Jews and by no means leave them at large there'.[31] This attitude stiffened with the outbreak of the war so that Jew and Turkish spy came to mean almost the same. The Ten, it seems, did not take exception when the governor of the galleys stopped a ship crossing from Pesaro to Salonica with '280 head of Jews counting old women and babes' and put the 70 able-bodied men to the oar. Better, he wrote, 'they serve Your Serenity in your present need of crews than with their lives and skills they serve our enemies and the enemies of the Christian faith'.[32]

From the moment, however, when the republic, worn out by the cost of the war and the blockade on trade with the Levant, signed a separate peace with the Porte, her policy towards foreign Jews underwent a radical change. Not a few misgivings must have accompanied the decision taken by the Council of Ten on 23 December 1573 that, following in the steps of 'the popes and other Christian princes', granted safeconducts to Venice for the next two years to 'Spanish and Portuguese Jews and any other Jewish stock living in the western [Ponent] regions and the territories of other Christian princes'.[33] A contorted formula, but the political intention was clear. The peace treaty had reopened the ports to Jewish Ottoman subjects, making no distinction between Oriental, Spanish, and Portuguese Jews, viz. Marranos.[34] Now the invitation was extended to Ponentine Jews living in Catholic countries, namely to the Marranos. No easy task for the ambassador in Rome to reassure the Pope about the impact of so open a

31. ASV, Consiglio dei Dieci, Parti secrete, reg. 8, fols. 114v-115v, 26 May 1568, to the *bailo*; fol. 122r–v., 12 June 1568, to the ambassador in Rome.
32. Ibid., Capi del Consiglio dei Dieci, Lettere di rettori e altre cariche, b. 278, fol. 281r–v, Lesina, 6 May 1571, Santo Tron to the doge. On 30 Aug. a quarter of the more than 200 passengers taken prisoners at Lesina (Hvar), were disembarked, after buying back their freedom, from four boats at Ferrara 'poor and wretched ... and sick from the suffering they say they have undergone': ASM, Archivio per materie, Ebrei, b. 2, Ferrara, 31 Aug. 1571, Gerolamo Montecucchi to the duke.
33. ASV, Consiglio dei Dieci, Parti secrete, reg. 10, fols. 159r–160r. The decision was approved by 21 votes to 5 with 5 abstentions. Actually the charter was never legally implemented, for it lacked the required ratification by the Senate; see B. Arbel, 'Gli ebrei e l'attività di Salomone Ashkenazi nella guerra di Cipro', in *Gli Ebrei e Venezia*, p. 181.
34. One of the clauses of the peace treaty between the Turks and Venice, signed at Pera (Constantinople) on 7 March 1573, was that 'in Venice, Turkish and Jewish merchants from Constantinople and other places be allowed to come and go freely with all their goods, and those of Venice be able to do the same in Constantinople and her dominions, without suspicion, in freedom and peace'. BAV, Urb. Lat. 1020, fol. 116v.

position, and Cardinal Gallio, the secretary of state, recapitulated a little later to the nuncio what made the Jews of Ancona different from the Jews of Venice: 'The Levantines there [in Ancona] are either Jews or have become Christians and live Christian lives without turning again to Judaism, while those Marranos [of Venice]...are baptized but live as Jews'.[35] In the 1570s rivalry flared up between the republic and the Papal States, for the route between Split and Venice, efficiently laid out by the 'Portuguese' Daniel Rodriguez, was emerging successfully as an alternative to that between Ragusa and Ancona. It so happened that, quite beyond the activities of the Holy Office, the republic itself laid its hands directly on some 'retagliato' who had confessed to being circumcised after baptism (as happened at Vicenza in 1575);[36] but now, at the end of the century, the route between Venice and the Levant was basically organized and run by Sephardim, whether they were Marranos or not.

It is worth comparing Venice's policy, vacillating between devotion and opposition to the Holy See, with the policy of two sovereigns on the other bank of the Po, who were unable to forget they were papal vicars. While the executions at Ancona were still ringing in the air, Ercole II explained to his ambassador at the Curia that the Pope could not treat him with the same haughtiness with which he treated the duke of Urbino.[37] In truth, anyone who could have read the minds of those two princes would have found them equally inclined to welcome the Marranos. In real terms, however, there was a difference, and it lay in the greater bargaining power the duke of Ferrara had compared with the duke of Urbino. And yet for all the weakness of his position, the latter held a card that the House of Este always lacked: a port, Pesaro, which, were it in fact opened to trade, might become a wonderful base for Jewish commerce.

In Ferrara the Sephardim were welcome right from the time when their wandering began. But the dukes stepped lightly: in 1520 communities of Italian and German Jews were authorized.[38] There was no Spanish community and the latter formed part of the Italian Jewish one. During the

35. ASVat, Nunziatura di Venezia, reg. 13, fol. 234r–v. A. Buffardi (ed.), *Nunziature di Venezia* Vol. 11, (Rome, 1972), pp. 139–40.
36. ASV, Consiglio dei Dieci. Parti secrete, reg. 11, fols. 45v–46r, 47r, 28 Jan., 25 Feb. 1575, to the *podestà* of Vicenza and Chioggia: an order that the three *retagliati* — the Spanish-speaking Andrea de Silva ('he has confessed to being born a Jew and baptized at Ferrara') and the two charged with spying for the Turks in Chioggia — be transferred to the prison of the Council of Ten.
37. Ercole II had written to his ambassador at the Curia: 'We cannot believe that we may be treated in this offhand way, as is the duke of Urbino, since we are not, thank the Lord, to be compared with him'. ASM, Ambasciatore Roma, b. 57, Ferrara, 3 Dec. 1555.
38. Simonsohn, *The Apostolic See: 1522–1538*, pp. 1613–14.

Congress of Bologna, the lord of Granvelle found time to stop the Este representatives and demand the arrest and extradition of a Flemish Marrano living in Ferrara, both for the reason that he was a Marrano and because he had fled after defrauding the merchants of Antwerp.[39] But then the sources fall silent and one is led to imagine that the duke took his agents' advice and helped the refugee get out of harm's way.

In the Este's dukedom we find a climate so favourable to Iberian Jews as to make the distinction between who is and who is not a Marrano extremely difficult. The loan-bank authorized in 1530 with outstanding franchises and the comforting support of a papal *tolleranza*,[40] opened the way for a long series of charters: a story rich with the achievements of the Abrabanels and (for some years) the Mendes, and documented by Jewish prints, both for the eyes of educated Europe and the pride of the Sephardim. In this period the ruler spoke out with a brutal honesty such as far more powerful princes have hesitated to use, and which he himself had to abandon soon enough. It was in 1550, with the clement Julius III in Rome, that the duke invited Jews 'of Lusitanian and Spanish nation' to take up residence in his lands, adding the specification 'even if, for whatsoever reason it be, they have said they are not Jews and call themselves Christians'.[41] By no other Christian prince was the meaning of the adjective 'Lusitanian' made plain with such outspoken and provocative clarity.

But the House of Este could not keep up this stance till the end of its much-disputed dominion over Ferrara, which it held by papal investiture. In November 1571 Abram Righetto, whom we have already seen declare himself a Jew, escaped from the dungeons of the Holy Office in Venice where he had been imprisoned as a Marrano, and the duke of Ferrara had to agree to his arrest and send him back at the request of the Spaniards for reasons not religious but military ('because, being kin with Gio. Michas, he could know something concerning the welfare of our League'). The Este government communicated the decision to the nuncio in Venice,[42] but it was acting more as an ally than as a devotee of the Holy Office, scourge of

39. ASM, Ambasciatore Roma, b. 33, Bologna, 30 Dec. 1529, 4 Jan. 1530.
40. The ten-year papal *tolleranza* granted to Leone Gedelia Juda Nigri, 'Portuguese' in Ferrara, his family and partners confirmed the banking charter awarded him by the duke, and entitled him to hold land, employ Christian wet-nurses in his home, resort to summary justice and be exempted from wearing the badge; dated 20 Nov. 1529, on 26 June 1530 it was prorogated for 30 years. See Simonsohn, *The Apostolic See: 1522–1538*, pp. 1761, 1794.
41. The charter was confirmed and extended on 24 Sept. 1559. See A. Balletti, *Gli Ebrei e gli Estensi* (Reggio Emilia, 1930), pp. 77–8; B. Polacco, 'Rafanellum', *Annuario di studi ebraici, 1963–64*, p. 60; Segre, *The Jews*, Vol. 1, p. 479.
42. ASM, Ambasciatore Venezia, b. 54, Venice, 31 Oct., 7 Nov. 1571, 16 Jan. 1572; b. 58, Ferrara, 7 Nov. 1571, 7 Jan. 1572; Carteggio principi esteri, b. 1410/156, Rome, 10 Nov. 1571, cardinal of Pisa to the duke of Ferrara. See also n. 8 above.

heretics, infidels, and apostates. By the end of the 1570s, however, the Holy See and the Inquisition at Rome had the dukedom well and truly in its clutches — this had not always been the case — and anything like the 1550 statement in favour of the Lusitanians had become unthinkable. Times had changed. Just what this change meant becomes clear when, in 1578, the Inquisitor General of Coimbra announced he had the testimony of New Christians returned from Ferrara, saying that they had seen their fellow countrymen receiving circumcision, going to the synagogue, and celebrating Jewish rites:[43] the Este dominions were not just a haven for fleeing Marranos, but a place where apostasy was publicly condoned. The evidence passed on by the Portuguese Inquisition to the Holy See was irrefutable, and continued to flow in. The fate of Joseph Saralvo and his two companions, extradited from the Duchy of Ferrara by the papal authorities and put to the stake in Rome in 1583,[44] marked the close of the good years for the Marranos of Ferrara, which was soon to fall completely under papal sway. Not only the 'Portuguese' left the city, but also a great many Spanish Jews — even the last members of the Abrabanel family — chose to set out for Salonica.

Though far weaker, the voice of the duke of Urbino was heard none the less, and he was well aware of the winds of change, for his diplomatic information service was efficient beyond what one might have expected in so small and frail a state. The moment of truth would come in 1556. Yet even before the Ancona tragedy, Portuguese Jews in Pesaro, with Emanuel Biciaccio as their constant leader, organizer, and financier, were at work opening the way to a safer life in the Ottoman lands.[45] For some time this little port was a centre for Jewish activity, and the powerful world of

43. J. Mendes da Remedios, *Os Judeus em Portugal* (Coimbra, 1928), Vol. 2, pp. 208–9; A. Baiao, 'A Inquisiçao em Portugal e no Brazil', *Archivo historico portugues*, 7 (1909), 233, 237, 238–9; 9 (1911), 477–8, 479. Information on the presence and activity of Marranos in the duchy continued to accumulate even in the 1580s. See, for instance, the testimonies presented to the Inquisitor of Ferrara by Cardinal Savelli in ASM, Archivio per materie, Ebrei, b. 2, c.1583–86.

44. Saralvo, alias Gabriel Enriques (1523–83) had emigrated in his twenties to Ferrara, where he worked as jeweller. C. Roth, 'Joseph Saralvo: A Marrano Martyr in Rome', in I. Elbogen *et al.* (eds.), *Festschrift zu Simon Dubnows siebziegstem Geburtstag* (Berlin, 1930), pp. 180–86. Of the six Marranos extradited to Rome in 1581, the brothers Giovanni, Diego, and Arias Lopez repented and were spared the stake. To prove their sincerity, they 'donated to the Holy Church' their wives and children, meaning they obliged them to follow suit. Giovanni Lopez, alias Samuel Abudente, is certainly no other than the trusted and powerful financial advisor whom we find a couple of years later in the Curia, under the pontificate of Sixtus V: ASM, Carteggio principi esteri, b. 1418A/168, Rome, 22 Aug. 1581, 26 Feb., 4 May, and 23 Aug. 1583, Cardinal Savelli to the Duke of Ferrara.

45. ASP, Notarile, Alessandro Allegrucci, filza 352, reg. 10–11, *passim*; M.A. Cohen (ed.), *Samuel Usque's Consolation for the Tribulations of Israel* (Philadelphia, 1965), p. 214.

finance in the Levant, in its attempt to boycott the papal docks, endeavoured perhaps to create a bridge between the Marranos grouped around the Mendes and the weak Italian-Jewish nation. The Della Rovere envoy at Rome indeed had to endure the signs of disdain and the humiliations that Ercole II of Este felt were incompatible with the dignity of a prince. Nevertheless, the little dukedom on the Marches still managed to maintain a certain degree of independence. During the months when they were taking in refugees from the Papal States, the Jews of Pesaro secretly contacted the duke of Naxos to secure the assistance of the Ottoman fleet against raids by Maltese galleys that were paralysing their trade.[46]

In Piedmont and Tuscany the Spanish Jews had a very different time of it. A comparison between the policies of the House of Medici and the House of Savoy shows how the former created an environment in which Sephardic immigrants were more easily able to put down roots. Matters started to take shape a quarter of a century earlier in Tuscany: Piedmont, under French occupation till 1559, only managed to reacquire its own political identity through the energetic rule of Emmanuel Philibert. In the court at Florence, Spanish was spoken quite widely: it was, after all, the language of Duchess Eleonora of Toledo, whom Benvenida Abrabanel had followed with her son Jacob. In 1547 the ducal government authorized them to run loan-banks anywhere in their dominions except Florence (in line with the Savonarolan principles of the republic, it was felt that the prohibition of Jewish money-lending stated half a century earlier should remain in force, though only in the capital).[47] In the first half of the sixteenth century the Jewish population in Tuscany had thinned out, and this sudden decision by the court in favour of a family recently banished from the Kingdom of Naples seemed to preclude any chance of revival for the small local communities. That this was in fact the case soon became apparent through a secret affair that puzzled even the Medicean chancery for over half a century.

In 1548 Duke Cosimo granted the representatives of the Portuguese Jews a charter containing a totally uncustomary clause stipulating that 'they be treated as nobles and citizens'. It included franchises to settle and trade, and the right to establish business premises, together with protection

46. D. Kaufmann, 'A Letter from the Community of Pesaro to Don Joseph Nassi', *Jewish Quarterly Review*, 4 (1891–92), 509–12.
47. The struggle for power in the world of Jewish finance between the two leading families (the Da Rietis, on one side, the Abrabanels and their partners the Da Pisas, on the other) is well illustrated by B.D. Cooperman, 'A Rivalry of Bankers: Responsa concerning Banking Rights in Pisa, 1547', in I. Twersky (ed.), *Studies in Medieval Jewish History and Literature*, Vol. 2 (Harvard, 1984), pp. 49, 64–5, 70.

against the Inquisition, excluding the Spanish- Portuguese procedure and incrimination for matters dating from prior to their arrival in Tuscany. The duke knew he was playing a dangerous game and gave the original deed to the Portuguese negotiators themselves, so that no trace of it remains in his archives. As a grand-ducal chancellor remarked at the beginning of the seventeenth century, Cosimo liked 'to have his cake and eat it' in these delicate issues, and for this reason 'never put pen to paper'.[48] He did not sign the charter, and the beneficiaries had to content themselves with a secretary's signature. This was an unusual procedure in the Italian states which foreshadowed, we feel, at least on a political level, the great Sephardic expansion in Tuscany, and especially in Livorno, half a century later.

This secret affair was not the only business of its kind. In 1551, for example, a licence was granted to the 'Oriental Jew' Servadio of Damascus to open a Levantine trading centre in Florence. The economic outcome appears to have been mediocre since the attempt to export Tuscan silks and woollen cloth to the Ottoman world came up against the already well-established Venetian competition. However, the entrepôt was still running 20 years later when another Levantine Jew took over the management from Servadio.[49] Cosimo was prepared to expose himself to censure from the Holy See only for the Sephardim: in fact he cared little for the 'old' Jews and was even ready to maltreat them. When desirous of papal investiture as grand duke, he agreed to concentrate them in a ghetto, where the Portuguese Jews, protected by their charter, were not enclosed. It was in line with this approach that 20 years later his son Ferdinand set the Livorno enterprise afoot, with such remarkable success.

In fifteenth-century Piedmont the Jewish population had been far more widespread and well fostered, but the preaching of the Franciscans and later the French occupation had made heavy inroads, and the surviving communities had been forced to gather in the narrow area along the Sesia which had remained under Hispano-Savoyan control. However, with the restoration of the House of Savoy, the Jewish population started to return. On 4 September 1572, Emmanuel Philibert signed a decree stating that the

48. For the unusual bureaucratic procedure, uncovered in 1602 by an investigation called for by the grand-ducal government and carried out by the fiscal auditor Vinta, see R. Toaff, *La nazione ebrea a Livorno e a Pisa (1591–1700)* (Florence, 1990), pp. 436–8. In 1573 Fernando Mendes held the text of the charter.

49. V. Bonazzoli, 'Gli Ebrei sefarditi del Levante e i Ragusei nel Cinquecento: dal commercio di cuoi e tessuti al profilarsi di nuovi equilibri mediterranei', in A. Di Vittorio (ed.), *Ragusa e il Mediterraneo* (Bari, 1990), especially pp. 168–70; Segre, *Nuovi documenti*, p. 132. In the preamble of the Piedmontese charter of 4 Sept. 1572 to the Levantines, including those of Sephardic origin, Emmanuel Philibert recalls the analogous charter, signed by Cosimo de Medici on 16 June 1551. See Segre, *The Jews*, Vol. 1, p. 479.

gates of his states were open 'to all the Jewish nation and its stock, Italian, German, Spanish, Portuguese, of Levant, Barbary, and Syria',[50] but he certainly did not mean to repopulate the Jewish communities with the usual run of small-time loan bankers, shopkeepers, and craftsmen from Provence and the Papal States, let alone from Germany and Bohemia. The newcomers were to be (unlike the 'old' Jews) exempted from sharing in the communities' financial obligations, protected against accusations and sentences dating from prior to their arrival, and guaranteed against forced baptism, allowed to keep slaves (who, should they wish to become Christians, must first buy back their freedom), to practise as doctors and treat Christian patients, and have cemeteries and synagogues of their own. What the duke aimed at was to attract wealthy and skilful Sephardic merchants and businessmen to his realm, as shown by the individual charters granted to a group of Portuguese 'of Jewish stock' soon after the solemn decree.[51] Moreover, they were authorized to trade in agricultural foodstuffs (cereals, rice, oil, and livestock) and above all to carry out financial operations on the international market up to the colossal sum of 20,000 *scudi*. Actually these privileges had effects that differed from area to area, and although, for instance, the Mazaods managed to establish a prosperous and long-lived bank in Savigliano (a little town mid-way between Turin and Cuneo), it was towards Nice that this policy looked, for the re-established Piedmontese dominion felt suffocated in the narrow space between the Alps and the Po, and wanted to open up a sea-port for the Levant. The duke did not aim at anything like Venice's huge international trade structure or the Medicis' scheme for uniting crafts, finance, and sea trade at Livorno, though to this latter element — sea trade — he gave a good deal of attention.

No sooner was the edict out of the ducal palace than the Spanish ambassador got wind of it.[52] King Philip II was not slow in complaining that the cousin who had served him so well on the battlefields of Flanders, was sheltering apostates and rebels in his territories and that, though the borders were already crawling with Huguenots, he was now allowing this new serpent to creep into the bosom of a Catholic state. At the same time the Portuguese Inquisition was claiming that he was granting safe-conducts to members of great New Christian families who had just recently emigrated in secret: once again, the Portuguese who kept flowing into Italy were true Marranos.

Emmanuel Philibert knew right from the outset how risky his plan was, and he knew he must negotiate. He gave the Pope a cautious answer: the

50. Segre, *The Jews*, Vol. 1, pp. 478–85.
51. Their names: Jacob Mazaod, Dr. Ruy Lopez, Paulo Lopez, Fernando Lopez, Antonio Rodriguez, and Antonio Hurtado; ibid., pp. 486–90.
52. Ibid., pp. liii–lvii

Jews had come — and it is not unlikely they were already there before the edict — 'and we can hardly go back on our word, even if they are Marranos, while to the best of our knowledge they are not'. If they turn out to be Marranos, it means they have abused his trust, and they will be 'punished'. There follows another decree stating that 'apostate Marranos' are excluded from the sovereign's favour, and the names of some holders of safe-conducts do in fact disappear without trace from the public documents. We may suppose that it is a question of the few known Marranos, and that, rather than undergoing punishment, they were sent away in good time.

Emmanuel Philibert's policy was to achieve its aim only partially and indirectly. Three-quarters of a century was to elapse before the free port of Nice could be opened and a modest maritime trade centre set up. The House of Savoy's only gateway to the sea did not have the same advantages as Ancona or Livorno: it was hemmed in by mountains, had no access by river, was far from the centres of production and consumption in Piedmont. None the less, Sephardim, Marranos or others, came and established numerous small communities during the 1570s. However, the next duke, Charles Emmanuel I (1580–1630), put the brake on Emmanuel Philibert's policy: his marriage with *infanta* Catherine, daughter of Philip of Spain, obliged him to renew his father's ban on Marranos. The words are the same but this time they are meant more seriously and the Sephardim who stayed on in Piedmont were compelled to erase any hint of apostasy from their lives.

After this general survey of policy in the Italian states towards immigrant Jews from the Iberian Peninsula, we can now examine the lives the new communities of Spanish and Portuguese Jews created for themselves on Italian soil.

One of the most significant aspects of the question is certainly privateering or, as the jurists put it, 'bellum omnium contra omnes'. Quite clearly it was not only the Jews who suffered the consequences, but their presence, along with their goods, aboard captured vessels was a constant subject for discussion and a good excuse. And, among Jews, the Sephardim were a particular target, as the cardinal of Granvelle, viceroy of Naples, reminded his sovereign: neither the duke of Savoy nor any other Christian prince should with impunity be allowed to give protection and safeconduct to Marranos since they are rebels against the Crown of Spain and when captured remain so and must be treated as such.[53] Giovanni Carafa, Duke of Paliano and nephew of Paul IV, was well known to have had fewer diplomatic qualms, but when one of his galleys stopped a

53. Ibid., pp. 497–8.

Venetian ship in 1559 and found five Marranos aboard, his captain took them as slaves and, as the Venetian ambassador (and later doge) Alvise Mocenigo wrote,[54] 'completely stripped them' not only of their goods but even of their clothes.

The highly respected patrician had a hard time of it explaining this affair in an audience with Paul IV (not an easy person to talk to at the best of times) since, he admitted, 'the laws of our most illustrious State forbid Marranos to sail in our ships'. He finds himself once again face to face with the old problem: are Marranos 'those who, once Christian, have become Jews, or simply those who though they live as Christians are intrinsically Jews': an important matter to clarify from the point of view of the negotiations though irrelevant in point of fact as (in Mocenigo's view), 'both are vilest and most loathsome breeds'. He managed, however, to have the prisoners released, comforted, and their goods ransomed. We get the end of the story not from the diplomatic correspondence, but from a notarial deed: the five captured Sephardim's attorneys paid back a loan given by the Venetian ambassador for the ransom demanded by the duke of Paliano and his captain.[55]

In one form or another this became normal procedure, without exception, for all the courts and princes who controlled privateering, either directly or otherwise. In 1578 the papal fleet stopped a ship with Sephardim aboard off Ragusa. It was naturally decided that they were Marranos and their release and the return of their goods were refused. In reply, the all-powerful Solomon Ashkenazi contacted the Venetian *bailo* in

54. The last lines of Mocenigo's letter to the doge give a glimpse of the horror a Marrano felt at the idea of capture at sea, especially by one of Carafa's triremes: 'I have no more to report you than that these Marranos or Jews claim they owe their lives to Your Serenity. Many from this country [the Papal States] who have come to the embassy to visit them, thank me for their deliverance as though I were almost a Messiah, and indeed, Very Serene Prince, it seems to those Jews, and in fact to everybody, wonderful and almost a miracle that the Pope has let them go free. Your Highness therefore may, in my humble opinion, know that His Holiness holds your illustrious dominion in highest esteem'. ASV, Senato, secreta, Dispacci ambasciatore Roma, filza 1, 14 Jan. 1559; BNP, MS Ital. 1146, fol. 16v.

55. The merchandise (cassia, carpets, and cloth) confiscated by the duke of Paliano's trireme, was donated to the Republic of Venice by the Pope and returned to the Marranos by the ambassador's secretary: a merchant from Trent received the goods for three of the Marranos (Moise Alalvo, Abram Colum, and Menachem de Stella), and the Abrabanels' agent Salomon Sarfati (see also n. 19) for the others (Davit Almedi, Josef Juda, and Abram Menachem), ASR, Notai Tribunale RCA, reg. 6463, fol. 80r (13 March 1559), 277r–v, (3 Sept. 1559). In this case the Duke of Paliano had to forgo the usual proceeds he derived from the ransom of the 'slaves': in the same period he sold a stock of them for 600 *scudi*, and it is not unlikely that most were Sephardim. ASR, Camerale I. Mandati camerali, Vol. 902, fols. 82v–83r, 7 July 1559. P.A. Guglielmotti, *La guerra dei pirati e la marina pontificia dal 1500 al 1560* (Florence, 1895), p. 260, distinguishes between private and state navy (run under the pontificate of Paul IV by his nephew, in his capacity as captain general).

Constantinople to arrange for the fleets of the Porte and Saint Mark to come together in a blockade off Ancona. The secret dispatch with which the Council of Ten sent its reply to the *bailo* seems almost to contain an impotent sigh of hope: the republic must act with great reserve, but would not mind at all if the Turks would go ahead on their own.[56]

With time it became a generally accepted principle that anyone who captured Sephardim shipping goods could decide they were Marranos and put them up for ransom. In 1582 the grand duke of Tuscany did not hesitate to demand 10,000 ducats from the Pope for the release of the Spanish Jews taken by the knights of Saint Stephen while travelling with a papal safe-conduct.[57] And it became not uncommon to see slaves being thrown into the 'bath' at Livorno by this knightly Order in the hope of a ransom from the local Jews, which was paid by means of an *ad hoc tax*, the *gabella*.[58]

So it was that the Sephardim lived in Italy, constantly looking towards the Levant, often ending up there, and always putting to good use their many, many contacts in Ottoman territory: a double-edged weapon this, at times their salvation, at times their ruin.

Till the early 1570s the two great Sephardic families, Mendes and Abrabanel played, the leading part as intermediaries between East and West. The Mendes were New Christians who had emigrated from Portugal to Antwerp, the Abrabanels Spanish Jews banished from Naples. The two families had, above all, a common triangular system of protection working between Venice, Ferrara, and — perhaps more for the Abrabanel than the Mendes family — Ancona. The whole business began with the almost mythical sum of 250,000 ducats that the eight-year-old Beatrice Mendes brought as her dowry when she moved with her family and following to Italy. She produced Venetian safe-conducts when the captain of the port of Pavia stopped her cortège, and made her way to the city of Saint Mark where she deposited part of her dowry in the Mint as a security. The Venetian government considered they had replaced the Spanish and Portuguese in the task of guarding their fabulously wealthy charge, and this is why in 1553 they banished the cousin who abducted her, João Miguez, later famous under the name of Joseph Nasi, duke of Naxos. But the Mendes were not to become a family of Italian Jews. In 1556 it was they who, from Constantinople, co-ordinated the retaliation to the burning of Marranos at Ancona. Notwithstanding the plethora of information, there is nothing to show that the two families ever saw eye to eye. In 1554 it was

56. ASV, Consiglio dei Dieci, Parti secrete, reg. 11, fol. 173v, 10 July 1578.
57. ASVat, Nunziatura di Firenze, reg. 7, fol. 475r, Florence, 8 Sept. 1582.
58. Toaff, *La nazione ebrea*, pp. 270–73.

the manager of one of the Abrabanels' banks in the Marches, Moise Basilea, rabbi of the Italian Jewish community — mostly refugees from the Kingdom of Naples — who regarded the Mendes with rivalry and suspicion.[59] And we certainly know how closely the Marranos were linked with the Mendes and how Basilea was in the forefront of the opposition to the boycott of Ancona.

The Abrabanels stand out as the finest example of Sephardim remaining as Jews in Italy. Their story spans some 30 years in Ferrara and reaches the apogee of its splendour in the 1560s. At one point the agent for the House of Este in Venice was angered by the audacity of 'a certain Don Lion the Jew' who came to the embassy with a patrician of the Pesaro family, claiming to have been asked by the duke of Ferrara to make an estimate of the house for a possible sale.[60] In reply, Alfonso II simply ordered his outraged representative to find a residence for Leone Abrabanel, who was under his protection: his respect for the Jew had perhaps led him to violate diplomatic form. By the end of the 1570s, however, even in Ferrara things had changed and the third child of Jacob Abrabanel, renamed Alfonso after his princely godfather,[61] had the duke's support *vis-à-vis* his two brothers and cousin who were trying to circumvent canon law in the division of inheritance with a Christian convert. It was the road to Salonica now for what was once the most illustrious Sephardic family in Italy.

Since the war of Cyprus (1570–73), the most prominent Sephardim of Italy no longer kept those close links with the great families that had developed freely in the middle years of the century. There were of course family ties with the Mendes for João Miguez who, after his unsuccessful abduction of the daughter of Doña Brianda de Luna, had married his other cousin Reina. But Miguez's politics were not family-oriented. Circumcised at Constantinople in 1553 and having gained power at the courts of Süleyman the Magnificent and Selim II, he was a source of suspicion and fear for the Serenissima of twenty years: the Venetians worried about his schemes to land Turkish troops in Italy, they believed he had engineered and run the war in Cyprus and, not without reason, they suspected he was financing the exodus of Jews to the Levant via the Italian ports.

59. Don Josef Abrabanel leased Rubi Moyse de Basilea his Arcevia loan-bank for five years at the considerable price of 2,550 *scudi*: ASA, Notarile, Antonio Manfredi, 754, 6 March 1554. Segre, 'Nuovi documenti', p. 152. For a biography of our rabbi, usually called Basola in Jewish sources, see *Jewish Encyclopedia*, Vol. 4, col. 310. It might be worth investigating whether it was a mere coincidence that in the same year the Synod of Ferrara issued the ordinance on community rabbis he fiercely opposed. See R. Bonfil, *Rabbis and Jewish Communities in Renaissance Italy* (London, 1990), pp. 140–43.
60. ASM, Ambasciatore Venezia, b. 52, Venice, 11 June 1565; b. 58, Ferrara, 14 June 1565.
61. BAV, Ott. Lat. 2452, fol. 124r.

Though of humbler origins and without Miguez's political influence, the 'Portuguese' Daniel Rodriguez stands out for the tireless and loyal service he gave the Venetian state. The mission the republic assigned him on 13 January 1574, namely to go onto 'Turkish soil' and ascertain the motivation and efficacy of the measures taken by the government to 'block trade goods' coming in from Venice, was not particularly important, as the modest sum of 100 ducats he received in payment indicates.[62] Two months later he was commissioned to examine the factors governing the establishment of the new frontier in Dalmatia. In the eyes of the Serenissima he was still a minor figure at this point, but a man to be trusted in delicate matters which only someone with the right contacts in Ottoman government circles could handle. He was even perhaps a man to be feared for, on 29 April 1574, the papal secretary of state gave orders to arrest him should he dare set foot in Ancona or elsewhere in the State of the Church.[63] Forty years of age, 'thin in the face and pale, dressed in Levantine style and wearing a yellow turban', he was guilty of 'having tried to stop traffic from the Levant coming to the port of Ancona'. Daniel was already setting his great project afoot, and would complete the planning stage in 1577. His idea was to direct trade from Bosnia to Split using Sephardic merchants from Serajevo and Monastir. The papal authorities got wind of the scheme while it was still in the pipeline, but once it came out into the open Daniel's name was prominent in the public and notarial documents for a considerable time.

Rodriguez figured not as a Levantine merchant but as an organizer, and we find other Jews working alongside him as Levantine agents. In September 1578 he is registered in the deeds of a Roman notary,[64] no longer as 'habitator Venetiis' but as 'de Zaffetta, territorii tribunali de Soria Damasceni' (namely from Safed), acting on behalf of the Levantine merchants of Venice and Ancona. However, perhaps as with Sixtus V things had changed in the Curia, he let another Venetian, Ventura Finzi, have his agency. This was in connection with business of a proposal to be made to the viceroy of Naples for the setting-up of a trade centre run by

62. ASV, Consiglio dei Dieci, Parti secrete, reg. 10, fol. 163r; reg. 11, fols. 4v–5r, 8r–9r, 16 March 1574.
63. ASVat, Lettere principi, reg. 38, fols. 332r–v, 355r–v, 396r–397r, Rome, 29 April, 27 May, 30 Oct. 1574, cardinal of Como to the governor of Ancona, Cristoforo Boncompagni. For an updating of the research on the issue, see R. Paci, 'La concorrenza Ragusa-Spalato tra fine Cinquecento e primo Seicento', in Di Vittorio (ed.), *Ragusa*, especially pp. 185–9.
64. Attached to the power of attorney of *dominus* Daniel Rodrigha to *dominus* Ventura Fintio is a list of 20 members of the Levantine nation in Venice and of 8 in Ancona who had given their 'consul' the task of negotiating charters with the Pope and all the 'lords and potentates of Christendom', with the aim of increasing trade with the Levant for the general good: ASR, Archivio dei 30 notai capitolini, Uff. 16, Bernardino Pascasio, reg. 11, fols. 371r–374r, 7 and 26 Sept. 1588.

Levantine Jews. This Finzi, so closely connected with Rodriguez, turns up again ten years later in Madrid in the final attempt to delay the banishment of the Lombard Jews from the last area of Spanish territory where they remained.[65] Not even Finzi has left any revealing evidence of himself in the mercantile transactions of the period, but he undoubtedly won the trust of the Levantine Jews as a representative in the corridors of power, especially in Spain. To get his letters patent for silk production (already approved by Sixtus V) extended to the Kingdom of Naples, Magino di Gabriele, another Venetian Jew, also called on the services of his influential cousin Ventura.[66] This is the same Maggino Meir di Gabriello Zarfati whom we find as consul of the Jews of Livorno, in the negotiations that lead to the 1591–93 charter known as Livornina.

In the documents of the period these same men turn up with a number of different origins and professions, and yet they are consistently related to one another and emerge as important figures in the Sephardic world. The authorities were becoming less concerned whether they were Marranos or not, and perhaps this is no longer really important from a historical point of view either.

Actually it was a far flung network that also included men of business, for whom ambiguity was more than just part of the game. Whereas the Venetian notarial deeds very seldom register the name Daniel Rodriguez, another Rodriguez appears prominently in the records: *magnifico* Diego Rodriguez *nobile portoghese* was the main partner of the 'Company Ruy Lopez, Diego and Bernardo Rodriguez in Venice'.[67] Of his associates and brothers, both *egregii mercatores lusitani*, Bernardo had lived for a while in Lyons — perhaps under the tolerant rule of Henry II (1547–59) — and Ruy in all likelihood was the doctor of that name we find living in Antwerp in 1572 and enjoying high repute the following year at the court of Emmanuel

65. R. Segre, 'Gli Ebrei lombardi nell'età spagnola. Storia d'un espulsione', *Memorie dell'Accademia delle scienze di Torino*, s. 4, 28 (1973), 123–4.

66. ASR, Camerale I, Diversorum del camerlengo, reg. 393, fols. 30r–31r; Archivio dei 30 notai capitolini, Uff. 16, Bernardino Pascasio, reg. 10, fols. 154r–v, 31 Jan. 1588; Toaff, *La nazione ebrea*, pp. 42–3, 109–12, 204. For a fuller portrait of Maggino more research on his international connections is required: in 1597 he secured trade privileges at Trier and letters patent at Nancy to establish an entrepôt of Levant merchandise (in his capacity as 'consul general of the Jewish and Levantine nation, be they Turks, Greeks, Armenians and other traders of Ponent and Levant'). The following year his attempt to create a Jewish trade enterprise in Württemberg met with less success, see H. Lepage, *Les archives de Nancy, ou Documents inédits relatifs à l'histoire de cette ville* (Nancy, 1895), Vol. 1, pp. 201–4; C. Pfister, *Histoire de Nancy* (Paris–Nancy, 1908), Vol. 3, pp. 312–13; G. Jaré, 'Abramo Colorni ingegnere del secolo XVI', *Atti della Deputazione ferrarese di storia patria*, 3 (1891), 274.

67. ASV, Avogaria di comun, reg. 3684; Notarile, Zuan Andrea Catti, *passim*, especially filze 3351, 3353, 3361; Ottavio Novello, *passim*, especially filze 10519, 10531, 10532.

Philibert at Turin. Their trade connections covered the Mediterranean, from the Catholic countries (Lisbon and Seville) to the Levant, where David Passi in Constantinople and R. Isach Saltiel in Ragusa looked after their business. But unlike Daniel Rodriguez, who was obliged to dwell in the ghetto, Ruy lived on the Canal Grande and Diego in a new residential district, right near Titian; and as a further mark of distinction they were allowed to employ Christian servants. The shroud of mystery in which, by choice or necessity, these men wrapped their lives means they must always remain partially hidden from our research.

Men of this stamp lived at the mercy of Fortune, yet most of the Sephardim dearly longed to see their children and children's children make a life for themselves in the Italy that had given them refuge. For them, to build a synagogue and bury their dead in their own cemetery with their own Spanish rites was their measure of success. In our view, however, a more essential symbol occurs when a prince gives a licence to publish in the Spanish language, as happened in Ferrara in the 1550s, and soon afterwards in Venice. To make the Torah and the books of prayer accessible in the Castilian vernacular meant that those Sephardic refugees with no Hebrew could have access to the religion of their ancestors. Some had been circumcised, others for prudence sake had to wait: whatever the case, when these books were published the Marrano had a living proof of his national and religious identity.

The Marrano was therefore a Jew who had refused to assimilate with Catholic society, and also had trouble assimilating with the 'old' Jews and for the most part apparently refused to do so. Where there were ghettos, the Sephardim (baptized or otherwise) almost always managed to avoid being confined to that area. Venice and Florence provide examples of this. The duke of Ferrara maintained that his protégés had the right as 'travellers' (namely non-residents in Venice) to live outside the ghetto, and some, as we have just seen, actually did live in the district of their choice.

It was in the realm of fiscal matters that the Sephardim most often fell out with the 'old' Jews (among the latter, German and Italian Jews sometimes joined forces in a way that otherwise would have been unimaginable). Fiscal exemption was urged as a way of attracting rich Sephardic immigrants, while the burden of taxation still rested on the local communities. Particularly in Venice the situation created friction, but it cropped up everywhere, and in Turin in 1573 it led the 'old' Jews to tell the nuncio that the duke was preparing his edict, so damaging to their interests.[68]

Rather than tension, it would be preferable to find evidence of

68. Segre, *The Jews*, Vol. 1, p. 509.

acceptance between the differing cultural backgrounds and religious traditions of the Italian and the Iberian Jews. So far research has revealed insufficient instances. When R. Azriel Trabot left Ascoli in 1569 under banishment from the Papal States,[69] he took with him an *aron* of Sephardic workmanship. This holy cabinet, wrought with carved figures and wooden lions, filled the Italian Jews of Pesaro with dismay.

As the century drew to a close the Levantine community at Livorno, now in full swing, broke away from the small and long-established mother community at Pisa, of which it was originally a coastal offshoot, and the fates of the two communities were linked with the progress of the young maritime port and the melancholy decline of the old port on the Arno.

Not a great deal is yet known about Jewish society in Italy in the seventeenth century, but from the rather scant information we can gather that for Sephardim and Italian Jews the way was fraught with problems and misunderstanding. Corfu, one of the outlying communities of Italy, furnishes, however, a revealing story. There were two Jewish communities, the one Italian and the other Greek (Romaniote), on the little island, military bastion of Venice at the edge of the Ionian Sea. In 1662 — a time of considerable military activity with the reopening of naval operations in the final phase of the war of Candia — the Greek Romaniotes complained that for some 60 years 'Ponentine, Spanish, and other foreign Jews' had been arriving with safe-conducts to swell the already outsized Italian community.[70] The privileges of the old and loyal Jews of Corfu were not to be given to all regardless, but should be reserved for the descendants of those families which had always inhabited the island and deserved the goodwill of the most serene doge: in a word, the Greek Jews. But the 'Italians' (a good many of them, it seems, speaking Spanish) came out on top by presenting a list of taxes for 1578 showing that, even though the Sephardim did not yet figure very strongly, the Greek were only a minority. Here alone, at the very gateway to the Levant, Sephardim and Italians managed to unite in the same cause.

69. D. Kaufmann, 'The Lions under the Ark in Ascoli and Pesaro', *Jewish Quarterly Review*, 9 (1896–97), 256–8. Giving testimony in Ferrara, Trabot's son Vitale (he, too, a well-known rabbi by his Hebrew name of Yeḥiel) recalled his family's wandering: 'The truth is that, as far as I go back with my memory, which may be for the last 30 years, in the synagogues at Pesaro, Ascoli, and Ancona I have always seen this to be the rite when there is controversy among Jews', ASM, Archivio per materie Ebrei, b. 2; Bonfil, *Rabbis and Jewish Communities*, pp. 113–14 (who calls him Yeḥiel of Ascoli to distinguish him from his grandfather, the elder Yeḥiel of Macerata)

70. ASV, Senato, mar, reg. 80, fols. 85r–86r, 29 July 1622 ('gli hebrei vecchi habitanti in Corfù, cioè delle case che da quaranta anni in su vi si ritrovano'); S.W. Baron, 'Jewish Immigration and Communal Conflicts in Seventeenth Century Corfù', in *The Joshua Starr Memorial Volume* (New York, 1953), pp. 175, 178–81; D. Kaufmann, 'Contributions à l'histoire des juifs de Corfou', *Revue des études juives*, 32 (1896), 226–31.

A Tale of Three Cities and their *Raison d'Etat*: Ancona, Venice, Livorno, and the Competition for Jewish Merchants in the Sixteenth Century

BENJAMIN RAVID

The term 'wandering Jew' usually conjures up the pathetic image of a group of poor and huddled men, women, and children wending their way with their meagre possessions on their backs as they searched for a new place of refuge where they could eke out a living. While this image all too often has been valid throughout Jewish history, nevertheless there were also certain occasions on which Jews, and especially Jewish merchants, were actively sought out by authorities who, in an attempt to strengthen economic and commercial conditions in their realms, offered Jews favourable charters as an inducement to settle in their territories.[1]

One of the most interesting and intense instances of competition for Jewish merchants took place in the sixteenth century, between various jurisdictions on the Italian peninsula. Ultimately, significant Jewish mercantile settlements emerged only in three ports: Venice, Ancona, and Livorno.

The immediate background for this competition consisted of several well-known concomitant factors. These were primarily the decline in the quantity of goods from the East reaching the Mediterranean Ottoman ports overland because of the discovery of the direct maritime routes around the Cape of Good Hope, the increasingly frequent presence of ships from the emerging northern European maritime nations in those

For reasons of space, I have presented some themes of this article — many of them first raised in my Ph.D. dissertation (Harvard University, 1973), by now largely published in expanded form in various publications referred to in the notes — very succinctly, and hope to return to them more extensively in the future.

1. See, for example, B. Blumenkranz, 'Germany, 843–1096', in *The World History of the Jewish People*, 2nd ser., Medieval Period, Vol II: C. Roth (ed.), *The Dark Ages: The Jews in Christian Europe* (Tel Aviv, 1966), p.163: 'It is no exaggeration to say that here was a kind of rival bidding among the various towns whereby they tried to entice them [the Jews] by offering the most attractive concessions'.

eastern ports to compete with the Italians for the decreased quantity of goods still available there, and above all the new phenomenon of the far-flung kinship network of Jews and New Christians spread out along the Mediterranean and Atlantic seacoasts. Many Italian authorities hoped that by attracting both Levantine Jews from the Ottoman Empire and judaizing New Christians from the Iberian Peninsula — or from those places on the Italian mainland to which they had already migrated — they could greatly enhance their maritime commerce and consequently also their general prosperity.

This sixteenth-century Italian competition is not only significant in itself, but also because its underlying considerations of *raison d'état* anticipated subsequent developments favourable to the Jews in Holland, England, southern France, and elsewhere. Additionally, it was profoundly reflected in early modern Jewish intellectual history. To ward off a potential expulsion of the Jews from Venice, the Venetian rabbi Simone Luzzatto wrote his *Discorso circa il stato de gl'Hebrei et in particolar dimoranti nell'inclita città di Venetia* (Venice, 1638), which stressed the great importance of maritime commerce to Venice and strongly defended the role of the Jews in that sector. Many of the arguments of the *Discorso* were repeated by Menasseh ben Israel in his *Humble Addresses* (London, 1655) and later by John Toland in his *Reasons for Naturalising the Jews in Great Britain and Ireland* (London, 1714). Thus ultimately the 'tale of three cities' not only serves as background for the readmission of the Jews to western Europe but also contributes to the subsequent debates over the 'amelioration' of their status and eventually also their emancipation.[2]

Within the framework of the general competition between Italian states to attract foreign merchants, and especially Jewish and judaizing New Christian merchants, initially the competition was sharpest between Venice and Ancona, reflecting the fact that then the Adriatic served as the frontier between East and West,[3] until the 1590s when, in a changing world, Livorno also became a serious factor. Both Venice and Ancona were located on the Adriatic, making them conveniently accessible to ships coming from the Levant. However, while Venice had the advantage of

2. See B. Ravid, *Economics and Toleration in Seventeenth-Century Venice: The Background and Context of the* **Discorso** *of Simone Luzzatto* (Jerusalem, 1978), especially pp. 96–8; also id., '"How Profitable the Nation of the Jewes Are": The *Humble Addresses* of Menasseh ben Israel and the *Discorso* of Simone Luzzatto', in J. Reinharz and D. Swetschinski (eds.), *Mystics, Philosophers, and Politicians: Essays in Jewish Intellectual History in Honor of Alexander Altmann* (Durham, 1982), especially pp. 171–4. On the background of the *Discorso*, see M.R. Cohen (trans. and ed.), *The Autobiography of a Seventeenth-Century Venetian Rabbi: Leon Modena's Life of Judah* (Princeton, 1988), pp. 143–6, and the historical notes of H. Adelman and B. Ravid, pp. 249–55.

being a better suited point of distribution for goods by land to northern Italy and Europe north of the Alps, since Ancona was located further south on the Italian coast, ships heading there did not have to cross the pirate-infested northern Adriatic. Also, because Ancona was situated across the Adriatic from the Dalmatian commercial centre of Ragusa (today Dubrovnik), it was in a better position to engage in extensive commerce with the Balkans than was Venice, and the fact that Ancona served as Florence's port for trade with the East was an additional factor in its favour.

Already in 1408, the Venetian Senate noted that as a result of certain Venetian restrictive legislation enacted against the Jews, many Jewish merchants from Apulia, the Marches, Romagna, and elsewhere who previously had brought various merchandise to Venice had ceased to come to the city; instead they went to Ancona, to the detriment of the city and the guilds which had worked with those imported goods, causing a loss of over 60,000 ducats annually; consequently, the Senate exempted those Jewish merchants from the restrictive legislation.[4] However, Jews were legally excluded from engaging in the lucrative Venetian Levant trade, which was permitted only to the upper two classes of Venetian society, the *nobili* and *cittadini originarii*. Legislation of 1476 reiterated that Jews, foreigners, and Muslim subjects of the Turks were not to engage in trade between Venice and the Levant, Lower Romania (namely Greece south of the Gulf of Corinth) and other places outside the gulf, and additional provisions for enforcement were enacted in 1483.[5]

The first Italian government to grant privileges to Levantine merchants was apparently that of Ancona. In January 1514, Ancona gave special customs status to merchants of Ioannina, Larissa, and Arta, three commercial centres in Ottoman Thessaly, in return for promising to send their trade primarily to Ancona. Within two months, they renegotiated their charter so that it applied to all Ottoman subjects, and four years later a more comprehensive charter was issued.[6]

Sometime prior to 1524, Venice relaxed slightly — *de facto* if not *de jure* — its restrictive policy regarding the Levant trade, for legislation of that year prohibiting Venetians from sending the goods of foreigners to the

3. See J. Delumeau, 'Un ponte fra Oriente e Occidente: Ancona nel Cinquecento', *Quaderni storici*, 5 (1970), 26.
4. See B. Ravid, 'The Legal Status of the Jews in Venice to 1509', *Proceedings of the American Academy for Jewish Research*, 54 (1987), 182–3.
5. See Ravid, 'The Legal Status', 189–90.
6. See B. Cooperman, 'Venetian Policy toward Levantine Jews and Its Broader Italian Context', in G. Cozzi (ed.), *Gli Ebrei e Venezia (secoli XIV–XVIII)* (Milan, 1987), p. 71. Cooperman points out that while the first charter was negotiated by a Greek and the second by a Muslim government representative, the third was arranged by a Jewish merchant of Valona.

Levant in their name and foreigners from using any kind of Venetian ships for such trade, exempted Venetian subjects living in the overseas *stato da mar* and also, reciprocally, Ottoman subjects 'as usual'.[7] In view of the previous restrictive Venetian policy, allowing Ottoman subjects to engage in trade between Venice and the Levant, now consolidated in the hands of the Ottoman sultan, no doubt was a concession Venice made in order to maintain its own privileges in Ottoman territory. This new exemption was increasingly utilized by Ottoman Jewish merchants, many — or even most — of whom presumably were originally of Iberian origin, and as time went by, of necessity increasingly of New Christian descent.

In 1534, shortly after Ancona came under the jurisdiction of the papacy, Pope Paul III issued a safe-conduct for 'all merchants of whatever nation, profession or sect, even if Turks, Jews or other infidels', allowing them to come with their families and goods and to stay and trade in Ancona and its Marches without having to pay any taxes other than the usual ones, and also exempting Jews and 'other infidels' from wearing any special distinguishing mark.[8] This safe-conduct was reissued in the following year, during which the Pope also took all Levantine merchants in Ancona, Christian, Muslim, and Jewish alike, under his protection.[9]

Next, in 1538 Duke Ercole II of Ferrara issued two *lettere patente*. The first, of May 30, promised 'all Spaniards and Portuguese and likewise all who speak Spanish or Portuguese who will come to live or trade in our dominion, state and lands' that they could continue to live as either Christians or Jews. The second, of August 13, ten weeks later, was more extensive. It was addressed to 'all Spaniards and Portuguese, Levantines, Slavs, Dalmatians, Greeks, Turks and [people] of every nation, whether Christians or infidels, who will come to live or trade in our lands, cities and any place of our state and dominion', and promised them that they could

7. Archivio di Stato, Venice (hereafter ASV), Senato, mar, reg. 20, fols. 113v–114v, 19 April 1524; text published in M. Sanuto, *I Diarii*, 58 vols. (Venice, 1879–1903), 36:238–40. I thank Benjamin Arbel for suggesting that there may not have been any formal revocation of the long-standing prohibition, but rather that during the course of time it came to be disregarded.
8. See S. Simonsohn, 'Marranos in Ancona under Papal Protection', *Michael*, 9 (1985), 234–67, especially 235, and document 1, 243, also published in S. Simonsohn, *The Apostolic See and the Jews*, 4 vols. to date (Toronto, 1988-), 3:1924, document 1690. I thank Renata Segre for giving me a copy of her article, 'Sephardic Settlements in Sixteenth-Century Italy', presented at the conference on 'The Mediterranean World after 1492' at Tel Aviv University (pp. 112–37 above) from which I obtained this last reference. On the Jews of Ancona, see generally V. Bonazzolli, 'Ebrei italiani, portoghesi, levantini sulla piazza commerciale di Ancona intorno alla metà del Cinquecento', *Gli Ebrei e Venezia*, especially pp. 732–49.
9. See Simonsohn, 'Marranos in Ancona', 235–6, and docs. 2 and 3, 244–6, also published in Simonsohn, *The Apostolic See*, 3:1960, doc. 1730 and 3:1989–90, doc. 1769.

continue in the way that they used to live.[10] Subsequently, in 1550, Ercole was to issue a safe-conduct explicitly for 'the Portuguese and Spanish Jewish nation', allowing its members to live in accordance with Jewish law and assuring them that they would not be molested if for any reason they had claimed that they were not Jews and passed themselves off as Christians.[11]

Venetian maritime commerce was disrupted by the Venetian-Ottoman war of 1537–40, especially since the Venetian Senate ordered the sequestration of the property of Turks, Jews, and other Turkish subjects in Venice,[12] leading to the diversion of the Adriatic trade of Venice to Ancona.[13] In an attempt to remedy the situation, in 1541 the Venetian Senate took certain steps. These included exempting all goods originating in Upper and Lower Romania, and also some from western countries exchanged in Venice for imports from the East, from all customs payments for the following two years.[14] Additionally, asserting that the merchants of Valona (today Vlore in Albania) were a very important factor in the sending of merchandise of all sorts from Romania to Ancona and other places, in order to attract them to Venice a requirement that they were obliged to pay a special customs deposit was eliminated. Above all, noting that the greater part of the merchandise coming from Upper and Lower Romania was brought by travelling Jewish Levantine merchants, as an encouragement for them to come to the city, their complaint that they did not have sufficient space in the Ghetto Nuovo was heeded, and they were given an enclosed area of 20 dwellings in the adjacent Ghetto Vecchio under the supervision of the Cinque Savii alla Mercanzia. There they were allowed to reside without their families and to engage only in maritime trade for a period of four months, which was soon extended to two years, the same length of time for which the customs exemption had been granted. Subsequently, the customs exemption was regularly extended for

10. Cooperman, 'Venetian Policy', pp. 72, 76; original text published in *Bulletin des Archives d'Anvers*, 7 (n.d.), 436–8.
11. See A. Balletti, *Gli Ebrei e gli Estensi* (Reggio Emilia, 1930), pp. 77–8 for further details.
12. See B. Ravid, 'The First Charter of the Jewish Merchants of Venice, 1589', *Association for Jewish Studies Review*, 1 (1976), 189–90.
13. See the observation of the Venetian Cinque Savii alla Mercanzia in 1540 that 'è manifesto ad ogn'uno [...] il gran corso che ha preso la mercadanzia in Ancona [...], al che non si facendo presta et gagliarda provision si pol esser certi che quella si habbi del tutto a deviar da questa nostra città'; quoted in A. Anselmi, 'Venezia, Ragusa, Ancona tra Cinque e Seicento', *Atti e memorie della deputazione di storia patria per Le Marche*, ser. 8, 6 (1968–70), p. 68.
14. Previously, in 1517, the customs on such goods had already been reduced by a half; see the report of the Cinque Savii alla Mercanzia in ASV, Cinque Savii alla Mercanzia (hereafter CSM), reg. 136, fols. 136r–137v, 8 Aug. 1579, also Senato, mar, filza 91, 17 July 1584.

two-year periods, but the extension of 1549 restricted the period for which the Jewish merchants could stay in Venice to one year, because of many unspecified alleged bad results of the length of their stay. However, this restriction, reiterated in 1551 and 1553, does not seem to have been enforced.[15]

A major step occured in 1544, when Pope Paul III issued for Ancona and the Papal States a charter preserved in Italian which, although officially addressed to 'Jews, Turks, Greeks, and all other merchants of the Levantine nations', on the basis of its contents seems to have been intended primarily for Jews. While generally the merchants were absolved of all punishment for misdeeds committed among themselves in the past, and no investigations could be made for they were to be regarded as if newly born,[16] a specific exception was made for 'offences in great scandal of the Christian religion or in "disservice" of the Pope'.[17] Interestingly, a confirmation of the charter issued a few days later, in Latin rather than Italian, omitted the clause excepting crimes against Christianity and the Pope.[18] Subsequently, in 1547, Paul III granted a safe-conduct inviting 'all merchants of whatever nation, faith or sect, even if Turks, Jews, or other infidels' to settle with their families and goods in Ancona and the Marches, reiterating that Jews and infidels would not have to wear any sign differentiating them from Christians.[19] Additionally, it was very significantly — indeed remarkably, for a document emanating from the papacy — stated that the terms of the safe-conduct were also to be valid for all persons coming from Portugal and Algarve, whatever their profession or status, even if they were 'of Jewish origin, called New Christians... or of the Jewish nation of whatever origin' and moreover no legate, vice-legate, judge, or other prefect was to bother them or interfere with their goods on grounds of heresy, apostasy, or blasphemy, or otherwise trouble them. That safe-conduct and other privileges were then reconfirmed in 1552 by

15. For further details, see B. Ravid, 'The Religious, Economic and Social Background and Context of the Establishment of the Ghetti of Venice', in *Gli Ebrei e Venezia*, pp. 222–8, and id., 'New Light on the Ghetti of Venice', to appear in the Festschrift in honour of Shlomo Simonsohn.
16. This expression, significantly, is used in Judaism *vis-à-vis* converts to Judaism.
17. See Cooperman, 'Venetian Policy', p. 73; Simonsohn, 'Marranos in Ancona', 237–8 and doc. 7, 249–53. Very unfortunately, lack of space precludes a discussion and comparison of the terms of the various charters mentioned in this article.
18. See Simonsohn, 'Marranos in Ancona', 238 and doc. 8, 254–8; cf. the texts on 252 and 257. Possibly this omission only reflects the incomplete nature of the text in question and should not be attributed too much significance.
19. See G. Saracini, *Notitie historiche della Città d'Ancona* (Rome, 1675), pp. 361, 431–5; Simonsohn, 'Marranos in Ancona', 238 and doc. 9, 258–62; Cooperman, 'Venetian Policy', pp. 73, 76. For the reconfirmation of 1549, see K. Stow, *Taxation, Community and State: The Jews and the Fiscal Foundations of the Early Modern Papal State* (Stuttgart, 1982), pp. 103–4, and Simonsohn, 'Marranos in Ancona', 239.

his successor Julius III.[20] Asserting that he desired in all ways to attract the Jews to the Catholic Church, outside of which there was no salvation, and that they should not betake themselves to those nations who did not recognize Christ the saviour,[21] Julius III took them under his protection and assured them that neither he nor his successors would permit any judges, whether secular or ecclesiastic, including cardinals and inquisitors, to investigate the Jews of the Papal States regarding their rites and ceremonies during the time they had previously lived as Christians or anything else concerning their religion.

Meanwhile, in 1551, as a result of the initiative of Jacob Abrabanel and a Greek Jew from Damascus named Servadio, Cosimo I of Tuscany issued a charter authorizing 'all you merchants, Greeks, Turks, Moors, Jews, Aggiumi, Armenians, and Persians' to come with their merchandise to trade in Florence and elsewhere in the Tuscan state on very favourable terms. While this invitation apparently did not lead to any significant results, it is of importance as an indication of the fact that one of the leading Italian states was concerned about its Mediterranean maritime trade and was willing to grant privileges to Levantine merchants, including Jews, to settle in its realm.[22]

In 1572, in order to develop his port of Nice, Emmanuel Philibert of Savoy issued a lengthy charter at the initiative of Vitale di Sacerdote and his son Simone 'for the entire Hebrew nation, of whatever rank, condition

20. See Saracini, *Notitie historiche*, pp. 435–8, dated 'die sexta Decembris 1552'; also M. Stern, *Urkundliche Beiträge über die Stellung der Päpste zu den Juden*, 2 vols. (Kiel, 1893–95), Vol. 1, pp. 108–12, misdated 5 Dec. 1553, as noted by H. Rosenberg, 'Alcuni documenti riguardanti i marrani portughesi in Ancona', *La Rassegna Mensile di Israel*, 10 (1955), 308–9; Simonsohn, 'Marranos in Ancona', 240–41; on the year as 1552, see also the references by Emmanuel Philibert and Sarpi, below. The subsequent vicissitudes of the Jews and former New Christians in Ancona, briefly alluded to below, are beyond the scope of this paper.
21. Whatever his real motivation, the Pope here seems to be paying at least lip service to the traditional Christian argument of tolerating the presence of the Jews in the hope that they would convert to Christianity, rather than advocating the commercial *raison d'état* later advanced by Sarpi; see below
22. See U. Cassuto, *Gli Ebrei a Firenze nell'età del Rinascimento* (Florence, 1918), pp. 89–90, 173–80, 409–13; (Jerusalem, 1967), pp. 71–2, 133–8, 336–40 (Hebrew), part of text reproduced facing p. 128. Earlier, in 1547 and 1548, Cosimo had issued two proclamations granting financial concessions and other privileges to settlers and foreign merchants coming to the ports of Livorno and Pisa; see L. Cantini, *Legislazione Toscana*, 31 vols. (Florence, 1800–1808), Vol. I, pp. 384–93; Vol. II, pp. 31–4. Although Jews or New Christians were not mentioned in those texts, one eighteenth-century author wrote that 'in 1548, he [Cosimo] invited the New Christians and Jews who had been persecuted and banished from Portugal to live there [in Pisa], offering them various exemptions and conveniences', including guarantees against any persecution by the Inquisition; B. Galluzzi, *Istoria del Granducato di Toscana sotto il governo della casa Medici*, 5 vols. (Florence, 1781), Vol. I, p. 151. See also R. Toaff, *La nazione ebrea a Livorno e a Pisa (1591–1700)* (Florence, 1990), pp. 34–7, and Segre, 'Sephardic Settlements'.

and language and for those of that descent, both Italians and Germans, Spanish, Portuguese, Levantines, and from Barbary and Syria'.[23] Most significantly, it stated that Vitale and Simone had shown Emmanuel Philibert previous *lettere patenti* of Paul III of 6 February 1553,[24] of the duke of Ferrara of 1559 ratifying all that his predecessors had done, and also of the duke of Florence of 1551. Relating that Vitale had requested that he grant the same terms as had the popes, Emmanuel Philibert cited at length from the safe-conduct of Julius III of 6 December 1552, asserting that it seemed reasonable to follow that example as well as he could, even for all crimes in matters of faith previously committed anywhere, and promised to observe inviolably everything mentioned in the papal documents. However, as is too well known to require elaboration here,[25] Pope Paul IV, after reversing the traditional papal policy towards the Jews with his harsh bull *Cum Nimis Absurdum* of 1555, in the following year revoked the promise of safe-conduct that his predecessors had made to the New Christians who had returned to Judaism in Ancona and had over two dozen of them burned at the stake for judaizing. Thus for Emmanuel Philibert to cite pre-1555–56 papal policy after that time was to ignore the new historical reality of the Counter-Reformation papacy.

As could be expected, this charter of Emmanuel Philibert ran counter to the interests and policies of both Pope Gregory XIII and of Philip II of Spain.[26] Philip II not only objected to the fact that New Christians could revert back to Judaism, but also among other things to the freedom of trade and residence permission granted to Turks, Moors, Persians, and others in Christian lands. Accordingly, he did all he could to hinder the travel of New Christians to Savoy, and pressured Emmanuel Philibert to rescind the privileges on the grounds that they were dangerous to Christianity and to Christian states, especially that of the duke, and consequently also to the Spanish holdings in Italy. The Pope also pressured Emmanuel Philibert, and in November 1573 the latter ordered all New Christians who had reverted to Judaism to leave Savoy within six months. However, the Spanish ambassador in Savoy pointed out that if Venice were to give refuge to the exiles, then the benefit from the Expulsion would

23. See E. Lattes, 'Documents et notices sur les Juifs d'Italie', *Revue des études juives*, 5 (1882), 223–8, 231–7.
24. Apparently a slip of the pen, since Paul III was Pope from 1534–49; on the other hand, the wording 'sommi pontifici Paulo Terzo' may indicate that something was omitted in the text as printed. See Simonsohn, 'Marranos in Ancona', 241–2 and document 12, 264–7, for privileges of Julius III of 17 Feb. 1553.
25. For a recent article with extensive bibliography, see R. Segre, 'Nuovi documenti sui marrani d'Ancona (1555–1559)', *Michael*, 9 (1985), 130–233.
26. This paragraph is based on H. Beinart, 'The Settlement of the Jews in the Duchy of Savoy as a Result of the Privilege of 1572', in *Scritti in memoria di Leone Carpi* (Jerusalem, 1967), pp. 72–118 (Hebrew).

be very slight, while Emmanuel Philibert complained to the papal envoy that it would not only harm him but also benefit Venice. The ambassador requested Philip II to urge the Pope to order Venice not to accept the exiles, and according to a subsequent letter from the ambassador to Philip II, the papal envoy told him that the Pope had decided not to consent to the settlement of former New Christians in Venice or elsewhere in Italy and had requested the duke of Savoy to implement the Expulsion. While a few Jews of Iberian origin were still in Savoy in 1575, clearly as a result of the combined pressure of emperor and Pope, Savoyan Nice never became the major commercial centre that its ruler had envisaged.

In the interim, in 1550 the Venetian Senate had issued a decree expelling the Marranos.[27] Nevertheless, the Venetian government almost never bothered New Christians who reverted to Judaism in the city and lived in the ghetto as Jews, and also allowed Levantine Jewish merchants to remain in the city with their families for extended periods despite the restrictive laws.[28] However, after the outbreak of war between Venice and the Ottoman Empire in 1570, all Turks, Levantine Jews, and other Ottoman subjects in Venice were detained along with their merchandise.[29] Out of concern that Ancona would take advantage of the situation and continue to hold its commercial gains after the war ended, the Venetian government sought to induce the Pope to prohibit Jews from trading between Ancona and the infidel Ottoman enemy on the grounds that it was not to the benefit of Christianity, but to no avail.[30]

After the war ended, new ideas began to be considered in Venice. First, in the spring of 1573, some of the leading Jews in Constantinople proposed to the Italian Jewish doctor Solomon Ashkenazi — who at great personal risk had served as intermediary between the Venetian *bailo* in Constantinople and the Ottoman grand vezir Meḥmet Sokollu and played a major role in negotiating the peace treaty of 1573 — to negotiate the terms for transferring all their commercial activities from Ancona to Venetian Chioggia at the edge of the Venetian lagoon. This aroused interest in the Venetian Council of Ten, which authorized the *bailo* in Constantinople to negotiate further with the Jews, but nothing resulted.[31]

27. See D. Kaufmann, 'Die Vertreibung der Marranen aus Venedig im Jahre 1550', *Jewish Quarterly Review*, OS, 13 (1900–1901), 520–32.
28. See P. C. Ioly Zorattini, 'The Inquisition and the Jews in Sixteenth Century Venice', *Proceedings of the Seventh World Congress of Jewish Studies, 1977: History of the Jews in Europe* (Jerusalem, 1981), pp. 83–92, and B. Pullan, *The Jews of Europe and the Inquisition of Venice* (Totowa, NJ, 1983), especially pp. 145–98.
29. See Ravid, 'The First Charter', 191–2.
30. See Pullan, *The Jews of Europe*, pp. 181–2, on what he felicitously termed 'parity of disadvantage'.
31. See B. Arbel, 'Venezia, gli Ebrei e l'attività di Salomone Ashkenasi nella guerra di Cipro', in *Gli Ebrei e Venezia*, p. 180; also id., 'Salomone Ashkenasi: Mercante e Armatore', in G. Todeschini and P.C. Ioly Zorattini (eds.), *Il mondo ebraico: Gli*

Shortly afterwards, in December 1573, the Venetian Council of Ten handily approved, with 21 members in favour, five opposed, and five abstentions, a safe-conduct to be issued in the name of the doge, Alvise Mocenigo.[32] Pointing out the general precedents of popes and other Christian rulers who had granted freedom and very extensive safe-conducts to merchants, including Jews of all origins, the doge stated that, prompted by their example, he was issuing for 'Spanish and Portuguese Jews and all others of Jewish descent living in the West and elsewhere under Christian rule' who wished to come to settle in Venice during the following two years, a very ample safe-conduct and inviolable security for their persons, property, and family. They could come to dwell in Venice and trade with anyone they wished and leave freely without molestation of any kind. Furthermore, it was to be firmly and inviolably observed that while they were residing in Venice in the place which was to be assigned to them,[33] living as Jews and wearing the yellow *baretta* or *sessa*,[34] they could not be accused or punished or in any way molested for the crime of apostasy or hypocrisy or any other error in matters of faith which they had committed in the past, nor for previously having lived as Christians. This was to be the case on the condition that while resident in Venice, they lived continuously as Jews without causing any scandal in religious matters. However, to become effective, the safe-conduct required the approval of the Senate, which for some reason was not forthcoming, and consequently it was never officially issued. Nevertheless, it is of significance in testifying to the attitude of the doge and one of the key Venetian assemblies towards merchants of New Christian origin living under Christian rule. And although the safe-conduct did not go into effect, *de facto* the Venetian government acted in its spirit, as it continued to be disinclined to proceed against those New Christians who broke completely with their Christian past and lived openly as Jews in the ghetto.

These developments regarding former Christian merchants from the West — and also no doubt the experiences of the Levantine Jews during the war of 1570–73 — presumably induced the latter, who as Ottoman

Ebrei tra Italia nord-orientale e Impero asburgico dal Medioevo all'Età contemporanea (Pordenone, 1991), pp. 111–28.

32. ASV, Council of Ten, secreta, reg. 10, fols. 159r-160r, 23 Dec. 1573. I owe the reference initially to Pullan, *The Jews of Venice*, pp. 183–4. As suggested by Arbel, 'Venezia', p. 180, quite probably this safe-conduct was the result of the initiative of Daniel Rodriga who had been in contact with the son of the Venetian *bailo* in Constantinople (on Rodriga, see below). See also the comments of the papal nuncio in A. Buffardi, *Nunziature di Venezia* (Rome, 1972), Vol. 9, pp. 136–7, 140–41. I plan to deal more extensively with this safe-conduct in the future.
33. Thus presumably they were to be required to live in the ghetto as were all the other Jews in the city.
34. See B. Ravid, 'From Yellow to Red: On the Distinguishing Head-Covering of the Jews of Venice', to appear in the memorial volume for Frank Talmage.

subjects would not have been covered by the proposed safe-conduct, to submit a petition to the government requesting that, in case of war, they could dwell in Venice and throughout the Venetian state and continue their business securely. Their petition was passed on to the Cinque Savii alla Mercanzia, who were instructed to report on it in writing.[35] The Savii responded that the request concerned matters of state, and therefore pertained to the consideration of the government, to which they would react, knowing that the benefit of the state would always be considered. It is tempting to interpret this vague response, to be repeated on other occasions, as indicative of a basic willingness to endorse the request of the Jews, but reluctance to innovate policy.

The course of maritime commercial developments in the Mediterranean was decisively changed by the indefatigable Daniel Rodriga.[36] Rodriga, presumably of Iberian origin, was certainly one of the leading — if not the leading — and most successful Jewish entrepreneurs of his century. For at least 30 years, he submitted numerous proposals and projects to the Venetian government. Although covering a very wide range, almost all in some way involved enhancing the maritime commerce of the city, augmenting its customs revenue, and improving the status of Jewish merchants. Many also included a calculated attempt to divert trade away from Ancona, and then later Livorno as well, after its port began to function. Certainly, this was true of Rodriga's two major interrelated projects first proposed in late 1576 or early 1577 but ultimately implemented only after 1589: the establishment of a free-transit *scala* at Spalato, a Venetian possession on the north-eastern Dalmatian Adriatic coast (today Split in Yugoslavia), and the granting of privileges to Jewish merchants.

When Rodriga first proposed his project to establish the *scala* of Spalato,[37] in support he pointed out that a new trade route between Venice and Spalato, further away from the harmful disturbances of the Uskok

35. ASV, CSM, reg. 136, fols. 33v–34r, 20 March 1574. Indeed, one of the clauses of the peace treaty between Venice and the Ottoman Empire had stipulated that Turkish and Jewish merchants from the Ottoman Empire could freely come and go in Venice with all their goods; see Segre, 'Sephardic Settlements', pp. 123–4 above.

36. On Rodriga, see R. Paci, *La scala di Spalato e il commercio veneziano nei Balcani fra cinque e seicento* (Venice, 1971); Ravid, 'The First Charter', 187–222; id., 'An Autobiographical Memorandum by Daniel Rodriga, *Inventore* of the *Scala* of Spalato', in A. Toaff and S. Schwarzfuchs (eds.), *The Mediterranean and the Jews* (Ramat Gan, 1989), pp. 189–213; id., 'Daniel Rodriga and the First Decade of the Jewish Merchants of Venice', in A. Mirsky, *et al.* (eds.), *Exile and Diaspora: Studies in the History of the Jewish People Presented to Professor Haim Beinart* (Jerusalem, 1991), pp. 203–23; also Segre, 'Sephardic Settlements'. I plan to deal further with Rodriga in the future.

37. For his memorandum, see ASV, Senato, mar, filza 69, 28 Oct. 1577, which also contains the legislation of the Senate, and Ravid, 'An Autobiographical Memorandum'.

pirates (namely to their north) would divert trade away from Ragusa and to Venice, to the very great benefit of its customs, guilds, and inhabitants. In return, he made several requests, including one that the privileges he had sought previously be granted to all merchants of whatever nation living with their families in the Venetian state. Although Rodriga referred to merchants of all nations and not specifically to Jews, since subsequently he was to concern himself only with Levantine Jews and Iberian New Christians, no doubt he primarily had them in mind.[38]

However, the granting of commercial privileges to non-Venetians ran counter to Venetian policy. While Ancona, Florence, Ferrara, and Savoy needed the help of foreign merchants in order to sustain their commercial life, Venice had a lengthy and successful native mercantile tradition and was not ready to make extensive unprecedented concessions that ran counter to its long-standing protective policies. It can also be conjectured that the reaction of many Venetians would be especially negative since among the beneficiaries, if not indeed the main beneficiaries, would be not only Levantine Jews, who were suspected of potential disloyalty, but also New Christians from the Iberian Peninsula who could revert freely to Judaism. While it was possible for the government to condone tacitly the presence of Jewish merchants, it was an entirely different matter officially to invite them to settle in the city. Otherwise, however, the proposal of Rodriga generally appeared attractive to the Venetian government. In October 1577 the Senate approved legislation providing for the establishment of the *scala* of Spalato, although subsequently nothing was implemented for well over a decade,[39] but took no steps regarding the privileges of the Jewish merchants.

Two years later, on or slightly before 20 June 1579, Rodriga submitted a lengthy memorandum to the Venetian government together with a separate brief introductory statement.[40] In that statement, he sought permission to bring to the city up to 50 families of Jewish merchants, each of whom he asserted would pay around 100 ducats annually in customs duties. Since Rodriga requested that these merchants be given those privileges already possessed by the visiting Levantine Jewish merchants and the resident so-called Tedeschi Jewish moneylenders,[41] presumably

38. Since to date no text of previously requested privileges has been found other than the above-discussed safe-conduct approved by the Council of Ten in 1573, presumably Rodriga was referring to it.
39. See Paci, *La scala di Spalato*, and Ravid, 'An Autobiographical Memorandum'.
40. ASV, Senato, mar, filza 91, 17 July 1584.
41. On the Tedeschi moneylenders, see B. Pullan, *Rich and Poor in Renaissance Venice* (Cambridge, 1971), pp. 476–578, and B. Ravid, 'The Socioeconomic Background of the Expulsion and Readmission of the Venetian Jews', in F. Malino and P. Albert (eds.), *Essays in Modern Jewish History: A Tribute to Ben Halpern* (East Brunswick, NJ, 1982), pp. 27–55.

they were not Levantine Jews but rather from the Iberian Peninsula. Additionally, he requested that as Venetian subjects they be guaranteed security of their persons and property in case of war. As precedent and justification, Rodriga pointed out that the government had granted a charter to the Jewish moneylenders who, he claimed, brought less benefit to the treasury than would these merchants who, if given the privileges, would be of enormous advantage to the market-place and satisfaction to the government.

The memorandum of Rodriga presented three specific reasons why the commerce of Ancona had increased in the past and indicated what he thought should be done in each case. The first reason was that the Pope had promised Levantine merchants that in all occasions of war, they would remain free and their trade would not be disturbed. This, Rodriga suggested, could be countered by granting the merchants the same privileges in Venice, as he had often requested. The second reason was that in Ancona the customs had been reduced and simplified and, Rodriga noted, in this matter the Venetian government had done that which he had recommended regarding their simplification, to the advantage of the customs and the merchants; thus, others had benefited from his labour and he requested that he be granted something.[42] Third and most important, *robbe grosse*, namely Cordovan leather, sheepskin, untanned hides, coarse blankets, fleece, and wool were going to Ancona rather than to Venice but since the Pope had just imposed a one per cent import tax at Ancona, the merchants, and especially the Ragusans, were resentful and would abandon Ancona if the Venetian government were to avail itself of the opportunity. Accordingly, Rodriga proposed eliminating the two and a half per cent Venetian export duties imposed on *robbe grosse*. Despite the resulting loss of revenue, he claimed that the city would benefit, since the *robbe grosse* would be accompanied by *robbe sottile* that was out of necessity going elsewhere, for although all the merchants wanted to bring it to Venice, they could not because if they used small ships, they ran the risk of corsairs and the Uskoks, while if they used large ships for such little volume, they could not pay the expenses. Accordingly, they were compelled to send the *robbe sottile* with the *robbe grosse* which was going to Ancona, from where it was then distributed throughout Apulia, Romagna, and the whole of Italy. Rodriga had no doubt that if Venice were to eliminate the two and a half per cent export duty, then merchants from all over Italy would come to buy in Venice rather than Ancona, because of the convenience and security of sailing by internal rivers rather than via the more dangerous Adriatic. Also, if all those goods came from the Levant to

42. On this point, which requires further investigation, see Ravid, 'An Autobiographical Memorandum', pp. 200–201 and n. 18.

Venice, without doubt goods from the West such as *cremise*, *carisee*, and *oltrafine* would no longer go to Ancona, since there would be no Levantine merchants or merchandise to trade for them in Ancona.

These documents of Rodriga were passed on to the Cinque Savii alla Mercanzia for their reaction.[43] They deferred commenting on the request regarding the privileges of the merchants because of its complexity and consequences, deemed commenting on the second unnecessary since legislation had been adopted regarding it, but discussed the third, which they termed very important, at great length. They concluded that reducing the customs would to a great extent take trade away from Ancona and therefore, since the increased volume would augment the general import and export duties to an extent significantly greater than the loss incurred by the decrease in the duties on the specific goods in question, they recommended that the government should try the proposal of Rodriga for at least two years and then decide whether it should be continued.[44]

Four years later, in 1583, Rodriga again attempted to secure residence privileges in Venice for the Jewish merchants.[45] Specifically, he asked that privileges be granted to Levantine and Ponentine (namely western, from *ponere*, the opposite of *levare*, referring to the setting and rising of the sun) Jews, enabling them to dwell securely with their families in the city on the basis of certain terms which he proposed. Of great significance is the fact that here for the first time, Rodriga was asking explicitly for privileges for both Levantine and Ponentine Jews. It was clear that many — if not most — Levantine Jews were of Iberian origin, either born there or descended from Jews born there who had subsequently moved to the Levant and now were coming to Venice as Ottoman subjects, and they had been recognized by the Venetian government as a distinct legal category since at least 1541. On the other hand, strictly speaking there was no such thing as a Ponentine Jew, since as of 1497 Judaism had been completely proscribed on the Iberian Peninsula. While the proposed safe-conduct of 1573, quite possibly formulated by Rodriga, had referred to Spanish and Portuguese Jews and all others of Jewish descent living in the West and elsewhere under Christian rule, now Rodriga only referred to Ponentine Jews. Although obviously the difference was basically only semantic, its significance should not be minimized.

43. ASV, CSM, reg. 136, fols. 136r–137v, 8 Aug. 1579; also Senato, mar, filza 91, 17 July 1584.
44. Some four years later, in 1584, the Senate eliminated for five years the two and a half per cent duty on Cordovan leather, sheepskin, untanned hides, and coarse blankets, while halving that on fleece and wool from five per cent to two and a half; see Ravid, 'An Autobiographical Memorandum', p. 201, n. 19. The report of the Cinque Savii also recommended reducing the duty on fleece and wool by that amount, so presumably Rodriga erred in including them in the category of goods like the skins and blankets, which owed two and a half per cent.
45. ASV, CSM, reg. 137, fols. 96v–97v, 26 Nov. 1583.

In view of the complex and ambiguous nature of the situation and the declining fortunes of Venetian trade, Rodriga had grounds for hoping that his request would be granted. However, the report of the Savii was very critical indeed.[46] At length it reiterated firmly that granting the sought-after privileges would be very bad both for individual Venetian merchants and also for the government. Its main point was that even though the customs might be increased a little, the harm which all the Venetian merchants would suffer would be incomparably greater, for if the Jews were allowed to trade freely, then they would be able to concentrate all the trade in their hands and control the prices; moreover, it was alleged, to achieve that end they also would try, in those ways that they knew well how to use when they wished, to find an opportunity to cause a war with the Turks.[47] The report concluded by pointing out that previous Savii had decided that the Jews could not stay in the city for more than four months, and had issued other regulations which it would be good to enforce so that the Jews could not live in the city for longer than that time; thus provision would be made for various difficulties that could arise because of their extended habitation, and accordingly the proposal of Rodriga should be rejected.

Such hostility on the part of the Cinque Savii was unusual, for as a rule they were favourably inclined towards the Jewish merchants out of general considerations of enhancing the declining commerce of Venice and the revenue derived from customs.[48] However, this report of the Savii constituted a firm reiteration of the traditional Venetian attitude towards the Jews, reflecting also the fear of the Jewish competitor at a time when the Venetian merchants were facing increasing difficulties in their commercial enterprises. When, in that report, the Savii conceded that granting the Jews privileges would lead to an increase in the customs revenue but then asserted that the resulting harm to the native merchants would outweigh this benefit, they were departing from their usual stand that increased customs revenues deriving from an upswing in trade led to a prosperity which rippled throughout the entire economy. In a behaviour

46. A substantial part of this report has been published in English translation in Ravid, *Economics and Toleration*, pp. 41–3.

47. Unfortunately, lack of space precludes further discussion of this theme of the Jews as a 'security risk'. Clearly, Joseph Nasi and the Venetian-Ottoman war of 1570–73 had not been forgotten. On Nasi and Venice, see B. Ravid, 'Money, Love and Power Politics in Sixteenth Century Venice: The Perpetual Banishment and Subsequent Pardon of Joseph Nasi', in the *Proceedings of the First Italia Judaica Congress, Bari, 1981* (Rome, 1983), pp. 159–81.

48. For other relatively rare occasions of hostility towards the Jews on the part of the Cinque Savii, see Ravid, *Economics and Toleration*, pp. 38–49.

pattern characteristic of groups threatened by new conditions beyond their control or understanding, they tended to place an exaggerated share of the blame on the Jews. The absence of any reference to the fact that Rodriga also requested permission for 'Ponentine Jews' to live in Venice as Jews (namely, to revert to Judaism with impunity) seems to indicate that the opposition of the Savii was primarily economic and political in nature rather than religious, even though in general they may not have been personally friendly towards Jews.

In the following year, 1584, Rodriga made yet another attempt to gain privileges for the Levantine Jews, apparently requesting that at least those Levantine Jews who were married and had settled in Venice be granted the safe-conduct that he had requested previously. The Cinque Savii of the previous year had been routinely replaced by new individuals, and perhaps that had encouraged Rodriga to try once more, albeit with a slightly more modest request. This time, the Savii briefly surveyed the history of the Levantine Jews in Venice since 1541, not citing the short time the Jews had been allowed to stay there in order to justify stern measures as had the previous Savii, but rather placing the Jewish presence in a much wider context in order to establish their positive role in Venetian commerce.[49] In the end, the Savii concluded that it could be seen that their predecessors had not only employed all possible efforts so that merchandise would be brought into the city, but also had made all suitable provisions for the Jews, giving them dwelling space without concern for any other considerations that could be raised. However, the Savii concluded very cryptically, they could not come to any decision because of the incident which was well known to the government, and therefore decided to defer until they could advise to their satisfaction whatever they then would consider to be for the public good.

Although the Savii were unable to come to a decision, significantly they did not seem to have any basic objection to granting the privileges Rodriga was requesting; they did not reject them or say that they were not competent to deal with the matter, but only deferred their final recommendation. Possibly they were willing to give a positive response, but were inhibited by various pressures compelling them to maintain the traditional policy of the Venetian republic.

Finally, when in 1589 Rodriga submitted the text of a charter granting extensive privileges to 'Levantine, Spanish, and other Jewish merchants living in Venice with their families', the Cinque Savii responded positively. They asserted that although in the past their predecessors had questioned the granting of such privileges, now 'because of the state of trade at present on account of the changes in the affairs of the world' they should be

49. ASV, CSM, reg. 137, fols. 135v–136r, 18 Dec. 1584.

granted, for there was no doubt that trade, which was the main support of the city, was strengthened by offering attractive terms to new settlers who could increase its commerce. Otherwise, they would be accepted elsewhere, as had been the Spanish exiles who, because they were not admitted into 'these parts', had gone to the Levant with their very substantial capital, augmenting the commerce of others to the great detriment of Venice. Although, the Savii observed, with the inclusion of Spanish and other Jews it appeared that the privileges were greatly broadened, actually it would not cause any greater harm since the merchants already could go to the Ottoman Empire and then come to Venice to enjoy the benefits of the law as Ottoman subjects. Accordingly the Savii approved the terms of Rodriga's proposed charter, only recommending a few relatively minor modifications. On the basis of those recommendations, the College *(Collegio)*, which was responsible for formulating official legislation for introduction in the Senate, somewhat reformulated the proposed charter and also added its own terms. Among other things, it changed Rodriga's 'Levantine, Spanish, and other Jewish merchants' to 'Levantine and Ponentine Jewish merchants'. Additionally, it introduced a time-limit of ten years and required the merchants to wear a special yellow head-covering[50] and to reside in the ghetto as did all the other Jews in Venice. Also, Rodriga's provision that the merchants could live securely as Jews without being subject to investigation by any office or magistracy, either ecclesiastic or secular, even if they had observed another religion elsewhere, was condensed to state merely that the merchants would be secure from being molested by any magistracy for reasons of religion. Then, introduced in the Senate on 27 July 1589, the charter was approved by the substantial majority of 110 senators in favour, with 11 opposed and 13 abstentions.[51]

The inclusion of Ponentine Jews in this charter was especially noteworthy. First, the granting of the immediate right to engage in maritime commerce with the Levant to non-Venetians, other than reciprocally to Ottoman subjects, on any terms, let alone those of equality with Venetian subjects and without the lengthy waiting period of 25 years, represented a concession completely unprecedented in the commercial history of Venice and serves as clear testimony to the Venetian perception of the importance of the Jewish merchants. Subsequent attempts on the part of Christian

50. For further details, see Ravid, 'From Yellow to Red'.

51. For further details, see Ravid, 'The First Charter', 197–222; also G. Cozzi, 'Società veneziana, società ebraica', in *Gli Ebrei e Venezia*, pp. 336–41, for the suggestion that after the Venetian Senate had not accepted the controversial Spanish offer to sell Venice all the pepper coming from the Indies in return for stopping to import it from the Levant (see F. Lane, *Venice: A Maritime Republic* [Baltimore, 1973], pp. 293–4), which had been supported by an influential pro-Spanish and pro-papal group in Venice, it granted the charter to the Jewish merchants in an alternative attempt to remedy Venetian commercial difficulties.

merchants from elsewhere in Europe to secure that privilege, on occasion citing the precedent of the permission granted to the Jews, were unsuccessful, although ultimately it was extended also to the Tedeschi Jewish moneylenders in 1634.[52] Second, these Ponentine Jews, to whom unprecedented commercial privileges were being granted, were former New Christians, who now were allowed to revert to Judaism with impunity upon arriving in Venice. This explicit authorization, over 30 years after the start of the hostile Counter-Reformation attitude towards the Jews and the reversion of New Christians to Judaism, was a potential provocation to traditional Christian circles,[53] and indeed was later challenged. However, Venice was not Savoy. Paolo Sarpi, the chief ideologist and spokesman for the Venetian policy, shrewdly identified Venetian commercial *raison d'état* with the greater good of western Christendom as he claimed that the permission for New Christians to settle freely in Venice as Jews was for the public benefit of Christendom, so that they would not take so much wealth and needed industriousness to the lands of the Turks.[54] Furthermore, he asserted, granting such a safe-conduct was lawful and legitimate and could not be condemned, since many popes had allowed Marranos to live in Ancona with greater privileges than those granted by Venice, specifically citing the charter of Julius III of 1552, but of course not pointing out that those papal permissions had been issued before the Counter-Reformation.[55] No doubt Sarpi agreed with the view expressed by one of the Venetian lay *assistenti* to the Venetian Inquisition in 1608, that it mattered very little that Paul IV had cancelled the Ancona privileges, since on one occasion the papacy actually had seen those privileges as good in principle.[56]

Events in Venice did not, of course, go unnoticed. The Medicean grand dukes of Tuscany sought to augment the maritime commerce of their

52. I plan to elaborate on these points; in the interim, see Ravid, *Economics and Toleration*, pp. 37–8.
53. A Vatican document of 1625 suggested that the charter had been issued in secret and not officially shown to the nuncio, presumably to prevent him from objecting; see Pullan, *The Jews of Europe*, p. 188; Cozzi, 'Società veneziana', p. 342, proposes that this may have meant without previous agreement with the papacy.
54. This argument did not originate with Sarpi, but presumably represented a trend within Venetian governing circles, for it had already been advanced in 1588 after a certain Jew passing through Venice had been arrested by the Inquisition on the grounds that he had posed as a New Christian in Flanders; see Pullan, *The Jews of Europe*, p. 186.
55. For further details, see Ravid, 'The First Charter', 207–11
56. See Pullan, *The Jews of Europe*, p. 193. Cf., however, the opposite view of the head of the Roman Inquisition in 1588 that 'these are different times, and circumstances have changed, and if the Pope did such things today we would consider proceedings against him' (ibid., p. 187), and later on of Cardinal Albizzi who asserted that 'the Pope himself could not permit such a thing without giving scandal to the Church', ibid., p. 198.

realm by creating a major trade centre at Livorno, which possessed one great geographic asset that both Venice and Ancona had to acknowledge. Livorno was conveniently located on the western side of Italy, directly across the Mediterranean from France and Spain. Thus it was easier for ships coming into the Mediterranean from the Atlantic ports of Spain, France, England, and Holland and from the Americas and Asia and wishing to trade with Italy, to sail to Livorno rather than to undertake the longer journey around southern Italy to Adriatic Ancona and Venice; similarly, ships sailing from Holland, England, France, and Spain bound for the eastern Mediterranean would also find it convenient to make Livorno their only Italian port of call. While these Westerners were unlikely to settle in Livorno, there was another group of people, very active in Mediterranean trade but not officially and securely allowed into Atlantic ports, who potentially could be attracted to Livorno, namely Jews and judaizing New Christians.

Accordingly, in the last decade of the sixteenth century, Grand Duke Ferdinand took the appropriate steps. In 1591 he issued a charter which was reissued — with slight changes — in its definitive form in 1593; known as 'La Livornina', it was basically to remain in effect until Napoleon, and then after him, again until the end of the Grand Duchy of Livorno in 1860.[57] Although formally inviting 'all you merchants of any nation, Levantine, Ponentine, Spanish, Portuguese, Greek, German and Italian, Jewish, Turkish, Moorish, Persian, and others' to settle in the city of Pisa and port and *scala* of Livorno, its provisions clearly indicate that it was intended primarily to attract Jews and judaizing New Christians; a nineteenth-century writer noted that this decree 'benefited more than any other the Jewish nation, which almost came to see in Ferdinand I the desired Messiah, and to find in Livorno another Jerusalem'.[58] 'La Livornina' contained the same basic privileges as the considerably briefer Venetian charter of 1589, with numerous very attractive additional ones. Most important, newcomers to Livorno were offered much greater security, since the provisions were to be in effect for a period of 25 years, subject to automatic renewal for another 25, unless notice was given 5 years in advance (paragraphs 1 and 36). Also, it was specified that no investigation, examination, denunciation, or accusation could ever be made against the newcomers or their families, even if in the past they had lived outside the jurisdiction of the Medici as Christians (paragraph 3).[59]

57. See Toaff, *La nazione ebrea*, pp. 15, 41–51, and pp. 419–35 for the integrated text of the two charters, showing additions and deletions; also A. Milano, 'La costituzione "Livornina" del 1593', *La Rassegna Mensile di Israel*, 34 (1968), 394–410.
58. E. Repetti, *Dizionario geografico fisico storico della Toscana*, 4 vols. (Florence, 1843–46), Vol. 2, p. 738.
59. On this third paragraph, see Toaff, *La nazione hebrea*, pp. 99–102, and especially n. 2.

The totally different attitude towards Jews manifested itself most clearly in the 29th of the 44 paragraphs of the charter. It granted the Jews the same right as Christian merchant citizens of Florence and Pisa to engage in all the *arte* as well as in all commerce except trade in second-hand goods (*strazzaria*; paragraph 31), which on the contrary was permitted to the Jewish moneylenders of Venice; thus the Jews of Livorno could, among other things, basically engage in retail trade, while the Jewish merchants of Venice were restricted to wholesale importation and exportation. That same paragraph also exempted the Jews of Livorno from having to wear any sign that would distinguish them from their Christian neighbours, as required in Venice, and permitted them to purchase real estate, which was forbidden in Venice. Also, the 42nd paragraph, added to the charter in 1593 at the request of the Jews, permitted them to employ Christian servants and keep Christian wet-nurses in their houses, as, it was asserted, was the practice in Ancona, Rome, and Bologna, viz. the Papal States; on the other hand, these relationships were prohibited to the Jews of Venice by the terms of their charters and allowed only occasionally by special permission.[60] In addition, no provision was ever made for the establishment of a ghetto in Livorno or Pisa, even though that institution had been required in Florence and Sienna, the two other places in Tuscany where Jews could reside, by legislation decreed some 20 years previously.

This succinct comparison of some key features of the charters given to Jewish merchants in Venice and Livorno demonstrates clearly that although identical considerations of *raison d'état* had indeed induced various authorities to grant privileges to Jewish merchants, nevertheless such rational commercial and economic factors did not exert a uniform effect on the conditions under which Jews lived. Their actual position depended on various additional factors that reflected the complex local religious, social, and political environment which differed substantially from one area to the next, and accordingly it is impossible to posit any simple economic determinism.[61]

60. Since the Venetian government recognized two distinct groups of Jews, the Tedeschi Jewish moneylenders and the Levantine and Ponentine Jewish merchants, and gave each group separate charters, any comparison of the legal status of the Jews of Livorno with that of the Jews of Venice must take into consideration both charters of the Jews of Venice, and also the additions made to them during the course of the seventeenth century, which I will trace in the future. In the interim, for the text of the charter of the Jewish merchants of 1589, see Ravid, 'The First Charter', 219–22, and for that of 1636, Ravid, *Economics and Toleration*, pp. 100–105, while that of 1598 is available in English translation in Ravid, 'Daniel Rodriga and the First Decade of the Jewish Merchants of Venice'; for the text of the charters of the Tedeschi Jewish moneylenders of 1618, 1624, 1629, and 1634, see id., *Economics and Toleration*, pp. 105–26.
61. On this theme, see Ravid, 'How Profitable the Nation of the Jewes Are', pp. 171–4, discussing the status of the Jews in Venice, Amsterdam, and London.

The course of events in Venice and Livorno no doubt induced the popes to maintain and even extend their conciliatory policy towards Jews in Ancona. Previously, in his bull *Hebraeorum Gens* of 1569, Pope Pius V had banished the Jews from all the Papal States except for Rome and also Ancona, because of their role there in the Levant trade.[62] Subsequently, Sixtus V in his bull *Christiana Pietas* (1586) had allowed the Jews to return to the cities, large castles, and lands of the Papal States, but not to the villages and hamlets.[63] However, this permission was revoked by Clement VIII shortly after his accession to the papal throne in his bull *Caeca et Obdurata* of 1593, which expelled the Jews from all of the Papal States, except for Rome, Avignon, and again Ancona.[64] This reiterated concession to economic reality was followed in the next year by a general bull 'in perpetuity', allowing merchants of all countries and nations who were trading with the Levant to come freely and securely to trade and live in Ancona and exempting their goods from all customs payments.[65] The Jews of the Levant and the existing Jewish community of Ancona — needless to say, on this occasion no mention was made of Jews from Spain or the West, or of New Christians — were specifically permitted to enjoy the privileges previously granted to them by Pope Paul III, with, however, certain unspecified modifications made a short time previously by the papal chamberlain; additionally, both groups of Jews were excluded from the previous decree of expulsion of 1593 and allowed to go freely anywhere in the Papal States in order to collect their mercantile debts.

This bull astounded the Venetian ambassador in Rome, Paolo Paruta, who observed that 'it is but a few days since the Pope complained that the Levantine Jews were received in Venice ... So the same things are differently regarded, not in accordance with their true nature, but in the light of the gains one hopes to make therefrom....'[66] Three years later, in 1597, the Cinque Savii singled out Ancona as 'one of Venice's greatest rivals', observing that the Pope had granted many exemptions to Levantine Jews in order to augment the commerce of Ancona to the very great harm of Venice.[67]

62. See *Bullarum, diplomatum et privilegiorum sanctorum romanorum pontificum*, 25 vols. (Turin, 1857–72), Vol. 7, pp. 740–42. Also, generally, see S. Baron, *A Social and Religious History of the Jews*, 18 vols. (Philadelphia, 1952–83), Vol. 14, pp. 49–56, and K. Stow, *Catholic Thought and Papal Jewry Policy, 1555–1593* (New York, 1977), pp. 28–32.
63. See *Bullarum*, Vol. 8, pp. 786–9.
64. See ibid., Vol. 10, pp. 23–8.
65. Latin text in ibid., pp. 104–8; Italian version in Saracini, *Notitie historiche*, pp. 379–83.
66. See Pullan, *The Jews of Europe*, p. 189.
67. See Anselmi, 'Venice, Ragusa, Ancona', 68; also 107.

After Rodriga secured his charter in 1589, he continued to submit to the Venetian government numerous proposals and requests, several of which aimed at enhancing the maritime commerce of Venice in the face of competition from Ancona and Ragusa. While many centred around lowering the customs duties, others involved providing for galleys to protect shipping in the Adriatic from the Uskok pirates, inaugurating organized shipping to Salonica under government auspices and granting the right of transit through Venice for Florentine wool and silk, which was rejected by the Cinque Savii alla Mercanzia since it was in complete opposition to the traditional protectionist policy of the Venetian republic.[68]

As the ten-year charter of 1589 approached its expiration, Rodriga proposed a new very similar one, with a few changes that experience had indicated were desirable. Introduced in the Senate with some further slight changes, it passed easily on 6 October 1598 by the vote of 94–6–16. Its preamble clearly indicated the attitude of the government: 'At this time, even greater are the reasons which on other occasions induced this Council to permit the residence of Levantine and Ponentine Jewish merchants in this city and their trade in the places of our state, principally for mercantile trade which is so important for the public service....'[69]

Those opening words of the Venetian charter of 1598 were not mere rhetoric. The long-range commercial trends leading to a decline in Mediterranean trade in general were having an increasingly detrimental effect on the commerce of Venice, while in the Mediterranean basin Venice was facing new challenges. Although the establishment of Spalato was leading to the importation of substantial merchandise to Venice to the benefit of the customs and the market place,[70] the trade itself was primarily in the hands of foreigners rather than in those of native Christian Venetians; also, even in the Adriatic the Venetian position was far from secure, for Ancona constituted constant competition. On the other hand, Livorno, located on the west side of Italy, was the first and often only Italian port of call for western ships entering the Mediterranean. The contemporary Venetians had every reason to perceive that as their commercial prosperity continued to decline, that of Livorno, favoured by the advantages of geography and commercial policy, would greatly increase. Unwilling otherwise to depart from its traditional protectionist policy of excluding all non-Venetians — except reciprocally Ottoman subjects — from trade between Venice and the Levant, the republic had to continue to attempt to attract Jewish merchants to its shores.[71]

68. For reasons of space, I will have to discuss these proposals elsewhere.
69. For details on this charter and a comparison with that of 1589, see Ravid, 'Daniel Rodriga and the First Decade of the Jewish Merchants of Venice', pp. 211–23.
70. See Ravid, *Economics and Toleration*, pp. 66–8, n. 64.
71. It is not yet possible to estimate the success of the Venetian policy. See Pullan, *The Jews of Europe*, pp. 195, 197, including the report of 1592 that 'the richest Jew with

The Jewish merchants residing in Venice well realized their importance to the city. At the time of subsequent negotiations over the renewal of their ten-year charters, they not only routinely drew attention to the merchandise they were bringing to the city and their payments to the customs, but also claimed that the renewal of their charter would induce additional Jewish merchants to come to the city, and generally the Cinque Savii alla Mercanzia endorsed their position. Thus, when in 1608 the Jewish merchants sought the renewal of the charter of 1598 for ten more years, they asserted that were it to be renewed, many additional families would come to the city and divert trade away from other ports. In support, the Cinque Savii affirmed that the Jewish merchants should in all ways be encouraged and not troubled, since they were being invited on liberal terms by other states, while one Savio added that many Jews in the Low Countries and elsewhere were waiting to see the government's decision regarding the renewal of the charter before coming in great numbers to live in Venice.[72] In 1624, as the merchants petitioned for the renewal of the charter of 1611, they claimed that during the intervening years, many families of Levantine and Ponentine Jews had come to Venice and the Venetian state, augmenting trade, promoting the general well-being, and benefiting the customs. The Savii reiterated this as they related that trade had increased considerably and was growing daily as families of Levantine and Ponentine Jews continually came to live in the city, and therefore the government ought to keep them satisfied and free from aggravation, since trade had for the greater part come into their hands.[73] In 1636, at the time of the next charter renewal, the Jewish merchants again asserted that many Jewish merchant families had come to live in Venice and the Venetian state, which increased trade to the advantage of all, and they hoped that in the future, trade would further increase with the arrival of many more who were expected to come in such numbers that it would be impossible to house them in the ghetto. Then when the Cinque Savii were asked whether they thought the Jewish merchants should be subjected to an annual tax, the Savii responded that any such step required careful consideration because of the possibility that if a greater burden were to be placed on the merchants, they might leave Venice, with the Levantines going to Ragusa and Ancona and the Ponentines to Livorno.[74]

the biggest business in Ancona' had recently arrived in Venice, and B. Ravid, 'The Third Charter of the Jewish Merchants of Venice, 1611' (forthcoming). In another forthcoming article, I will discuss the subsequent charters of the Jewish merchants of Venice.

72. See Ravid, 'The Third Charter'.
73. ASV, Senate, terra, filza 276, 24 Sept. 1625.
74. Ibid., filza 387, 10 July 1636.

The Tedeschi Jewish moneylenders on two occasions also evoked such arguments as they sought official permission to engage in overseas maritime trade, which apparently they were doing without authorization.[75] Specifically, in 1628 they claimed that such permission not only would give them hope of economic relief but also would benefit the government by greatly increasing the customs revenue because it would attract to Venice many rich Jewish families from other lands who had not come to the city and would not come because they were not allowed to engage in any activity other than the very meagre and declining second-hand goods trade.[76] In 1634, when they repeated this argument, it was endorsed by the Venetian magistracy of the Cattaveri.[77] Indeed, the Venetian government so greatly desired to attract additional merchants that in 1633, at the request of the Jewish merchants already in Venice who claimed that additional very wealthy Jewish merchants would settle in the city if they had room to live, the Senate authorized the expansion of the ghetto into what immediately became known as the Ghetto Nuovissimo.[78]

The Venetian rabbi Simone Luzzatto, who in his *Discorso* so well presented and defended the commercial and financial activities of the Jews in the wider context of the Venetian economy, also clearly articulated the long-standing competition between Italian states for Jewish merchants and placed it in a wider perspective.[79] In the course of discussing how to prevent wealthy foreign traders from retiring to their homeland with the wealth that they had amassed in the city of their sojourn (Venice), he pointed out the obvious solutions of either forbidding them to engage in trade, or of hampering their activities by imposing higher customs duties or by adopting other measures. After citing the successful experiences of the English in breaking the power of the Hanseatic League in English ports because England was a well-situated island whose rulers had complete control over its many ports, he observed that this would not so easily succeed in Italy, for since the ports of Italy were in the hands of many different rulers, should any one of them make conditions difficult for foreigners and exclude them, they would immediately be welcomed elsewhere with very extensive privileges. This indeed was the situation on

75. See n. 52 above.
76. ASV, Senato, terra, filza 306, 8 Sept. 1629. Presumably these Jews were not Levantine or Ponentine merchants and therefore not eligible to settle in the city under the terms of the charter of 1625.
77. Ibid., filza 368, 29 Dec. 1634.
78. See B. Ravid, 'The Establishment of the Ghetto Nuovissimo of Venice', in *Jews in Italy: Studies Dedicated to the Memory of Umberto Cassuto on the 100th Anniversary of his Birth* (Jerusalem, 1988), pp. 39–54.
79. Luzzatto, *Discorso*, 12r–14v; see Ravid, *Economics and Toleration*, p. 61.

the sixteenth- and seventeenth-century Italian peninsula, where the authorities sought to attract Jewish merchants as the easiest way to enhance their maritime commerce which was declining because of numerous factors beyond their control.[80]

80. The attempt to attract foreign merchants to Venice continued well into the eighteenth century, as I hope to show in the future. In the interim, see L. Poliakov, 'Un tentativo di Venezia per attirare gli Ebrei di Livorno', *La Rassegna Mensile di Israel*, 23 (1957), 291–5. On the other hand, see Balletti, *Gli Ebrei e gli Estensi*, pp. 79–80, on Jewish immigration, primarily from Iberia, to Reggio in the mid-seventeenth century; a list of 60 such families arriving between 1652 and 1657 included 29 from Livorno and 6 from Venice.

Venice between Jerusalem, Byzantium, and Divine Retribution: The Origins of the Ghetto

ELISABETH CROUZET-PAVAN

The Expulsion of 1492 marked the end of the *Reconquista*; yet, at the same time, it revived it on another — internal — front in Christian Spain. In this article, I propose to discuss the dynamics of a parallel reconquest: that which wrought upheaval in Italy during the late fifteenth and early sixteenth centuries and which, in one city, Venice, culminated in the establishment of the ghetto.[1] My intention is not to study the effects of the expulsion of the Jews from Spain on sixteenth-century Venetian history.[2] However, when considering that period, it is evidently more significant to bring out the Italian side of what seems to have been a common structure of European imagery, manifested in Venice by a turning point in the collective attitude towards the Jews.[3] With the formation of the ghetto a style of cohabitation developed, a kind of spatial segregation which broke with former socio-urban reality, and which was to persist well into modern Europe, although having taken shape within a chronological sequence marked by the outbreak of hypermedievalism.

Thus, in exploring the very conditions of the ghetto concept, several themes must be taken into consideration. Firstly, the image of fifteenth-century Venice as the city of Providence, city of God, shapes the portrayal of that city in its own eyes and those of the world at large, as a place of sacred glory and serene inhabitants. Yet, secondly, that image was blown out of proportion and set off balance by the prophetic and eschatological

This article was translated by Sharon Neeman

1. For a study of the more strictly socio-economic conditions giving rise to the ghetto, see B. Ravid, 'The Religious, Economic and Social Background and Context of the Establishment of the Ghetti of Venice', in G. Cozzi (ed.), *Gli Ebrei e Venezia. Secoli XIV–XVIII* (Milan, 1987), pp. 211–60.
2. B. Pullan, *The Jews of Europe and the Inquisition of Venice, 1550–1670* (Oxford, 1983), pp. 170–75, for the first evidence of Jews — or rather, Portuguese Jews — in Venice.
3. The last centuries of the Middle Ages are discussed in D. Jacoby, 'Les juifs de Venise du XIVème siècle au milieu du XVIème siecle', in H.G. Beck *et al.* (eds.), *Venezia centro di mediazione tra Oriente e Occidente (secoli XV–XVI). Aspetti e problemi* (Florence, 1977), Vol. 1, pp. 163–216; P.C. Ioly Zorattini, 'Gli ebrei a Venezia, Padova e Verona', in *Storia della cultura veneta*, III/1, Dal primo Quattrocento al Concilio di Trento (Vicenza, 1980), pp. 537–76.

tension which rocked Italy starting from the last decades of the fifteenth century. For Venice, the Time of Tribulations had arrived, and the ghetto was to emerge from that eschatological universe. Thirdly, the collective fantasies of the guilt-ridden Christian population in search of its own reformation were freely taken out on those driven by military events to take refuge in large numbers in the city: the Jews. The religious element involved in the foundation of the ghetto has already been discussed in Robert Finlay's study.[4] Still, it seems to me that the understanding of the Christian perception of the world must encompass more than the mere set of circumstances, in order to bring out the scope and duration of what amounted to a process of torment. That process was tolerated, though kept in check, by the Venetian public authorities. By creating the ghetto, by locking up those considered in the collective concept as a stain on society, the political powers of Venice took charge of achieving the mythical Christian city, while at the same time imposing the braking force of their rationality on a pathologically anti-Semitic culture.

A CITY UNDER THE BENEVOLENT HAND OF GOD

Thus, at the starting point of our analysis — that is, up to about 1480 — the collective image of Venice was one of great serenity and inner peace. Diving into the murky waters of medieval chronicles concerning the city, one cannot but note to what extent its history is initially depicted as providential. Everything in Venice is described as serving the will of God and in accordance with His order.

With the transfer of St. Mark's relics to safekeeping in the ducal palace began the city's mystical process of identification with its patron saint.[5] The testaments of Doge Giustiniano Partecipazio and Bishop Orso of Olivolo[6] bear witness to the growing awareness of the unique nature of Venice. The legend of St. Mark's preaching at Aquilea, which dates from the late sixth or early seventh century, is repeated, but corrected to refer to Venice. The city may be viewed in that context as illustrating the concepts of predestination and the absolute. Popularized in texts by Marco and by Martino da Canale,[7] the theme of St. Mark's dream that the lagoon would be the place of his eternal rest was structured — and restructured — to

4. R. Finlay, 'The Foundation of the Ghetto: Venice, the Jews, and the War of the League of Cambrai', *Proceedings of the American Philosophical Society*, 126 (1982), 140–54.
5. G. Fascoli, 'Nascita di un mito', *Studi storici in onore di G. Volpe* (Florence, 1988), Vol. 1, 450–52; H.C. Peyer, *Stadt und Stadtpatron im mittelalterlichen Italien* (Zurich, 1955), pp. 8–24.
6. R. Cessi, *Politica e economia di Venezia nel Trecento* (Rome, 1952), 157.
7. A. Niero, 'Questioni agiografiche su San Marco', *Studi veneziani* (1970), 3–27.

create a prophetic suggestion that the city's very foundation was miraculous. Similarly, Andrea Dandolo's account of the angel appearing to St. Mark at Rialto places Venice within the order of divine creation.[8] This self-awareness, concretized out of its own history, permitted the city to consider itself as born to preserve and defend the faith.

This textual predestination reached its zenith during the thirteenth and fourteenth centuries. Even after that theme began to fade in narrative tradition, at the end of the fourteenth century, the Venetians' own faith continued to affect the history of the Holy Republic.

The chronicle of Giovanni Diacono evokes the image of the *aurea Venecia*, describing the construction of both sacred and profane edifices.[9] Later narrative texts maintain the trend of listing catalogues of relics, recalling processions, enumerating churches — not so much in evidence of the city's material prosperity as of the piety of its inhabitants. Venice in their day, by its very nature, abounded with holy sites and objects. Throughout the complexity of its diplomatic and military history, Venetian sources have tended to emphasize this essential aspect: the faith of the Venetians. True, the city was also often depicted in those years as cruel, arrogant, and avaricious; yet the defence of the faith transformed its military and commercial adventures into crusades. Indeed, that theme seems to have been intended as an explanation of Venice's very existence.

Starting from the fifteenth century, this same analysis of the city's providential history became extended and transformed by an exaltation of the unique nature of its site. Whereas medieval chronicles had listed this site among the *mirabilia*, later historical accounts not only improved its geographical description, but often linked it with political considerations. Analyses of the city's location led to a glorification of its independence. Built amid the waters by the will of God, the city was considered indestructible; that indestructibility guaranteed Venetian freedom. As of the second half of the fifteenth century, it was primarily this commendation of the site which led to the image of Venice as the city graced by God. The admirable location of the city dictated its admirable nature. According to the collective concept shaped by or reflected in the writings, Providence itself — divine inspiration — presided over Venice's rise from the briny waters. The Republic was of God: a creation, the most amazing one within Creation.[10]

8. *Andresse Danduli Chronica per extensum descripta*, ed. E. Pastorella, R.I.S. (Bologna, 1942), Vol. 12, Pt. 1. p. 10.
9. R. Cessi (ed.), *Origo civitatum seu Venetiarum* (Rome, 1933); G. Fasoli, 'I fondamenti della storiografia veneziana', in A. Pertusi (ed.), *La storiografia veneziana fino al secolo XVI. Aspetti e problemi* (Florence, 1970), pp. 11–44.
10. Bernardo Giustiniani's treatise illustrates the outcome of one such reflection: *De origine urbis Venetiarum rebusque gestis a Venetiis libri quindecim*, ed. J.G. Graevius, *Thesaurus antiquitatum et historiarum italiae* (Lugduni Batavorum,

As a result, various fifteenth-century chronicles and treatises — though in slightly different terms — stress the city's providential birth, miraculous growth, and unrefuted liberty. The other characteristics of the urban system, in one way or another, remain of less importance than these original definitions.

An additional motif belonging to this system of portrayal, invoked by Venice in the late fifteenth century to reaffirm its identity, was the triumphal image of two cities: Jerusalem and Byzantium.

Venice, even before Rome, sought to attain the somewhat mystical status of the 'centre of the world'. Its situation as a port city led it to organize the voyage to Jerusalem — a move more or less in line with the theme of predestination. Though aspiring neither to supplant nor to equal the Holy City, it was henceforth to herald it. Venice was the 'ante-City', the portal of the Holy Land. Pilgrims, celebrating (for example) the Feast of the Eucharist, sanctified the Venetian domain. Most notably, Venice strove to reproduce within its own borders some of the sites or monuments of the Holy City at the centre of the terrestrial sphere, thus actually distorting the image of Jerusalem in favour of its own. The building of San Sepolcro[11] on the bank of San Giovanni in Bragora was a reconstruction of sacred topography. A *translatio* had been initiated — and was to take its full effect when, facing the image of the king of France, leader of the Crusade, designated by Jean Michel and Guilloche de Bordeaux as the man who would retake Jerusalem,[12] Venice claimed to include the very presence of Jerusalem within its own limits. In the same way, following the conquest of Constantinople, Venice declared itself ready to take up that city's heritage, and thus to conclude a *translatio* undertaken long since. In the second half of the fifteenth century, various relics were transferred from one city to the other.[13] By this very process, Venice was transformed into a chosen city, a spiritual focus.

The concept of divine protection assuredly also explains this rescue mission. During those years, Venice fed on the spoils and relics of fallen cities, enclosing them within itself, and thereby implementing a reconse-

1722), Vol. 1, cols. 1–172.

11. Archivio di Stato di Venezia (hereafter ASV), C.R.S., San Sepolcro, B. 1, *pergamene*, for the testament of Elena Vioni, who bequeathed a house in which the hospice was to be established. After having housed oblates of the Order of St. Francis, San Sepolcro became a convent of the Poor Clares in the early sixteenth century.
12. *La Prophecie du roy Charles VIII par maître Guilloche*, publ. Marquis de La Grange (Paris, 1869); *La Prophécie vision et revelation divine révelée par le très humble prophète Jehan Michiel* (n.p., n.d.).
13. I shall restrict myself to mentioning, moreover, that Bessarion's gift may be quite neatly placed within the framework of these transfers, and that an incontrovertible neo-Byzantinism is evident in Venetian architecture between 1468 and 1509.

cration — or an appropriation. By their grace, the city was protected and could survive. Sanuto, on the eve of Holy Saturday 1515, recalls Venice's quasi-ontological search for relics. Evoking that impressive heritage, preserved in the lagoons for all of Christendom, he concludes: 'Molte vi sono. E queste e quelle che mantien la cità nostra, ch'è senza muraglie'.[14] This passage is no more than a reprise of an already abundantly illustrated theme: the preservation of Venice.

This, then — the continuance of Venice — is the third theme in this interplay of images. A large part of the narrative tradition written during the period in question still links this remarkable preservation to divine grace. Yet, in some of the first humanist treatises of that time, the constituent parts of the myth are already assembled into a different shape. For their authors, this theme of the city's continuance, like the related ones of its liberty, equilibrium, and concord, revolve around a common axis: the Venetian institutions.

Although the dynamics of the image of Venice in the late fifteenth century tend to expose additional components in focussing the praises of that city on the political, I find it of definitive importance to show how powerful the themes of Venice's providential history and mythical accomplishment still were. True, for certain authors of the time, the city's history began to centre on a new accomplishment, that of the State. Yet, throughout this interweaving of images, the incessantly enriched image of the predestined and miraculous city remained triumphant. Created and ordered amid the waters, charged with the Christian heritage of the cities which it claimed to succeed, Venice aimed to offer, for its own consumption and that of others, the spectacle of a city 'under the benevolent hand of God'. Yet that very Christian concept of a human order achieved by God in the conformity of human life to God's will was suddenly to go awry.

MIRACLES, SIGNS, PROPHECIES: *UNA TANTA RUINA ALA REPUBLICA VENETA*

The turning point marked by the 1450s in the history of Italian religious sensitivity has already been noted.[15] Similarly, the converging signs which, starting from the 1480s, indicate the dramatization of this spiritual climate

14. Marino Sanuto, *I Diarii*, 58 vols. (Venice, 1879–1903), Vol. XX, col. 99.
15. For the activity of prophesying preachers throughout the peninsula, see O. Niccoli, *Profeti e popolo nell'Italia del Rinascimento* (Rome–Bari, 1987); C. Vasoli, 'L'attesa della nuova era in ambienti e gruppi fiorentini del Quattrocento', in *L'attesa dell'età nuova nelle spiritualità della fine del Medioevo* (Todi, 1962), pp. 340–432; G. Miccoli, 'La storia religiosa', in *Storia d'Italia* (Turin, 1974), Vol. II, 1, pp. 968–75.

have also been discussed. The prevailing historiographic image of Venice, however, remained that of a Signory which took pains to protect the city from ferment, disorder, and religious panic. After 1496, Sanuto's *Diarii* fortunately counterbalanced this unilateral focus of historians' interest on the official and apparently dominant forms of religious ritual within the borders of San Marco.[16] In any event, Venetian chronology does not diverge from this pattern, and the same break is marked by the 1480s. Let us, then, prepare the ground for the examination of that period.

Admittedly, the history of Venice is punctuated by wonders, by miraculous apparitions. None the less, one cannot ignore the drastically increased outbreaks of miracles in the late fifteenth century. On 2 July 1480, the first of a series of Marian miracles took place in the *contrada* of S. Lio. Another Virgin launched Her own miracles on 23 August 1480 in the parish of Santa Marina. Although the two phenomena have never been studied in parallel, their relationship seems rather close. The same lineage of *cittadini*, the Amadi, owned both of the miraculous images. In both cases, the Franciscans, supporters of the Immaculate Conception cult, trained (or, at least, directed) the local devotion, leading to the establishment of two new churches: Santa Maria della Fava and Santa Maria dei Miracoli.

These two Virgins may thus be considered as one of the first manifestations of this increased sensitivity to the sacred. Yet, in Venice of the 1480s, miracles were still capable of dissipating anguish. The years to come were to plunge the city into crisis. A new dimension of penitence and expiation — reflecting the mounting tension of the emotionally convulsed atmosphere — was to characterize the forthcoming apparitions of the Virgin: in Padua in 1500,[17] Chioggia in 1508,[18] Venice in 1509.

As early as the 1480s, in both Italy and Germany, a multitude of handbills and flyers, doubly armed with rhetoric and imagery, circulated accounts of wonders and prognostications. Until that time — and, indeed, well into the 1500s — Florence had played a major role in the publication of ancient prophecies translated into the vernacular.[19] Now, however, Venice rose into ascendancy with these handbills and poems, written in support of a new prophecy, whose extremely wide circulation — according to O. Niccoli's research — is a proven fact. These publications provided an essential medium for the propagation of eschatological anguish and

16. Certainly worth mentioning at this point is E. Muir's sound analysis, *Civic Ritual in Renaissance Venice* (Princeton, 1981).
17. *Apparizione della beata Vergine addolorata del Torresino in Padova* (Padua, 1864).
18. For the subsequent construction of the sanctuary, see Sanuto, *Diarii*, Vol. VII, col. 575.
19. See O. Niccoli, 'Profezie in Piazza. Note sul profetismo popolare nell'Italia del primo Cinquecento', *Quaderni storici*, 41 (1979), 500–539 (esp. Appendix, pp. 535–9).

contributed to the obsolescence of the serene vision of earlier decades. Divine retribution seemed imminent; and an increasing number of signs convinced the general population of the portents of these prophecies, giving modern-day researchers a precise impression of the instability of the collective mind. Such events as the great planetary conjunction of 1484 and the birth of various monsters[20] presaged an age of misfortune. Both revelations and astrological predictions announced a future fraught with tribulations and called for atonement, for reform.

Thus prophesied, the plunge indeed took place. Wars, plagues, famines, and heavenly portents: all of the chronicles, following Charles VIII's first expedition into Italy, give the same apocalyptic account of Italian misfortunes. The crisis was soon to touch Venice directly: Sanuto's description of the year 1509 begins with the huge fire which ravaged the world, Italy, and the Venetian Signory relentlessly pursued by the coalition of kings.[21] And the terrible blaze which wiped out the Arsenal brought the first signs of that *gran combustione* into the city itself. Earthquakes and explosions rocked the very ground, fire swept through homes and public buildings: one of the safest foundations of Venetian power had crumbled.[22] The collapse of the domination under which the Republic had held the *terra ferma* marked the nethermost point of the slide into disaster.

The most gripping vision of this divine punishment inflicted on Venice is presented by Girolamo Priuli.[23] The fate of Venice in Priuli's day is compared with those of the biblical cities laid waste by the will of God. The fall of *Veneto Imperio* is attributed to one sole cause: its sins. The fact that these *Diarii*, like those of Sanuto, bear the stamp of a violently conservative tradition, eager to invoke its new, bad customs in the denunciation of the city's entire recent socio-political evolution, explains nothing. On the contrary: it must be noted that Priuli's words express the same *angoisse* which had already begun to grant legitimacy to the councils' rulings on sodomy or to the condemnation of luxury. The ruin of Sodom and Gomorrah is similarly noted in the proceedings of the Council of Ten, to explain the urgent need for atonement and repression of evil: 'For similar acts, cities and men have perished'.[24] In addition to the obvious political concerns, the eschatological mainstay of the creation of Venice lay open to attack. Divine retribution, in 1509, was well in progress.

20. Such as those sent from Venice to Florence in 1489, see L. Landucci, *Diario fiorentino dal 1450 al 1516*, ed. J. del Badia (Florence, 1883), p. 57.
21. Sanuto, *Diarii*, Vol. VIII, col. 1.
22. Ibid., cols.17–19.
23. *I Diarii di Girolamo Priuli (aa 1499–1512)*, ed. R. Cessi, R.I.S. Vol. 24, Pt. 3 (Bologna, 1936–38), Vol. 4, p. 30.
24. ASV, C.d.X, *Miste*, reg. 17, fol. 127v.

The incredible shock of the events of 1509 may thus be seen to have two explanations. Notwithstanding all the signs and portents, despite the overall destruction, Venice proved its capacity for resistance. In that year, the city abruptly left the realm of legend, to re-establish itself in that of time and history. Yet the collapse of the Venetian myth must also be noted in another context. Venice ceased to be 'a city under the benevolent hand of God' and became no more than a city of sin, doomed to calamity. It lost its unique nature, blending into the guilt-ridden system of images which characterized Italy in those years. The two mainstays of the Venetian myth, jointly represented in contemporary portrayals, collapsed simultaneously; and with them, the 'superb' city of Venice, 'first of sins against Christian law'. It was submission to the laws of Christianity which, in the eyes of medieval chroniclers, had explained the success of the city. The first humanist historiographers had perceived the rationalized nature of the State and the excellence of its institutions as guaranteeing its longevity. With the fall of the *dominium*, both rationales went bankrupt.

Priuli, for whom Venice was first and foremost a Christian city, foresaw a single remedy: atonement and penitence. Venice, indeed, immersed itself in mass repentance, warding off the wrath of God with what the chroniclers describe as a ritual of supplication. The proud became humble, the magnificent took off their silk and scarlet, the hedonistic renounced their lascivity, the women left their jewelry at home. No feasts, no parties, no plays, no balls; only a profound penitential tension. Prayers, fasts, tears, alms and penances ... In response to the *angoisse*, to the immediate quest for a gesture of mercy such as that accorded by the Almighty to Nineveh, the Venetians — spontaneously and universally — sent their vanities to the stake.

Although this first wave of penitence has already been discussed in literature, as has the second wave in 1516,[25] I feel it necessary — if only so as not to favour conjectural explanations — to emphasize the duration and depth of that panic-ridden movement. The study of the turning point of the 1480s aims to reconstruct the long chronology of this process. Let us examine its forms and its main expressions after the start of the Time of Tribulations.

The first reality encountered was that of the Word as preached by men of God. Sermons were heard throughout the city at Christmas, during Lent, at Easter. Naturally, preaching was organized in the basilica, before the doge and the Signory.[26] Yet other preaching took place as well, in the larger churches. Sanuto's *Diarii* provide us with a long list: San Francesco della Vigna, San Polo, Santa Maria Formosa, Santi Giovanni e Paolo, Santa Maria dei Frari, Santo Stefano, Santa Maria dei Servi, Santa Maria

25. Finlay, 'The Foundation of the Ghetto'.
26. Sanuto, *Diarii*, Vol. I, cols. 836–7.

dell'Orto, San Moisè, Santa Croce ...[27] During Holy Week of 1507, four preachers, at the four corners of the city limits, spoke from the pulpit.[28] On each of these occasions, Sanuto notes that crowds were present. For some of the more inspired of these ministers of God, the faithful were too numerous to be contained within a church, and the Word was brought out of the sacred confines.

Some of the themes of these sermons have come down to us, as has the identity of some of the preachers: Fra Timoteo de Lucca, Fra Egidio, Fra Rufino Lovato — Minorites of the Observance or Hermits of St. Augustine. The dynamics of their sermons revolved around penitential atonement. Preaching against the Conventual Brothers[29] and the women's monasteries — real 'brothels'[30] — they denounced a Church degraded, perverted. Against the sodomites, they described the spread of the most abominable sin[31] throughout the city. Against the blasphemers, they impugned a society which ridiculed the name of God.[32] And, in the name of that very God, they challenged public action, criticized the *cosse di Stado*.[33]

In early sixteenth-century Venice, these were extreme manifestations of that other Christian *Reconquista* initiated, as we have been able to observe, by the Brothers of the Observance in fifteenth-century Italy. Yet, within that movement, the prophetic and eschatological dynamics were now increasingly severe — to the point of transforming it entirely. The wrath of God was on the move.[34]

In the late fifteenth and early sixteenth centuries, the conversion of the Jews was said to herald the end of the Time of Tribulations.[35] Palm Sunday of 1505 marked what appears to have been the first public baptism of a Jew in Venice. Attended by a great assemblage, the ceremony was performed by Fra Egidio, the preacher of Santo Stefano.[36] Four years later, in April 1509, a second baptism — announced in advance by Fra Rufino Lovato — took place in the *campo* of San Polo. Three Venetian noblemen served as godfathers to the new Christian.[37] Another baptism, also in a *campo*, was

27. Ibid., cols. 61, 836–7; Vol. II, cols. 493, 546; Vol. V, col. 1037; Vol. VI, col. 141; Vol. VII, col. 40; Vol. VIII, col. 78.
28. Ibid., Vol. VII, col. 40.
29. Ibid., Vol. I, col. 61.
30. Ibid., col. 836.
31. Ibid.
32. Ibid., Vol. VIII, cols. 70–71; Vol. XVIII, col. 131.
33. Ibid., Vol. XIX, col. 460.
34. See, in this connection, Fra Helia de Brexia's sermon: Sanuto, *Diarii*, Vol. XIX, col. 462.
35. In this connection, let us recall the influence of Joachim de Flore on the Franciscan Friars of the Observance; see M. Reeves, *The Influence of Prophecy in the Later Middle Ages: A Study in Joachimism* (Oxford, 1969), pp. 229ff.
36. Sanuto, *Diarii*, Vol. VI, col. 141.
37. Ibid., Vol. VIII, col. 88.

held in March 1513.[38]

The Jews were also a favourite target for the ire of these same preachers. At San Marco, on Christmas of 1497, befor the doge, Fra Timoteo da Lucca cited usury among the innumerable sins of Venice.[39] Although the sermon explicitly mentioned the Piazza di Rialto and does not, at first glance, appear to refer to the Jewish moneylenders — who, in any event, were forbidden to reside in the city for periods exceeding fifteen days during the period in question — echoes of the struggle maintained by the Minorites of the Observance against Jewish moneylending are no less apparent in the text. On Good Friday of 1509, Fra Lovato's sermon was even more explicit, stating that it was 'legal to take the Jews' money and not to let them live.'[40] Two years later, on Passion Sunday, this preacher addressed the same theme, albeit a bit more gently. After having despoiled the Jews, he now stated, it was acceptable to expel them.[41] The campaign for the institution of *monti di pietà* disputed not only the Jews' financial activity, but the very possibility of their existence in the midst of Christian society.[42] From that point on, preaching in Venice became much more radical. All the looting and pillaging which marked the conquest of the *terra ferma* — as was more generally the case throughout the history of the Italian wars[43] — were given not only legitimacy, but encouragement, by men of God. Throughout the period in question, preachers' words and gestures on the subject of Jews revolved around this double theme: convert them or get rid of them. As such, this theme became a major motif of prophetic and penitential dynamics. The winds of prophecy continued to fan the flames of *angoisse*. In 1507, '... a preacher in Servi threatened Venice with plagues and wars against the Turks, and he claimed to be a prophet, and there began to be a great crowd of people there.'[44] In 1509, according to Agnadello, the crowd poured into the basilica, in an attempt

38. 'Frate Rafael de Uberti, el qual in campo...', ibid., Vol. XVI, col. 53.
39. For an analysis of this brother's activities, see M. Luzzatti, *La casa dell'ebreo. Saggi sugli ebrei a Pisa e in Toscana nel Medievo e nel Rinascimento*, (Pisa, 1985), pp. 179–202.
40. Sanuto, *Diarii*, Vol. VIII, cols. 70–71.
41. Ibid., Vol. XII, col. 121.
42. Worthy of mention here are several recent studies, representative of an impressive bibliography: S. Boesch-Gajano, 'Il comune di Siena el il prestito ebraico nei secoli XIV e XV: fonti e problemi', *Aspetti e problemi della presenza ebraica nell'Italia centro-settentrionale (secoli XIV–XV)* (Rome, 1983), pp. 171–220; A. Toaff, *Il vino e la carne. Una comunità ebreica nel Medioevo* (Bologna, 1989), pp. 85–6, 152–4; or Pt. IV of the proceedings of the colloquium *Gli ebrei a Venezia: Prestatori ebrei e banchi di pegno a Venezia e nel Dominio*, pp. 565–700. For a more general analysis, see D. Owen Hughes, 'Distinguishing Signs: Earrings, Jews and Franciscan Rhetoric in the Italian Renaissance City', *Past and Present*, 112 (1986), 3–59.
43. Let us recall that, in 1495, a synagogue and some houses of the Jewish quarter in Rome had been sacked by Charles VIII's troops.
44. Sanuto, *Diarii*, Vol. VII, col. 40.

to decipher the properties of its mosaic paving — which Sanuto as well as a whole tradition attribute to Joachim de Flore,[45] and in which Jean Lemaire de Belges discovered the causes of decadence in Venice.[46] On Palm Sunday of 1513, in a packed church, Fra Girolamo of Verona counted the generations in the Old and New Testament[47] in order to prophesy 17 years of schism and suffering.

To appease the wrath of God — 'e per placar l'ira di Dio' — these same preachers opened the way to penitence. Sanuto goes on to describe precisely this wave of faithful seized by the fear of divine retribution. Piety took the form of a penitential procession in which children, clad in white, marched at the head, directly behind the cross. This event differed in all aspects from ordinary Venetian processional ritual: the participants — women and children; the geographical setting — the Cannaregio area; and the timing — not a single event, but repeated throughout Lent.[48]

In such a context, the power of portents was redoubled; their interpretation could instil fear, or quell it for a certain time. There was, for example, an earthquake in April 1511. The *contrade* held processions. A Jewish astrologer in Treviso had predicted that a terrible earthquake would take place on 2 April. On the Venetian *terra ferma*, that timely example illustrated certain elements of the role played by the Jews — some of whom, no doubt, were impatiently waiting for the Antichrist to come and avenge oppressed Israel — in the mounting tide of eschatological hopes and fears. The entire country fled.[49] By contrast, the Sign of the Cross which appeared in the moon on the night of Holy Thursday, 1507, marked a respite. After months of drought, rain began to fall; and the doge, for the first time in years, attended Vespers on Easter Day. Similarly, in late March 1509, the fire which swept through the city spared an image of the Madonna in the *Merceria*; given that miracle, the Venetians hoped that God had not totally abandoned them.

There can be no doubt that some of the high points of this tension had to do with military events. Yet it is just as certain that the tension flared up again each Lent,[50] reinforced even further by its aim — precisely within a temporal and spatial framework saturated with human sin — to restore the

45. Ibid., Vol. VIII, col. 326; see Reeves, *The Influence of Prophecy*, pp. 96–9.
46. Jean Lemaire de Belges, *La Légende des Vénitiens ou autrement leur cronique abbrégée par laquelle est démonstré le très juste fondement de la guerre contre eulx* (Paris, n.d.).
47. According to calculations made by Joachim de Flore: Niccoli, *Propheti e popolo*, 135; Sanuto, *Diarii*, Vol. XVI, col. 53.
48. Sanuto, *Diarii*, Vol. XIX, cols. 460, 462; Vol. XX, col. 20.
49. Ibid., Vol. XII, cols. 98–9.
50. The Venetian example does not differ in this regard from the situation observed in other cities during the Lenten weeks, starting from the fifteenth century.

city to its own predestined order, that of the New Jerusalem. The city was profoundly aware that it had entered into a time of divine portents; a time eminently appropriate to the creation of the ghetto.

THE HOLY REPUBLIC

Generally mentioned along with the activity of Franciscan preachers on the *terra ferma* and the opening of *monti di pietà* in Padua, Verona, and Treviso is the refusal of the Venetian authorities to open such an establishment within their city.[51] This fact attested to the relative moderation of the Signory following the Trent affair, when the cult of the Blessed Simon was spreading through Venice along with other accusations of ritual murder.[52]

As from that time, one may discern two different aspects of Venetian politics. In Padua, for example, preachers had been threatening that city with divine judgment for some time. In 1488, the fear of an outbreak of plague led the Paduan ruling élite to take what it considered as real prophylactic measures. The city, under Venetian domination, sent ambassadors to the Signory, to petition for the closure of moneylending *banchi* or the expulsion of the Jews from Padua. The response to each of these petitions was affirmative.[53] None the less, within Venice itself, Jewish moneylending continued in its traditional forms.[54]

The stability of the Jews' position in Venice was undoubtedly determined more by economic factors than by political considerations. As shown in R. Mueller's analyses, the variations in Venetian politics concerning the Jewish community were closely linked to the evolution of the credit market.[55] Sanuto's comments, to the effect that wheat was less scarce in Venice than money, bear witness to a situation of real monetary deprivation in the first decades of the sixteenth century. Economic reality,

51. For example, P.C. Ioly Zorattini (ed.), *Processi del S. Uffizio di Venezia contro ebrei e giudaizzanti (1548–1560)* (Florence, 1980), p. 23.
52. The spread of such accusations outside the well-known Venetian framework is discussed, *inter alia*, in S. Simonsohn (ed.), *The Jews in the Duchy of Milan* (Jerusalem, 1982), Vol. 1, p. xxii.
53. A. Ciscato, *Gli ebrei in Padova* (1300–1800), repr. (Padua, 1401), pp. 51–2, 54–5.
54. Its legislative and judicial framework is well known. In 1502–3, the Senate renewed the *condotta* of the Jews of Mestre. The *capitoli* of 1508 concerned the Jewish communities of the *dominium* in general. In July 1513, on its expiration, the pact of 1503 was renewed. See B. Pullan, *Rich and Poor in Renaissance Venice: The Social Institutions of a Catholic State to 1620* (Oxford, 1969); R.C. Mueller, 'Charitable Institutions, the Jewish Community and Venetian Society: A Discussion of the Recent Volume by Brian Pullan', *Studi veneziani*, 14 (1972), 37–82.
55. R.C. Mueller, 'Les prêteurs juifs à Venise au Moyen Age', A.E.S.C., 3 (1975–76), 1277–1302.

then, played a considerable role in explaining this tolerance, often noted in ancient historiography.[56]

Venice was by no means isolated from news, nor from dominant cultural and religious trends. A miracle occurring in June 1497 was discussed within the city for a long time thereafter. A poor peasant from Piove di Sacco was forced to go to a Jewish moneylender in order to buy seed wheat. The crop came up longer and more beautiful, with fuller heads of grain. Brought to Padua, the peasant celebrated God's aid to His faithful against the Jews.[57] Another example: when the cult of the Blessed Simon was organized in Trent and his body borne in processions, Venice was notified immediately.[58] And, like any other city, Venice was overrun by collectors raising contributions to the work of God — namely funds for the construction of churches and *monti di pietà*.[59]

The Councils apparently remained faithful to their traditional politics. Preachers cited the Jews as one of the causes of divine retribution. The Signory managed the affairs of the Jewish community and heard its representatives. 'The Jews were reminded that they had to pay this week what taxes were left for them to pay.'[60] In an understandable defence of primarily Christian interests, the Council of Ten posted small boats at Mestre to guard houses belonging to Jews which contained pledges.[61] Once the first storm on the *terra ferma* had passed, in order to conserve supplies, Venetian authorities demanded that all the Jewish refugees leave the city. This measure, however, also applied to all of the *forestieri*.[62] Moreover, although the various magistrates' courts in the city were informed of the order of expulsion, there are very few recorded instances of its implementation.[63] In an exact counterpart of the pillage which accompanied the fall of the *terra ferma*, the Venetian troops retaking Padua sacked *li banchi di Zudei*. The Signory received a delegation of complaint.[64] Sixty German soldiers were taken prisoner at Serravalle. One single soldier — a Jew — was paraded before the population in the *Merceria*, along the same route as those used by condemned men going to their execution. None the less, several days later, his return to Germany was negotiated.[65] And, in the hours which followed the most violent sermons, notables of the Jewish community used to come and bare their fears to the heads of the Council of

56. For example: C. Roth, *Gli Ebrei in Venezia* (Rome, 1933).
57. Sanuto, *Diarii*, Vol. I, col. 653.
58. Ibid., Vol. VII, col. 275.
59. Ibid., col. 293.
60. Ibid., Vol. VIII, col. 91.
61. Ibid., cols. 355–6.
62. Ibid., and col. 406.
63. Ibid., Vol. XII, cols. 110–11.
64. Ibid., Vol. VIII, cols. 523, 527.
65. Ibid., col. 550.

Ten. As a result, various preachers were admonished, reminded of the necessary moderation 'acciò non segui contra zudei qualche cossa'.[66]

The evidence attesting to this moderate government policy is manifold. Venice, in its reconquered *dominium*, ordered its officials to ensure public order for *all* of the population;[67] within the city, even at the worst moments of crisis, the councils continued to meet with the principal representatives of the Jewish community, the bankers. This dialogue, in fact, lasted until the weeks immediately prior to the settling of the Jews inside the ghetto.

Faced with the call for pillage and crimes against the Jews, which invoked anathema and misfortune on a State which had ceased to respect Christian precepts, the councils made efforts to restrain the preachers, to confine them to the spiritual sphere alone. Yet the various spheres were by no means autonomous: the political, in the years under study, can barely be distinguished from the religious.

Of primary importance is the fact that the prophetic and eschatological tension which tormented the city spared none of its inhabitants. Sanuto himself went to hear the preachers. True, he criticized them, and at times even rejected them;[68] none the less, starting from April 1515, his entries bear the stamp of contemporary preaching. The Jews, too numerous, mingled with the Christians. During Holy Week, instead of hiding decently away, they exhibited themselves. And, since the wars had made their economic activity necessary, there was now no way to limit them.[69] As early as May 1509, Sanuto had consulted Fra Piero Nani, 'zentihomo nostro'. Nor was he alone in so doing: a number of patricians had gathered around this prophet[70] — 'dice molte cosse, quale le traze de profetie'. Niccoli notes the two social levels of prophetic traffic: the nobles around the noble Piero Nani, the masses in the basilica deciphering the mosaics. Yet the anxiety in both classes was more or less identical.[71] Similarly, the Easter sermon of 1511 at San Marco clearly appears intended to liberate the Venetian ruling élite from its eschatological fears: '... the Republic will not disappear; in the end, God will help it'.[72] The Venetian authorities' reproof of the preachers should not be allowed to cloud a fact of vital importance: in the basilica, before the doge, even the most violent of these preachers were still allowed to speak.[73]

66. Ibid., Vol. XII, col. 98–9.
67. Ibid., Vol. VIII, col. 548.
68. Ibid., Vol. XVI, col. 53.
69. Ibid., Vol. XX, col. 98.
70. Ibid., Vol. VIII, col. 326.
71. Niccoli, *Propheti e popolo*, pp. 37–9.
72. Sanuto, *Diarii*, Vol. XII, col. 122.
73. The example of Fra Timoteo da Lucca has already been cited. The sermons given at San Marco by Fra Egidio and Fra Rufino Lovato, baptisers of Jews within the city, were documented in 1503 and 1511; Fra Rafael di Uberti, disciple of Bernardino de Feltre, preached there in 1512.

From that time onwards, the Republic truly made efforts to be holy. It ordered prayers to be said in all the monasteries for the success of its armed forces, and worked towards the reformation demanded by the preachers. The city's prostitutes repented and were converted.[74] Under direct pressure by Fra Rafael, reform of the monasteries run by the two bishops of Venice and Torcello was ordered in 1502. The Patriarch himself had to visit the monasteries in order to achieve a prompt implementation of this reform.[75] The following year, the opening of the hospital of Gesu-Cristo was viewed as an act of propitiation in the framework of the quest for *divina gratia*.[76]

As a direct consequence of the many sermons addressing the issue of blasphemy, new instruments were adopted in the struggle against that sin. The Council of Ten undertook a programme of repression. A decree issued in 1500 empowered them to handle the more serious offences, leaving the *Signori di notte* (Lords of the Night) in charge of minor derelictions only. In 1505, the latter, more ancient institution was relieved of its last powers.[77] The imposition of extremely harsh penalties, as evidenced by the numerous examples reported in the *Diarii*, can leave no further doubt. It seems to have been part of a general outbreak of repression, a process of purification in which the old normative body was revived, expanded, and vested with real powers. In those years, public authority extended beyond the domain of the social system. Its object was also to restore the city to the Messianic order of God's will, by making it conform to that will. Along with the struggle against luxury, blasphemy, sodomy, and all other sins, guided by the preachers and their incitement, a preventive effort at reformation was undertaken and set in motion by the political arm. The State, in short, served as a means of restraining the fantasies of the inspired Word.

This long process of religious *angoisse* and penitential activity was the framework for the act adopted by the Senate on 20 March 1516: 'Li zudei stano male in la terra, sicome li predicatori predicano le perversità dil Stado vien da questo'. The proposed law, as formulated by Zacaria Dolfin, stated that the misfortunes of Venice had no other cause but the presence of the Jews.[78] The Councils gave in, as they had given in regarding all the other offences against divine majesty[79] — especially given the military situation in March 1516, which raised the general anxiety to unprecedented heights.

74. Sanuto, *Diarii*, Vol. II, col. 493.
75. Ibid., Vol. IV, cols. 237–8.
76. Ibid., cols. 810–11.
77. R. Derosas, 'Moralità e giustizia a Venezia nel '500–'600; Gli esecutori contro la bestemmia', in G. Cozzi (ed.), *Stato, società e giustizia nella Repubblica veneta (sec. XV–XVIII)* (Rome, 1980), pp. 431–528.
78. Sanuto, *Diarii*, Vol. XX, col. 72.
79. Ibid., col. 86.

In the weeks which followed, the situation reached its denouement. The establishment of the Jewish quarter, as Finlay has stated, was equivalent to an act of prayer and thanksgiving, a tangible sign addressed to God by Venice, a gift made to Him by His city. The decades of prophesy and penitence leading to the formation of the ghetto were certainly among the factors contributing to that event, which seems less inspired by a sudden set of circumstances than by the flood of anguish which had engulfed Venice since the 1480s. The creation of the ghetto is no more than an example of the long, dull, and sometimes discordant dialogue between the inspired Word and the political decision-makers, the end of an ambivalent search for the way back to that lost serenity which had shaped the ancient concept of the city.

In 1387, Venice had promised the Jews with whom it had concluded the *condotta* a quarter 'ita quod potuerunt stare simul pro commodo eorum'.[80] A similar tendency — aimed at preserving Christian society from danger, stain, or contamination — had become increasingly evident throughout the cities of Italy, especially as from the late fourteenth century. Thus, for example, we see the repeated edicts concerning 'distinguishing signs'. The campaign undertaken by the Franciscan Friars of the Observance gave a real boost to this dynamic process,[81] which was further accelerated in Venice by the prophecies and misfortune of the times. Finally, the imposition of spatial segregation on the Jews was instituted in the name of the mythical tradition of the Holy Republic.[82] The debates on the subject, according to Sanuto, were quite controversial. An initial proposition to

80. Mueller, 'Les prêteurs', 1285. This attempt at spatial cantonment did not succeed. In 1397, the Jews were forced to relocate to Mestre. This expulsion, despite the fifteenth- century regulations lengthening the compulsory intervals between visits to Venice, was never totally effective (witness the nuclei of Jews who continued to live within the parishes, especially near the Rialto). Despite an official ban and numerous discriminations, the cantonment in Mestre, throughout that entire period, was only imperfectly achieved; see Jacoby, 'Les Juifs', 163–216.
81. Although there do exist some earlier attestations of Jews' being obligated to wear a distinguishing sign, even according to the rulings of Lateran IV (thus in Sicily since 1221; see Owen Hughes, 'Distinguishing Signs', 21), it was only in the fourteenth century that certain cities passed regulations in this regard. In any event, the widespread adoption of such signs is clearly linked to the preaching of the Franciscan Friars of the Observance in the fifteenth century: D. Pacetti, 'La predicazione di S. Bernardino in Toscana', *Archivum Franciscanum historicum*, 30 (1940), 282–318; R. Manselli, 'Bernardino da Siena', *Dizionario Biografico degli Italiani*, Vol. 9 (Rome, 1967), pp. 215–26; and, for several case studies, see A. Toaff, 'Gli ebrei a Città di Castello dal XIV al XVI secolo', *Bollettino della Deputazione di Storia Patria per l'Umbria*, 72 (1975), fasc. 2, 1–105; id., *Gli ebrei a Perugia* (Perugia, 1975) (Deputazione di storia patria per l'Umbria, Fonti per la storia dell'Umbria, 10); id., *The Jews in Medieval Assisi, 1305–1487* (Florence, 1979).
82. Prototypes for such segregation may be found in the last decades of the fifteenth century in cities like Bari, Cesena, and perhaps Ravenna.

that end, made the preceding year, had not succeeded. In the text finally adopted, the concerns of public order were mixed up with those of Catholic order. The enclosure of the Jews on the edge of the city, at Cannaregio, is in line with the hierarchization of Venice from the centre out to the edges — the latter having then been viewed as places of danger and pollution. After all, the trend in urban politics had for some time been to exile to the extremes of Venice any activities which could threaten the city: the glass industry because of the fire involved, the slaughterhouses for reasons of hygiene ... This process of relegation was perceived by Venice as guaranteeing good order.

And indeed, automatically as it were — whether due to historical evolution alone, or perhaps to a certain liberation arising from the enclosure of that which the Christian mindset held to be a source of human corruption and divine retribution — the dangers facing Venice did subside. In 1516, Lateran V denounced apocalyptic preaching; yet the structure of the ghetto remained, as if in witness of the collective anxiety attack on Venice — an attack which historians, caught up in the myth of the City Triumphant, have deprecated or ignored. This crisis may well give rise to another image of Venice: a city which, only after having internalized the theme of the Holy City and the consequences of that theme to the most extreme degree possible, could now develop another set of dynamics — that of a rationalized State; that of the Venetian Republic.

Owners, Houses, Functions: New Research on the Origins of the Venetian Ghetto

ENNIO CONCINA

The settlement of Jews in the Ghetto Novo began at the end of July 1516, and was preceded by negotiations which were far from easy, especially with regard to property ownership,[1] while the complexity of the situation was compounded bv social, political, and cultural factors.

Initially, several property owners in the ghetto, among them Giacomo da Brolo, Cristina, widow of Alvise, and Marco Michiel were summoned by the authorities.[2] Some of the owners agreed to the demands, but reactions varied. There was some resistance by those who were living on their property and had invested money in it. Doubtless the relationship between the republic and the ghetto property owners continued to be satisfactory due to the economic measures in the decree. The Jews had to pay a rent increase of a third, and this revenue was exempt from taxation. It must be stressed that the decree legitimized existing procedures and conditions. We know of cases, dating to 1513–14, where owners who let properties to both Christians and Jews made the latter pay more than the 'good people' and did not have to pay tax on the difference. The fiscal archives of the republic confirm that in 1521 these laws were already enforced. Sometimes the owners themselves would notify the office of the *Savi Sopra le Decime* of the sum of money exempt from tax, 'that which paid no tithe'.[3]

These documents, together with information from other sources, enable us to reconstruct the essential characteristics of the very early stages of organization, from 1516–21, when the Jews first inhabited the ghetto. The area was regarded as 'very spacious' by the government, but the Jewish banker Anselmo claimed that the area was too small. His estimate proved to be correct: already then about 700 heads of households lived in the city

1. For a complete bibliography on the ghetto of Venice, see E. Concina *et al.*, *La città degli Ebrei. Il Ghetto di Venezia, architettura e urbanistica*, 1991, especially Pt. 1, Concina, 'Parva Jerusalem'.
2. Archivio di Stato, Venice (hereafter ASV), Savi alle Decime, Redecima 1514, b. 44, cond. 11, 16.
3. M. Sanuto, *I Diarii* (Venice, 1879–1903), Vol. XXII, col. 86; ASV, Compilazione delle Leggi, b. 189; ASV, Savi alle Decime, Redecima 1514, b. 44, cond. 76, 77.

and, due to the crowded conditions, many left it during the spring and summer.[4]

Anselmo's opinion is backed by information found in documents dealing with the housing situation in a group of buildings owned by the da Brolo family just before and soon after the Jews came to live in the ghetto.[5] We find references to 'developments carried out in ghetto houses', and these were clearly extensive. Seven habitations were divided into sixteen dwellings; ground floors were split into *casette* (small houses). The characteristic Gothic architecture of the ghetto was modified or eliminated. Many architectural features which replaced the ancient late Gothic style in the Ghetto Novo date to the first half of the sixteenth century.

The documents provide additional information. Simon, a school teacher, lived in a wooden hut, which would indicate that there was a lack of sufficient space from the start, despite the fact that many had left. The problem is further illustrated by the fact that a certain clerk from Anselmo's bank lived in a garret, but paid more rent than Simon the school teacher.

It seems that only three of the houses belonging to Giacomo da Brolo, which were inherited by his son Giovanni and his brothers, preserved the functional characteristics which had marked them in 1514, or were even renovated. One was inhabited by the banker Vita, and another by Lazzaro, a physician, probably the same person against whom the monk, Brother Giovan Maria, heatedly preached on Good Friday in the Frari Church. The third house, which apparently remained intact, was inhabited by Scaramella and Salamon *partioro* (an assessor of the quality of gold), or had been inhabited by them a short while beforehand.

The details of the original housing plan are clear. The ground floor was taken up by the humbler low-rental dwellings inhabited, for example, by a widow, two Jews who made bonnets — Donna Bella and Abram Spagnol — and Lazzaro, a butcher.

The upper floors were usually divided into more spacious dwellings, occupied, for instance, by Damian the physician, and Aron the innkeeper. Moisè Luzzatto, his brother Salamon, and companions, rag merchants who had previously owned a shop at the Rialto, lived together in a house near the bridge in the northern part of the ghetto islet. Jacob da Cesana, of unknown profession, also lived there.

From these bare facts, it is quite clear that, by 1521, some essentials of community life had been reorganized, including the inn which was the only one in the city authorized to serve Jewish travellers. No documentation has

4. Sanuto, *I Diarii*, cols. 108–9.
5. ASV, Savi alle Decime, Redecima 1514, b. 44, cond. 11, 16, 76, 77.

been found concerning places of worship, whose possible existence before 1527 was alluded to by Gabriele Moro.[6]

A very interesting observation, backed by documentation, derives from the comparison between Giacomo da Brolo's property in 1514, and that of his son seven years later. The income from the rented property was much higher than it would have been if only one-third had been added on to the basic rent.

There is evidence that, during those years, a different type of work was underway in some nearby buildings. In 1522 Marco Michiel informed the authorities that he had added one storey to a building in order to create storage space for keeping pawned objects on behalf of Anselmo, the famous banker. This is the oldest existing documentation regarding the ever-increasing phenomenon of adding floors to the ancient buildings in the ghetto. Above all, Michiel's declaration provides the first scraps of interesting information on building work commissioned by the owners, documented in greater detail by others later on. Three houses that were owned by Piero Zuane da Modon underwent drastic building alterations and modifications and in fact were turned into twelve habitations, among which was the large opulent residence belonging to Salamon dal Banco, details of which were sent in 1537 by his widow Caterina to the office of the Dieci Savi (ten wise men) at the Rialto.[7]

These testimonies clearly indicate that, the space allotted to the Jews being very limited, investing in property in the ghetto or renting it to the Jews was a highly lucrative business, and speculation was brazen and uncontrolled.

In 1566 Massimo Valier, and in 1582 Bertuzzi, both new owners of a da Brolo house in the Ghetto Novo, expressed the view that if the Jews left Venice, profits would be negligible because the houses were small and a lot of money would have to be spent to make them fit for Christian habitation.

The situation regarding buildings also applied to open spaces. The Campo (square), which was once a field on the small island, remained in the possession of the property owners and their families. They constantly confirmed their rights in great detail, as can be seen from a document written on 9 August 1581 by Bernardo da Brolo, in which he declares that 'he and his brothers have exclusive rights on the whole Ghetto Novo Square where the Jews live, including three wells and anything else in the above-mentioned square, between one house and another, in any direction and between one front door and the next.' This meant that the slaughterer had to pay 'for the place where he cut meat in the square', and sites for

6. Sanuto, *I Diarii*, XLIII–XLIV, col. 305.
7. ASV, Savi alle Decime, Contente Reg. 1=b. 1231, fol. 194, 1.7.1522, 7.8.1522; reg. 2=b. 1232, fol. 72r., 5.5.1527; Redecima 1537, b. 93 cond. 748; Redecima 1582, b. 158, cond. 981.

butchers' chopping blocks for beef and veal were leased under formal contract.

Water from the wells also yielded an income. At first, possibly throughout the sixteenth century, the wells were entrusted to a water-vendor, but from 27 February 1605 — a date which marks the most ancient contract of its kind — the wells were leased to the Talmud Torah Brotherhood of the Jews in the Ghetto Novo.[8]

The available information only provides a partial picture regarding the formation and organization of Jewish settlement in the Ghetto Novo. Despite the fact that some of the oldest documents have been lost and the resultant lack of information, it is certain that the Jews were not only housed through direct agreements with the owners. During the first decade of Jewish occupation in this area, a public body — the Office of the Cattaver — also assigned houses to the Jews, though there is some doubt when this actually began.

From the 1530s on, once the contract once between owner and tenant had been formalized, building work was carried out directly by the new Jewish residents. It was not easy for the Venetian Jews to create their own living and working premises and for some time this was a cause for controversy. On the one hand, this arose partly due to the fact that developments were carried out by the house owners, while, on the other, the new tenants themselves made great improvements to the buildings. From that moment, the history of the ghetto is one of never-ending building and change.

During the years 1531–36, the Serenissima Signoria was summoned, first at the request of Consiglio and Angelo and their Jewish Companions, and then by Salamon dal Banco, to reach a decision regarding building alterations carried out by the Jews in the ghetto. This decision was of particular importance.

The Signoria recognized that, once the houses and working areas had been assigned by the official organization of the republic, the work done on the buildings by the tenants themselves was of direct relevance not only for living conditions but also for the initiation and development of commercial activity in the area as a whole and in particular for the tenants themselves: 'They have greatly improved those houses ... and it is not right to do so just to suite other people's convenience.' The Jews could dispose of their property any way they wished as long as the ground rent had been paid to the owners.[9]

8. ASV, Savi alle Decime, Redecima 1566, b. 133, cond. 684/5; Redecima 1582, b. 162, cond. 202; Procuratia di San Marco de Ultra, Commissaria da Brolo, b. 57,16/23 and b. 57,20.
9. ASV, Collegio, Notatori, reg. 23, fols. 87v.–88, 31.1.1536 m.v.

During the time of Doge Andrea Gritti, the resolutions passed by the Venetians had a series of very important consequences. The Venetian Jews were guaranteed a form of possession in the ghetto which did not imply actual ownership of the property; all building and architectural interventions were recognized, and conditions were stipulated regarding income the Jews obtained from the property.

This did not mean that the 1423–24 *comminatorie* (warning), stipulating that Jews were not allowed to purchase or hold as security any buildings, estates, or freehold property had been abrogated. It simply implied a new approach to the issue which was not only seen from the legal angle.[10]

This period was significant for the change in attitude of the Venetian Jew towards his property. The reciprocal effects of such an attitude, their different interpretation depending on the various people and opinions involved, and of course the inevitable conflicts which ensued, were very important during the three centuries of institutional life in the Venetian ghetto and were a determining factor in its building and architectural history. In reality, the protection granted by the republic to rights granted on the basis of occupation of houses in the Ghetto Novo and recognition of the rights deriving from *miglioramenti* (housing improvements) acquired by Venetian Jews basically implied acknowledgment of the Jewish legal concept of *ḥazakkah*. To a certain extent this concept was also reflected in the relationship between the Jewish community and the regulations which governed the city-state itself.

As is explained in Marco Ferro's eighteenth-century legal dictionary, the Jews were not allowed to actually own property in Venice according to early medieval laws, but they could enjoy special contracts regarding 'the houses where they lived'. In fact the Venetian dialect absorbed the Venetianized form of *casacà* from the Jewish term to define hereditary leaseholding, 'a contract which gives someone hereditary rights to possession in return for annual payment in money, fruits or services'.[11]

The *casacà more hebreorum* issue did not follow a path of linear development, either from the judicial point of view or with regard to mentality, as explained below. Certainly, the fact that negotiations to sublet were officially recognized and any housing improvements carried out were to the advantage of those who had undertaken the work soon meant that the Jews were integrated into the Venetian system of taxation concerning income on property — just like any other resident or citizen in the town.

The Dieci Savi, in their Rialto office, were responsible for stipulating the following declaration, dated 7 November 1583: 'Jews must inform the

10. ASV, Compilazione delle Leggi, b. 188, 16.2.1423, 26.9.1423, 30.12.1424.
11. M.Ferro, *Dizionario del Diritto comune e Veneto*, Vol. V (Venice, 1779), p. 4; G. Boerio, *Dizionario del dialetto veneziano* (Venice, 1856), s.v. 'casacà'.

above-mentioned office regarding income obtained from any let houses, shops, rooms, warehouses or other places in the Ghetto Novo or in the Ghetto Vecchio, or if there is any financial gain from buildings, whether improvements have been made or not'.[12]

A number of Jewish 'owners' of buildings came forward and responded to this clear and direct request, listing possessions and income, lodgings and floors of buildings added for their own use or sub-let to tenants. Some Jews spontaneously presented their declarations before the Dieci Savi specifically ordered them to do so. The patricians who were in charge of the office, having no cause for objection, recorded the details and imposed taxation. Meanwhile, their register of *Traslati*, in which changes of ownership and income were noted, listed buildings, parts of buildings and 'buildings ... greatly improved'. Thus our first documentary evidence that an intensive building programme was in progress, one which led to to the 'purchase' by Levantine Jews of land for building purposes in the Ghetto Vecchio, dates to 1560.

During the second half of the sixteenth century it is evident that there was a tendency and desire among the Jewish community to integrate with the Venetians, and this was facilitated by the newly-won rights. This change is attested in moving declarations like the one written by Davit de Florio who, in March 1582, wrote to the Savii: 'I live in the Ghetto Vecchio in my own house which has to be estimated by Your Excellencies'.[13]

Serious difficulties and heated arguments over these issues began during the first half of the seventeenth century. One controversy lasted five years — from 1636 to 1641 — first passing through the courts and then through the political institutions. The case was brought before the court by Marco da Brolo versus the ghetto community, in connection with the income from and rights attached to his 46 houses. The accusations by this descendant of the founders of this complex were both harsh and significant, the main argument being that 'it is not right that the Jews can use my property as a commodity' or that 'they merchandize houses'.

The reasoning presented was highly relevant. Marco da Brolo affirmed, for example, that for 15 habitations situated in the original buildings the tenants paid little more than 241 ducats, but then they sub-let them for more than 952 ducats. He added that from another group of houses, where no alterations had been done, he received only 230 ducats while 'a very advantageous estimated 947 ducats are received [for them] by the Jews'. Even from the butchers' *stazi* 'the tenants receive an income of 199 ducats

12. ASV, Savi alle Decime, 'Sommario et pratica generale delle Leggi', b. 8, fols. 148, 167; Capitolare II, b. 2, 1463–1685, fol. 154r., 7.11.1583.
13. ASV, Savi alle Decime, Redecima 1582, b. 163, cond. 720; b. 183, cond. 780; b. 162, cond. 25; b. 162, cond. 1131; b. 162, cond. 29 and 87; b. 164, cond. 806; b. 164, cond. 1229; b. 162, cond. 45; Traslati, reg. 1251 (1580)–1346 (1807).

but they only pay me 30'. And he continued at length, singling out cases of very low rents, work undertaken, lists of expenses, and giving the names of tenants cited by the owner to back up his argument: Isach Luzzatto, Mandolin Baldosa, Jacob Luzzatto, Mazo Mai, Moisè Giona, Rosa Sacil, and others.[14]

It should be noted that at the beginning of the seventeenth century a particular type of transaction developed, which was legally registered with Venetian notaries. The Jewish owners of *miglioramenti*, on property owned by Minotto and Brolo in the Ghetto Novo and Ghetto Vecchio, ceded part of their rights of possession to Venetian nobles, receiving them immediately afterwards as 'emphyteusis'. The interpretations of rights relative to *ragioni utuli* (income and usufruct) and *miglioramenti* therefore covered a wider field, sometimes with economic consequences.[15]

According to Brolo, it was necessary to discuss the whole matter afresh:

> - The Jews have no right of dominion in the houses ...
> - The Jews should not be able to sell their houses; if they no longer wish to live there, the house should be handed back to the owner; if there have been any improvements, the owner or his descendants should pay what they are worth at the time of giving up the house;
> - Great care should be taken when estimating rents, making a distinction between houses where no improvements have been made — and the original structures are still intact — and *miglioramenti*.

It was Simone Luzzatto who retorted, in reply to the accusation of 'upstart Jewish interests' regarding the houses in the Ghetto Novo. He supported the legitimacy of 'possessing houses we live in' and he tried to define the limits of the concept of possession. He clearly underlined the symmetry between Venetian legislation and Jewish traditions regarding 'the developments and improvements called *casacha*'. He stressed the great economic resources which the Jewish community had poured into the building enterprises in the Ghetto Novo, estimated at being 'thousands and thousands of ducats'. After hearings, appeals to the doge and government, and inspections, the Senate nominated a committee of 20 members to solve the problem. On 23 March 1641, it tried to clarify the issue by making a clear distinction between buildings that had remained in their original state, *fondi vecchi*, and buildings that had been rebuilt, altered, refurbished, and so on, which lay at the root of so much controversy. To all appearances, Marco Brolo had won a legal victory.[16] In reality, after this difficult issue had been reviewed, the Jews' rights to buildings in the ghetto were confirmed and consolidated.

14. ASV, Procuratia di San Marco de Ultra, Commissaria da Brolo, b. 56, 10 and 11.
15. Ibid.
16. Ibid., 10, fol. 14.

At this point in developments the situation is clear. The notaries registered negotiations, conveyances, and purchases which they defined as follows: 'tutte le ragioni, utili sive miglioramenti o siano inviamenti et casacod more hebreorum'.

The history of housing is also the history of ideas with regard to stability, permanence, continuity, and the various symbolic values involved, as indicated by some important examples. A contract signed between Samuel Baruch and the patrician Cappello family about 1660, allowed the former to construct a small palace. In fact, Samuel Baruch owned several houses which were to be converted or utilized as he wished, with the prerogative and right guaranteed at a public and private level. He could also carry out any work that he wished on the property. 'Naturally', were the Jews banished from Venice, he would have lost any rights on the house he was building, as was stipulated by medieval laws. But — and this is important — if after banishment the Jews were allowed to return to Venice, then it was guaranteed by contract that Samuel Baruch could regain his house and other property.

This is not an isolated episode. During the same period a Venetian noblewoman who owned property in the Ghetto Vecchio added to the compulsory clause 'unless the Jews are banished from Venice', in an official housing contract, that she hoped 'God would not wish such a thing'. And an inhabitant of one of the buildings in the Ghetto Vecchio speaks about his house with the deep devotion that the 'house of his ancestors' deserved.

Throughout the second half of the seventeenth century, first in the case of the register of property let by the Minotto family in the Ghetto Vecchio, then in Pietro Venier's notarial documents, and finally, at the end of the seventeenth century with the notaries Carlo Gabrieli and Angelo Marice Piccini, all transactions were recorded as 'livello perpetuo et enfiteutica investitura' (emphyteusis). This showed the marked tendency — cultural if not legal — to regard the 1424 law as being obsolete. In this way, in the seventeenth century, the term *casacà* often became synonymous with a building or house, in fact a building defined in several ways.[17]

This development meant that in the mid-seventeenth century the Jews and the income obtained from their property were regarded with less suspicion and doubt from both a political and social point of view.

17. ASV, Notarile, A. Calzavara, b. 2967, fols. 718v.–720, 18.11.1654; C. Gabrieli, b. 7073, fols. 15v.–22, 12.3.1712; F. Giberti, b. 6930, fol. 254, 1.9.1681; A. Calzavara, b. 2990, fols. 334v.–335v., 14.6.1663; b. 2994, fol. 662v., 9.9.1664; Archivio Marcello-Grimani-Giustinian-Querini, fondo Querini, b. 4; Notarile, P. Venier, b. 13933, fols. 161–2v., 27.2.1665 m.v.; C. Gabrieli, A.M. Piccini, b. 7065, fols. 252 v.–254v., 25.8.1698.

The register of proprietors, tenants, and rents compiled in 1661 for the Jewish quarter, as was the case regarding all the districts in Venice, enables us to define the situation shortly after the clash between the land owners and the interests of the residents/possessors mentioned above.[18] In spite of some lack of clarity, the situation is most interesting.

In fact, in the Novo, Vecchio and Novissimo ghettos, 138 Jewish heads of households lived and worked on their own 'property', paying only the ground rent for houses and shops 'for their own use'. All the other inhabitants lived in rented property which yielded an annual income of little more than 14,600 ducats. Out of this sum only 1,269 ducats went to non-Jews as compared to 13,392 ducats earned by the Jewish owners of *miglioramenti* or *casacà*.

The 4,000 ducats' ground rents paid to the Brolo, Minotto, and Malipiero families, their heirs and minor property owners should certainly not be forgotten. And the flow of gifts specified in the contracts, which poured annually out of the ghetto to enrich the cellars and storerooms of the patricians and Venetian citizens should also be remembered: sugar loaves, fine quality candles, sausages made of *lengual*, and raisins, tasty *botarga* of tuna, 'belle et onorate', fine fat geese, and capons for carnival time.[19] Nevertheless, the Jews were entitled to most of the income deriving from property in the ghetto. According to the *redecime* twice carried out by the republic, once in 1712 and then in 1740, the proportion continued to increase.[20]

Going back to the 1660s, 275 owners of *casacod* shared the income we have already mentioned. It is useless to try to find any large property owners among these people — at the most they possessed two or three apartments from which they received between 100 and 200 ducats in rent.

However, there are some exceptions. First and foremost there was R. Moisè Coen Spilimbergo, with an income of 650 ducats from 12 houses. Marco Baldosa, Iona Tobi, and Jacob Leoncini in the Ghetto Vecchio stood out from other possessors as did the Cividal Zemel brothers, Anna widow of Simon dal Banco, Benetto and Abram Mazo, Aron Volterra, and Anselmo Gentili in the Ghetto Novo.

18. ASV, Savi alle Decime, b. 425, Ghetto Vecchio; b. 426 Ghetto Novo e Ghetto Novissimo, 1661; Redecima 1661, bb. 219,221,222, *passim*.
19. See n. 18, and ASV, Savi alle Decime, Condizioni aggiunte, reg. 382, n. 3279; Notarile, A. Calzavara, b. 2967, fols. 718v.–720v., 18.11.1654.
20. E. Concina, *Venezia nell'età moderna. Struttura e funzioni* (Venice, 1989), pp. 202–3, 216.

It is also possible to single out family group interests, such as those owned by the Jesurum family in the Ghetto Novissimo, the Luzzatto and Saraval families in the Ghetto Novo, and the Dal Medico, Namias, Pappo, and Penso families in the Ghetto Vecchio. The names changed during the eighteenth century, but the tendencies — at least up to the mid-1700s — were already well-defined.

The Role of Ragusan Jews in the History of the Mediterranean Countries

IVANA BURDELEZ

It is a well-known fact that, after expulsion from Spain (1492), and later from Portugal (1498), Sephardic Jews travelled along various routes in search for a new homeland. One of the countries they were directed to was the Ottoman Empire, at that time developing into a powerful state which welcomed and tolerated the banished Jews, who brought with them their vast knowledge and culture.

On their way to the Turkish lands a large number of Sephardim crossed the Adriatic sea from Italy and arrived at the port of Ragusa, modern Dubrovnik, where they hired transport and the caravan continued on its way. Some of those refugees stayed on in Ragusa and soon formed one of the most successful and most important Sephardic communities on the southern Adriatic coast. This was the start of the Ragusan Jewish community. There had been some Jews in Ragusa beforehand (according to the surviving documents, from the thirteenth century on),[1] but these were merely visiting the town for business reasons or temporarily practising medicine there. As those Jews did not represent any larger group, only after the arrival of the Sephardim in the Balkan peninsula can we follow, for several centuries, their continuous presence in the Republic of Ragusa where they played an important role in its economic, cultural and social life.

From the early period of their arrival in Ragusa two contracts for transporting Jews to Salonica and Skopje (1501, 1502)[2] were stipulated and registered in the notarial books still to be found in the archives of Dubrovnik, and among the numerous documents we can also find descriptions of the attacks on the ships transporting Jews and their goods.[3]

1. G. Cremosnik, *Spisi dubrovacke kancelarije* [Documents of the Ragusan counting-house] (Zagreb, 1951).
2. Historijski archiv Dubrovnik (hereafter HAD), Diversa Cancelariae, Vol. 95, pp. 64–5; Consilium Minus, Vol. 27, fol. 4.
3. J. Tadíc, *Jevreji u Dubrovniku do polovice XVII st.* [Jews in Dubrovnik up to the mid-seventeenth century] (Sarajevo, 1937).

The Ragusan government often intervened and helped the endangered parties, and even gave them protection when needed.[4]

For those who decided to stay, the Republic of Ragusa seemed to offer many attractions: a description of the town at the end of the fifteenth century (more precisely, in 1481) was given by an Italian Jew, Meshullam de Volterra, who stopped there on his way back from a visit to the Holy Land:

> ... Ragusa is a very pretty town and I have never seen such splendour and such beauty. I dare say it looks like Florence. Ragusans are very rich and there are also many immensely rich merchants dressed like noblemen ... Ragusa is situated on the right side of the Adriatic coast and is a free state like Florence.... But there are no Jews there ...[5]

However, only a few decades after Meshullam's visit, there was already a large number of Sephardim in the very town which by the beginning of the sixteenth century was a powerful maritime, commercial and cultural centre, playing an important role in trade between East and West. Moreover, it was an active transit port, a centre for commerce throughout the sixteenth and the first half of the seventeenth century.

Actually, the first half of the sixteenth century could be called the 'period of adaptation' of Ragusan Jews to their new surroundings, but equally it was the period of adaptation of the Ragusans to the newcomers. In a short time the Sephardim of Ragusa fitted into the commercial life of the republic due, in the first place, to the fact that they were closely connected with the other Sephardic communities in the Mediterrranean countries and with those further inland. Thus the maritime and the caravan commerce formed a complex network of firm bonds between various regions, connecting large western trade centres with undeveloped eastern regions which were rich in the raw and primary materials so badly needed by western industry. That vivid movement of people, ships, caravans, and cargo represented the basis on which the Braudelian concept of a Mediterranean world was created. This world was a vibrant entity with its own rules and methods, in whose chain the Jews, with their intensive economic activities, formed one of the most important links.

From the inland areas of the Balkan peninsula (Bosnia, Serbia, Bulgaria, Macedonia, et cetera) Ragusan Jews were exporting all kinds of skins, leather, wax, fish, wine, and spices to the western countries, while they imported silks, ready-made clothes, textiles, paper, jewellery, and

4. B. Krekić, 'Gli Ebrei a Ragusa nel Cinquecento', in G. Cozzi (ed.), *Gli Ebrei e Venezia (secoli XIV–XVIII)* (Milan, 1987), pp. 835–44.
5. J. Ramón Magdalena Nom de Déu, *Relatos de viajes y epistolas de peregrinos judíos a Jerusalem*, Ausa edn. (Madrid, 1987), p. 41.

various other goods to the East. By paying customs and other taxes Sephardim enriched the republic's budget, but, nevertheless, they were not always welcomed by the Ragusan citizens. In fact, in 1502 several of them were sentenced to death for a presumed ritual murder,[6] and in 1515 the Senate decided to expel all Jewish families from the territory of the republic.[7] Although this was a serious threat to the Jews the situation did not last long, and they soon continued with their impressive commercial activities, bringing goods from the Balkan hinterland to the port of Ragusa, from where Ragusan ships transported them to the ports of Venice, Ancona, Sicily, Livorno, Genoa, Patras, Alexandria, and elsewhere. As the participation of Jews was becoming more and more significant for the flourishing economic life of Ragusa, the Senate decided to enable them to reside permanently in the town. On 25 February 1546 a ghetto, which included a synagogue, was finally established.[8]

Ragusan commerce developed in particular during the periods of the First and the Second Wars of the Holy League (1537–40 and 1570–73), when sailing across the sea was rather dangerous. Besides enjoying a favourable geographical position and skilled diplomacy the Republic of Ragusa succeeded in maintaining neutrality, and in those years of crisis was able to hold the role as sole mediator between the Christian and the Muslim world.[9] Ragusa became the centre for merchants and traders and if one compares the surnames of the Jews working there to those in Ancona or Venice and then to those, for example, in Sofia, it is immediately apparent that all of them belonged to the same families or groups. The intensity of trade in Ragusa is most clearly indicated by the data registered in 1572, which show that of 50 commercial agents, the so-called *sensali* who worked in Ragusa, 30 were Jews.[10] According to the numerous archival documents, Jewish trade reached its peak between 1590 and 1645. During that period the Balkan Jews exported a total of 1,602 bundles of *giambelotti*, 77,202 bales and 267,384 pieces of *cordovani*, 135,881 pieces of oxhide, and 4,872 bundles of other hides.[11]

6. L. Münster, 'Proces zbog ritualnog ubistva u Dubrovniku 1502 i tragicna smrt Moše Maralija' [Proceedings on the ritual murder in Dubrovnik in 1502 and the tragic death of Moshe Maralija], *Zbornik Jevrejski Istorijski musej* (1971).
7. HAD, Consilium Rogatorum, Vol. 33, 160–61.
8. Ibid., Vol. 47, 249.
9. V. Foretić, *Povijest Dubrovnika do 1808* [The history of Dubrovnik up to 1808], 2 vols. (Zagreb, 1980); R. Samardžić, *Veliki vek Dubrovnika* [The great century of Dubrovnik] (Belgrade, 1962); T. Popović, *Turska i Dubrovnik u XVI veku* [Turkey and Dubrovnik in the sixteenth century] (Belgrade, 1973).
10. HAD, Consolati I. 1572–75.
11. Sneška Panova, *Die Bulgarischen Kaufleute im 17 Jahrhundert* (Sofia, 1980); Zdenko Zlatar, 'Export Trade of Balkan Jews via Dubrovnik in XVI and XVII Centuries', *Zbornik Jevrejski Istorijski musej* (1979).

The data resulting from research on trade relations have already proved useful though still not complete, but there is one further aspect of the multiple Jewish activities that has not been sufficiently stressed and is worth mentioning. Besides being good tradesmen, commercial and legal agents, Jews also played an important role as spies both for the Christians and for the Muslims, and from the documents assembled from the archives of Spain, Venice, and Dubrovnik it becomes evident that Ragusa was one of the crucial meeting-points for these spies; from there all the information brought from the Ottoman Empire was sent on to Venice, Ancona, Naples, and Madrid. The *avvisi* were written in encoded letters and were especially frequent during the Second War of the Holy League. One of the Jews working for the Venetians in Ragusa, David Passi, even passed on information about the war with the Turks.[12] Nevertheless, the Venetians suspected him of being a double spy, while Spanish documents reveal that he was working for the Spanish king.[13] Guzmán de Silva, Spanish ambassador in Venice, wrote an encoded letter to his king requesting a Spanish passport for David Passi in a reward for his services to the Spanish Crown. He described him as '... an intelligent person who knows a lot about the Levant and, what is more important, he is a good friend of João Miches ...'[14] One other letter on the same subject came from a spy in Ragusa: '... a Jew named David Pas left for Venice to give you all the information. He was working as a commercial agent here in Ragusa. His uncle is a doctor at the Turkish court, he has a wife in Ferrara and his father lives in Salonica ...'[15] In August 1573 de Silva added more information about him: 'A Jew by the name of David has come from Ragusa; he is the nephew of a Jewish doctor. He speaks Spanish and considers himself to be a native of Spain ... He is dressed in a Greek manner and is very rich....'[16] Another Jew had arrived in Venice from Ragusa in 1572 and immediately a letter was sent from Ragusa: '... a Jew from Constantinople, Cain Suruco, came here and embarked a Venetian galley. It seems that he is an important person since they paid him great attention. He is trying to negotiate peace between the Ottoman Empire and Venice....'[17] De Silva informed his king in 1573 about one *judío* by name of Cain Suruco who came from Constantinople and wanted to pass on some information: '... he could be a double spy' continued de Silva in his letter, 'and is not to be trusted....'[18] Again in 1573 a Spanish agent in Ragusa

12. Archivio di Stato, Venice (hereafter ASV), Consiglio dei Dieci, Parti secrete, R.9 cc. 89–90, fol. 101 v.
13. Archivo General de Simancas (hereafter AGS), Papeles de Estado, Legajo 1334.
14. Ibid., 1332, fol. 78.
15. Ibid., 1331, fol. 233.
16. Ibid., 1332, fol. 110.
17. Ibid., 1331, fol. 236.
18. Ibid., 1332, fol. 6.

wrote about Lazaro Hebreo who brought him information about many Jews who were working for Venetians. He also mentioned a certain Leon Barcon who was sending his letters to Ragusan Jews via Skopje.[19]

On another occasion the same agent from Ragusa asked for some money in order to pay Cayn de Salin, a Jew who had provided useful information and who had faithfully served His Majesty the King of Spain.[20] In 1576 the ambassador of Florence asked the Spaniards to grant safe-conduct (*salvacondotto*) to some Levantine Jews and their goods, transported by French, Venetian, and Ragusan ships: called Sechiel Zaccur and his son Mordochay, they lived in Constantinople, and had volunteered to forward all the information they had about the Turkish plans.[21]

On the Muslim side, the centre of espionage was the house of João Miches, a famous Marrano who became the Duke of Naxos. He had a wide net of agents, mostly Slav and Jewish merchants, working for him in Corfu, Dalmatia, and even Venice. Although it is still not very clear it would seem that he was a double spy working for Spain as well.[22]

In those periods of 'secret wars' all kinds of tricks and weapons were used. Thus the Christian spies, knowing how easily the Jews were able to travel throughout the Ottoman Empire, used to disguise themselves as Jews. Caballero Alvar Mendez travelled dressed as a Jew, assumed the name Abraham and pretended to be a merchant. He even took along many silver candles for the synagogues in the Levant.[23]

Poison was very often used as a means to eliminate enemies. In 1654 a Jew, Zacuti, promised the Venetians that he would poison one of their renegades in Constantinople[24] and a Jewish doctor, Michel Angelo Salomon, offered to prepare a special poison contaminated with the plague, with which they could impregnate clothes and textiles destined to be sold in Turkey.[25]

The success of Jewish espionage would not have been possible without the knowledge of several languages. And indeed the Sephardim showed a great aptitude for foreign languages, in addition to their use of Ladino which served as the lingua franca no matter where they lived. Unfortunately, little research has been done on this topic in the Historical Archives of Dubrovnik although there are numerous documents written in Judaeo-Spanish and in Latin letters.[26] The content of the documents mentioned is

19. Ibid., fols. 167–8.
20. Ibid., 1346, fol. 2.
21. Ibid., 1449, fols. 102–6.
22. P. Preto, 'La guerra segreta, spionaggio, sabotaggi, attentati', in *Venezia e la difesa del Levante: Da Lepanto a Candia* (Venice, 1986).
23. AGS, Estado Leg. 1341, fol. 22.
24. ASV, Inquisitori di stato, b. 417.
25. Ibid., b. 274.
26. B. Kalmi, *Dokumenti o Jevrejima u Dubrovačkoj državnoj arhivi* [Documents on the

mostly of a commercial character, due to the fact that all of them belong to a single archival series and represent a pattern,[27] but there are also many other private letters, testaments, legal documents and *ketubbot* which testify to the language spoken by the Ragusan Sephardim several centuries earlier. It should be noted that the Spanish or rather Castilian in these texts is almost classical at the beginning while it becomes corrupt towards the end of the seventeenth century, a fact which has a socio-linguistic explanation. A closer study reveals two types of documents: on the one hand, those that are basically in Castilian (as it was spoken in fifteenth-century Spain), with the addition of words borrowed from Italian, Slavonic, and Turkish, and on the other hand those basically in Portuguese. This division leads us to believe that two main groups of Sephardim coexisted in Ragusa, the one of Spanish origin and the other from Portugal. There is no doubt that they mixed and lived in the same ghetto but their language was obviously not the same, and they probably had other differing customs originating from the Iberian Peninsula. The work done so far on these texts is in fact a pioneer study of languages which developed in close contact, research which began only a year ago, and it is now clear that there is a need for the help of Hebrew and other scholars in order to completely decipher the available texts. The introductory section preceding each of the texts was written in Latin and notaries would usually stress the fact that they had translated the letters from Hebrew ('...scriptura translata de hebraico in linguam Italicam ...')[28] or from Spanish ('...Hispanico idiomate ...'),[29] and very often the Jews themselves acted as notaries and translators. Thus Salomon Oef wrote a letter which was translated by Joseph Fermi, and about it Fermi writes: 'Jo Joseph Fermi ho tradotto la sopraditta coppia dal originale proprio presso il sudetto Achiua barmanoach esistenti da Caratere leuantino de lingua Ebraicha et spagnola a lingua Italiana e caratero Vuolgare questo giorno 9 maggio 1653 in Ancona ...'[30] Recently I found corroboration for Fermi's translation. Among the numerous Turkish documents in the Historical Archives of Dubrovnik several letters registered as Turkish are in fact written in Rashi script, and they are still awaiting translation.[31]

Jews in the State Archives of Dubrovnik] (Zagreb, 1931); A. Kovačec: 'La lengua de los Sefardíes de Ragusa', *Bollettino dell' Atlante linguistico Mediterraneo*, 13–15 (1971–73); *Un texto judeoespañol de Dubrovnik*, Studia Romanica et Anglica Zagrabiensia, 33–6 (Zagreb, 1972–73); I. Kunchev, *Archaisms and Innovations in the Phonetic System of the Spanish-Jewish Speech in Bulgaria, Annual*, 11 (Sofia, 1976).

27. HAD, Diversa de Foris, 1600–1650.
28. Ibid., Vol. 1, 1593.
29. Ibid., Vol. 53, 1633.
30. Ibid., Vol. 83, fol. 223.
31. HAD, Acta Turcarum, ser. E., fasc. II.

The Jewish names more frequently mentioned in the texts are as follows: Semuel Maestro, Aron Coen, Daniel Campos, Israel, Josef, and Samuel Franco, Abraham Ferro, Salamun Luzena, Isaach Naamias, Rafael Penso, Salamon Leui, Isach Lima, Abram Pardo, Jacob Abenun, Salamon Oef, Mose Pappo, Abram Atias, Joseph Baruh, et cetera. These and many others established the trade routes and continued the work of their ancestors in the seventeenth century. The proceedings against Isach Jesurun in 1622 endangered their position but only partially, which leads us to believe that, apart from several persecutions of Jews, the Republic of Ragusa was rather friendly towards them.[32] The best proof of this is that Jews lived there for several centuries. Many of them, for example, had respectable jobs as physicians or even as consuls of the republic to Albania or Algeria,[33] and they in turn valued the place where they made their home, as can be seen from the beginning of one *ketubba*: '... Marcordi a 2 del mese di Sinan [*sic*] l'anno 5342 della creation del mondo qui in Ragusa uicino al mare, e ambondante di aque, di fontane, e di cisterne ...'[34] To this poetic description of Ragusa we should also add the fact that by publishing the works of two great figures, Amatus Lusitanus (a famous physician)[35] and Didacus Pyrhus (a great writer and poet),[36] the Sephardic community of Ragusa made a considerable contribution to the flourishing cultural life of the republic. And finally, besides being able to reconstruct the Ragusan Jews' way of life by analysing the archival documents, many aspects come very much alive. For example, among the newly discovered documents there are several dating from the beginning of the eighteenth century and sent from Ragusa to Florence or *vice versa*. In those letters the Ragusan merchants ordered various kinds of textiles from the Florentine manufacturers and they sent them the samples together with the measures. As those samples are very well preserved we can still enjoy the magnificent colours of silk and brocade entwined with golden thread. Their agent to whom the materials had to be sent was a Jew from Ancona, Raffael di Moisé Coen.[37]

32. A.G. Kaznačić, *Processo di Isach Jesurun israelita di Ragusa nel 1632* (Ragusa, 1882).
33. B. Hrabak, 'Jevreji u Albaniji od kraja XIII veka do kraja XVII veka i njihove veze s Dubrovnikom' [The Jews in Albania from the end of the thirteenth to the end of the seventeenth century and their links with Ragusa], *Zbornik Jevrejski Istorijski muzej* (1971).
34. HAD, Acta Turcarum D XXIII/6; I. Mitíc, *Konzulati i konzularna služba starog Dubrovnika* [Consulates and consular service in the Republic of Ragusa] (Dubrovnik, 1973).
35. M. Rešetar, 'Amat Luzitanac, dubrovacki ljekar XVI vijeka' [Amatus Lusitanus, Ragusan physician in the sixteeenth century], *Brankovo kolo*, 6 (1900).
36. Dj. Körbler, *Zivot i rad humanista Didaka Portugalca, napose u Dubrovniku* [Life and activities of the humanist Didacus Portuguese, especially in Ragusa], RAD JAZU 216, 1917.
37. HAD, Miscellanea, Masa Negrini.

This brief survey covers some of the topics awaiting serious research on the Jews, based on the archives of the Mediterranean and inland countries alike. Each of these can add to the picture, and only by examining all the parts will we be able to understand to what extent the role of Ragusan Jews and other Sephardic communities was important in the history of the Mediterranean countries.

Between the Eastern and Western Mediterranean: Sephardic Music after the Expulsion from Spain and Portugal

EDWIN SEROUSSI

Significant interest in the music of the Sephardic Jews has been generated, particularly in Israel and North America, following field work[1] and the publication of music anthologies in the twentieth century.[2] These printed and recorded collections possibly give the impression that the study of Sephardic music has greatly advanced. However, it would not be far from the truth to state that, despite the rich documentation at hand, a definitive scholarly evaluation of the historical development of the Sephardic musical culture is still far from sight. The present state of Sephardic music research can be more properly evaluated if compared with the achievements in other fields of Sephardic studies to which music is closely linked, most notably that of oral literature.

1. The most voluminous sound collections of Sephardic music are to be found at the National Sound Archives in the Jewish National and University Library, Jerusalem (which also holds the early collections of A.Z. Idelsohn, R. Lachman, L. Levy, and E. Gerson-Kiwi); the Ladino and Folklore departments of Kol Israel, the radio station of the Israel Broadcasting Authority; the Departmento de Estudios Sefardíes (Instituto Arias Montano) at the Consejo Superior de Investigaciones Científicas in Madrid; the private collections of Professors S.G. Armistead, J.H. Silverman, and I.J. Katz in the United States.
2. Anthologies of Sephardic music issued after the Second World War, like those of Levy and Attias, are based on materials collected from Sephardic immigrants in Israel. See I. Levy, *Antología de litúrgia judeo-española*, 10 vols. (Jerusalem, 1965–80); id., *Chants Judeo-Espagñols*, 4 vols. (Vol. 1: Paris, 1957; Vols. 2–4: Jerusalem, 1970–73); M. Attias, *Cancionero Sefaradi*, with musical transcriptions by Y. Mazor (Jerusalem, 1972). Serious efforts made prior to the massive displacement of the Sephardic Jewry from North Africa and the eastern Mediterranean countries to Israel, western Europe, and the Americas prove that interest in this music has roots in the early twentieth century. Worth mentioning are the unpublished collections of Judaeo-Spanish songs by M. Manrique de Lara (1863–1929) in Morocco and the eastern Mediterranean (1910–17), and A. Hemsi (1896–1975) in Turkey, Alexandria, Rhodes, and Salonica (1920–38), and the monumental collection of liturgical music by Abraham Z. Idelsohn (1882–1938) in Jerusalem (c.1910–20): *Hebräisch-orientalischer Melodienschatz*, Vol. IV: *Gesänge der orientalischen Sefardim* (Jerusalem–Berlin–Vienna, 1923). For a critical summary of all major and minor collections of Judaeo-Spanish songs published

As many Mediterranean societies, the Sephardic communities that sprung up around the Mare Nostrum after the Expulsion from Spain and Portugal were exposed to manifold cultural influences. This fact is reflected in their music too, which indeed consists of a complex web of superimposed layers originating in different sources and periods. Thus, complex questions of historical nature have to be tackled in Sephardic music research, such as those concerning the Spanish component, the stylistic identity of the music tradition of each Sephardic community around the Mediterranean, and the cross-relations between these sub-traditions. This article is an attempt to summarize in a critical manner some of the basic premises underlying the scholarly treatment of these questions by reviewing the relative extant literature.

The presence of a Spanish component in Sephardic music is one of the most controversial issues in the research on Sephardic music and raised several hypotheses.[3] One can reduce this issue to two basic questions: 1) To what extent were characteristics of medieval Spanish music retained by the Sephardim after 1492? 2) Did Spanish music reach the Sephardic communities after the Expulsion?

The idea of Iberian elements incorporated to a greater or lesser extent in Sephardic music has for a long time captured the imagination of several authors. Research on folk literature, which examines the closeness of orally-transmitted Sephardic literature to medieval Spanish sources, played an important role in the application of the same hypothesis to music research.[4]

Although existing data concerning the music of the Jews in Spain prior to the Expulsion is almost nil, we possess some scattered references to the role of Jews in the musical life of medieval Muslim and Christian Spain. For example, Jews were acquainted with the theory of Arabic music and used melodies of Arabic *muwashshaḥat* when singing Hebrew poems. They

prior to 1972, see I.J. Katz, *Judeo-Spanish Traditional Ballads from Jerusalem: An Ethnomusicological Study*, 2 vols. (New York, 1972–75). No parallel study of Sephardic liturgical music sources has been compiled to date. For a comprehensive list of these sources see E. Seroussi, '"Schir Hakawod" and the Liturgical Music Reforms in the Sephardic Community in Vienna, ca. 1880–1925', Ph.D. dissertation (University of California, Los Angeles, 1988), pp. 380–83.

3. For elaboration on this issue, see M. Sánchez, 'Música sefardí: (¿Música castellana?', *Las tres culturas en la Corona de Castilla y los sefaradíes* (Salamanca, 1990), pp. 127–35; E. Seroussi, 'El legado español en la cultura musical sefardí: Estado de la cuestión', to be published in the proceedings of the conference *La proyección histórica de España en sus tres culturas* (Medina del Campo, 16–18 April 1991).
4. See I.J. Katz, 'La música de los romances judeo-españoles', in S.G. Armistead and J.H. Silverman (eds.), *En torno al romancero sefaradí (hispanismo y balcanismo de la tradición judeo-española)* (Madrid, 1982), pp. 243–53; id., 'The Musical Legacy of the Judeo-Spanish Romancero', in J. M. Sola-Solá, *et al.* (eds.), *Hispania Judaica: Studies in the History, Language and Literature of the Jews in the Hispanic World*, Vol. II, *Literature* (Barcelona, 1982), pp. 45–58.

also occupied steady positions as *juglares* at the courts or performed sporadically for members of the emerging Christian aristocracy.[5] However, this scant information does not enable us to evaluate in any detail the musical life or repertories of the Jews in Spain. From the data available it emerges that music of the non-Jewish environment influenced all forms of Jewish music, including that of the synagogue. R. Profiat Duran, for example, in his work *Ma'aseh Efod* (1403), warns against the practice of singing religious poetry and other sacred texts with metrical melodies adopted from the non-Jews. The attitude of Duran echoes similar concerns already implied some three centuries earlier by Judah Halevi in a famous passage on music in his *Kuzari*.[6]

The meagre references to the music of the Jews in Spain leave researchers with one viable option: to extrapolate any traces of an assumed Iberian heritage from present-day Sephardic music traditions. This hypothetical component was sought in two contexts: in Judaeo-Spanish folksong, particularly the *romancero*, and in the liturgical music. The first case rests on the assumption that Sephardic romances, and to a lesser extent lyric songs, with parallel textual versions recorded in medieval Spanish sources, also keep elements of their original music.[7] A related idea postulates that if the same Judaeo-Spanish song is found in the repertory of communities at both extremities of the Mediterranean basin, then its origins may be Iberian. This is the case of several Judaeo-Spanish romances and wedding songs with variants found in Turkey and northern Morocco.[8]

Supporters of the proposition concerning liturgical music allege that melodies common to the diverse Sephardic traditions at the present are remnants of a Pan-Iberian synagogal music repertory. This argument, found for the first time in the mid-nineteenth century in the writings of cantor David Aharon de Sola from London, was later reworked by several modern authors.[9]

5. Most of what is known about the music of the Jews in medieval Spain appears in H. Anglés, 'La musique juive dans l'Espagne médiévale', *Yuval*, 1 (1968), 48–64.
6. I. Adler, *Hebrew Writings Concerning Music* (Munich, 1975), p. 129 (for Duran), p. 199 (for Halevi). See also A. Shiloah, 'Melody and Meter in the Kuzari', *Tazlil*, 6 (1966), 5–8 (Hebrew); N. Allony, 'Melody and Poetry in the "Kuzari"', *Yuval*, 3 (1974), 7–17 (Hebrew section).
7. See, for example, E. Gerson-Kiwi, 'On the Musical Sources of the Judeo-Spanish Romance', *Musical Quarterly*, 50 (1964), 31–43. For an attempt to find links between Sephardic and Spanish sources during the era of the Expulsion see, H. Avenary, 'Old Melodies to Sephardic *pizmonim*', *Tesoro de los judíos sefaradíes*, 3 (1960), 149–53 (Hebrew).
8. For a case of a song probably of Spanish origin found in the eastern Mediterranean and North Moroccan Sephardic traditions, see S. Weich-Shahak, 'A Dance-Song in the Sephardic Repertory', *Jahrbuch für Volksliederforschung* (in press).
9. De Sola expressed his views in his important introductory study to E. Aguilar and D.A. de Sola, *The Ancient Melodies of the Liturgy of the Spanish and Portuguese Jews*

Opposition to the 'Spanish heritage hypothesis' in all its variations stems from the conspicuous lack of hard-core musical evidence to verify it. This position was vehemently stated by I. Katz in relation to the music of the *romancero*.[10] A more balanced view on the music of the romances which seeks to tackle this issue on the basis of oral and written sources was lately advanced by J. Etzion and S. Weich-Shahak. They argue that links between Spanish romance melodies in fifteenth- and sixteenth-century polyphonic compositions and contemporary Judaeo-Spanish romance tunes can be shown through a sophisticated analysis that seeks to uncover the deep structures underlying the melodies used for variants of the same text in both repertories.[11]

In the field of liturgical music the medieval Spanish hypothesis encounters even more serious difficulties. First, it is based on the assumption that all the Jewish communities in Spain shared a common musical repertory in their synagogues. Second, it supposes the stagnation of the Sephardic liturgical repertory for over five centuries.[12] In face of these two problematic propositions, the researcher has to find alternative routes to explain the similarities and differences between the liturgical music structures or performance practices of the eastern and western Sephardic communities. One possible explanation of the shared heritage of liturgical music of geographically distant communities is to follow the wanderings of *ḥazzanim* from one location to another. For example, we were able to trace the eastern origins of some aspects in the Spanish-Portuguese liturgical tradition of Amsterdam to one of the first *ḥazzanim* serving in that city, R. Joseph Shalom Gallego, who immigrated from Salonica in 1614.[13]

(London, 1857). See E. Seroussi, 'The "Ancient Melodies" of Cantor David Aharon de Sola of London: An Inquiry into their Ancientry', *Pe'amim*, 51 (Hebrew, in press). For recent studies where this question is discussed, see H. Avenary, 'Persistence and Transformation of a Sephardic Penitential Hymn', *Yuval*, 5 (1986), 181–232; A. Shiloah, *Continuity and Change in the Sephardic Musical Heritage* (Jerusalem, 1988). For a critical discussion of several positions on this subject, see the concluding chapter in E. Seroussi, *The Spanish-Portuguese Liturgical Repertory in Western Europe according to German Sources* (Jerusalem, forthcoming).

10. I.J. Katz, 'The "Myth" of the Sephardic Legacy from Spain', *Proceedings of the Fifth World Congress of Jewish Studies: Jerusalem 1969*, Vol. 4 (Jerusalem, 1973), pp. 237–43.
11. J. Etzion and S. Weich-Shahak, 'The Spanish and the Sephardic Romances: Musical Links', *Ethomusicology*, 32 (1988), 1–37; id., 'The Spanish "Romances Viejos" and the Sephardic Romances: Musical Links across Five Centuries', *Atti del XVI Congresso della Societá Internazionale di Musicologia. III: Free Papers* (1989), pp. 7–16.
12. Such ideas can be found in A.Z. Idelsohn, *Jewish Music in its Historical Development* (New York, 1967), pp. 49–50; E. Werner, 'Music', *The New Grove Dictionary of Music and Musicians*, Vol. 9, p. 627.
13. E. Seroussi, 'R. Joseph Shalom Gallego: A Cantor from Saloniki in Seventeenth-Century Amsterdam', *Asufot* (Hebrew, in press).

Here it has to be pointed out that the romantic view of Spanish musical elements from the pre-Expulsion period surviving in contemporary Sephardic traditions was also fostered by extra-musical factors. Advocates of the re-establishment of ties between Spain and the Sephardim in the twentieth century, pursuing any argument that would provide evidence of the fidelity of these Jews to their ancient Iberian heritage, heartily embraced this idea. For example, the composer and field researcher A. Hemsi championed this path, encouraged by his friend and mentor, the noted Spanish composer and investigator José Subirá.[14]

To complete this discussion on the Spanish heritage in Sephardic music, it should be noted that the musical ties between the Iberian Jews and the peninsula were never totally discontinued. Sporadic contacts occurred, either due the flow of *conversos* who fled Spain and returned to Judaism after the sixteenth century or through direct contacts with Spanish merchants, diplomats, and adventurers. Proof of such post-Expulsion musical contact is found in some titles of Spanish songs serving as melodic instructions for singers in collections of religious Hebrew poetry in Turkey and Greece from the sixteenth century on. For example, the renowned poet R. Israel Najara (c.1555–1625) based his poem *Nidḥei Israel tekhanes* on the melody of the sixteenth-century Spanish song *Morenica que te pones*.[15] Finally, in the Sephardic repertory today one can find items originating in contemporary Hispanic music such as zarzuela, flamenco, and tangos. These recently adapted songs reflect the renewed musical influence of

14. J. Subirá, 'El orientalismo hispanista del compositor A. Hemsi', preface to A. Hemsi, *Coplas Sefardíes*, Op. 7 (Alexandria, 1932), reprinted subsequently with minor additions in Op. 8 (1933), Op. 13 (1934), Op. 18 (1935), and Op. 22 (1938).
15. See E. Seroussi, 'R. Israel Najara: Moulder of Hebrew Sacred Singing after the Expulsion from Spain', *Asufot*, 4 (1990), 285–310, especially n.12 (Hebrew). This phenomenon is known in the musicological literature as contrafacts. The earliest evidence for the use of melodies of songs in Spanish by Jews before the Expulsion from Spain appears in a fifteenth-century manuscript of liturgical dirges. See J. Yahalom, 'Poetry as an Expression of Spiritual Reality in the Late Sephardic Piyyut', in A. Mirsky, *et al.* (eds.), *Exile and Diaspora: Studies in the History of the Jewish People Presented to Professor Haim Beinart on the Occasion of his Seventieth Birthday* (Jerusalem, 1988), pp. 337–48 (Hebrew). For musicological studies of Judaeo-Spanish contrafacts, see H. Avenary, 'Cantos españoles antiguos mencionados en la literatura hebrea', *Anuario Musical*, 25 (1971), 67–79; A. Bahat, 'Les contrafacta hébreux des romanzas judeo-espagnoles', *Revista de Musicología*, 9/1 (1986), 141–68; I.J. Katz, '"Contrafacta" and the Judeo-Spanish Romancero: A Musicological View', in J.V. Recapito (ed.), *Hispanic Studies in Honor of Joseph H. Silverman* (Newark, DE, 1986), pp. 169–87; J.R. Cohen, 'Musical Bridges: The Contrafact Tradition of Judeo-Spanish Songs', in F. Gerson and A. Percival (eds.), *Cultural Marginality in the Western Mediterranean* (Toronto, 1990), pp. 121–7; E. Seroussi and S. Weich-Shahak, 'Judeo-Spanish Contrafacts and Musical Adaptations: The Oral Tradition', *Orbis Musicae*, 10 (1990–91), 164–94. See also the seminal study by S.G. Armistead and J.H. Silverman, 'El antiguo cancionero sefaradí: citas de romances en himnarios hebreos (siglos XVI–XIX)', *Nueva Revista de Filología Hispánica*, 30/2 (1981), 453–512.

Spain on the Sephardim which started in the late nineteenth century.[16]

The rejection of the Iberian heritage hypotheses places *ipso facto* great emphasis on the opposite view, namely, that the Sephardic music revealed to collectors and scholars in the twentieth century was mostly crystallized after the Expulsion from Spain, heavily drawing on the music traditions of the Muslim societies of North Africa and the Ottoman Empire.

Musical evidence to substantiate this hypothesis is overwhelming. Muslim influence is not limited to the adoption of single Arabic or Turkish melodies in all genres of Sephardic music, but it also includes the absorption of modal systems, new musical genres, and instruments.[17] Sociologically this process is confirmed by the substantial documentation available concerning leading Sephardic performers and composers who held important positions as musicians outside the Jewish community. Such was the case of the renowned Isaac Fresco Romano, known in Turkish sources as Tamburī Izak, who was chief musician at the court of Sultan Selim III (1789–1807).[18]

The impact of Muslim music on Sephardic music after the Expulsion largely spread to their liturgy too. The constantly growing emphasis placed on performances based on Muslim music in Sephardic synagogues, particularly in paraliturgical rituals, raised strong opposition from rabbinical authorities. R. Mordechai Atzban, for example, in his *Zobeaḥ Torah* (Constantinople, 1733, fol. 7v) complains about the *ḥazzanim* who concentrate on the practice of the Turkish *makamat* (musical modes) for the pure pleasure of music during the singing of *baqqashot* in early morning vigils.[19] R. Eliyahu Ha-Cohen of Izmir, in his *Shevet Musar* (Constan-

16. See E. Seroussi, 'The Growth of the Judeo-Spanish Folksong Repertory in the 20th Century', *Proceedings of the Tenth World Congress of Jewish Studies*, Division D, Vol. 2 (Jerusalem, 1990), pp. 173–80. Similar processes can be detected in the traditional oral poetry; see I.M. Hassán, 'Última (?) poesía tradicional sefaradí de la zona del Estrecho', *Congreso Internacional 'El Estrecho de Gibraltar'* (Ceuta 16–19/11/1987) (Madrid, 1987), pp. 301–13.
17. Turkish, Arabic, and Greek melodies were adopted since the sixteenth century as contrafacts (see n. 15 above). For the adaptation of entire musical systems by Sephardic Jews after the Expulsion from Spain see, for example, A. Shiloah 'La Nuba et la celebration des bakkashot au Maroc', in M. Abitbol (ed.), *Judaisme d'Afrique du Nord* (Jerusalem, 1980), pp. 108–13; E. Seroussi, 'La música arábigo-andaluza en las *baqqashot* judeo-marroquíes', *Anuario Musical*, 45 (1990), 297–315; id., 'The Turkish Makam in the Musical Culture of the Ottoman Jews: Sources and Examples', *Israel Studies in Musicology*, 5 (1990), 43–68; id., 'R. Israel Najara'.
18. On leading Sephardic musicians in Turkey, see A. Galante, 'Les juifs dans la musique turque', repr. in *Historie des juifs de Turquie* (Istanbul, 1985), Vol. 8, pp. 66–73. On Morocco see I. Ben-Ami, 'Jewish Performers and Ensembles in Morocco', *Yahadut Morocco* (Jerusalem, 1986), pp. 223–8 (Hebrew).
19. The opposition of Sephardic rabbis to the use of elaborate music in the liturgy was primarily based on the distraction it caused, deviating the attention of the worshippers from the contents of the sacred texts. A thorough compilation of the opinions by Sephardic rabbis on this matter after the Expulsion from Spain is still a *desideratum*. Meanwhile, see Y. Ratzhabi, 'The Foreign Tune in the Song and the

tinople, 1712, fol. 77v), recommends the *hazzanim* to avoid wasting time by singing songs with nonsense syllables, a practice which is characteristic of several Turkish genres. We may add here that the Muslim influence on the Sephardim was not at all exclusive. In certain areas of the Ottoman Empire, we find that the Sephardic Jews were even influenced by the music of the eastern Orthodox Church.[20]

In summary, the musical legacy of the Sephardim can be defined as a musical culture in flux. The sound and printed collections assembled in the twentieth century present us with a complex conglomerate of styles of the most diverse historical and geographical origins. Residues of melodies of medieval Spanish romances may be sung in a single breath by the same informant followed by a Judaeo-Spanish translation of a Turkish *sharki* or an Argentinian tango. A cantor, in the same service in the synagogue, may combine venerable old melodies which might carry vestiges from a pre-Expulsion repertory with tunes adopted from the dervish or Orthodox Church rites in Turkey, or from the Andalusian secular repertory in Morocco.

This apparent eclecticism in the Sephardic musical repertory does not have to be interpreted as symptomatic of a lack of creativity or as evidence of cultural inconsistency. Music connotates meanings on an abstract level, and in this differs from other aspects of culture. Thus the evolution of this repertory is a process which reflects and symbolizes, in patterns of sound, the history of these communities on the edge of the Mediterranean during the past five centuries. The stock of melodies in the Sephardic repertory evolved incessantly throughout the ages, but the society that produced it remained faithfully attached to its Hispanic heritage in many other spheres of its culture, particularly its language. While texts, such as romances or the liturgy, remained unchanged, their music was transformed, symbolically representing the dialectic development which ensured the survival of the Sephardim as a cohesive society in so many different environments. A sober appreciation in this direction was recently expressed by J. Cohen:

> ... The very concept of Sephardic music as a separate phenomenon owes much to the mass media and while one may justly mourn the passing of the old repertory ... one must admire the flexibility and

Piyyut', *Tazlil*, 6 (1967), 8–13 (Hebrew); E. Seroussi, 'A Hassidic Exemplum in a Sephardic Homily of the Early Nineteenth Century: A New Source on "Secular Music" in the Synagogue', *Jerusalem Studies of Jewish Folklore*, 11–12 (1990), 121–38 (Hebrew).

20. The most important passage in this respect is the testimony by the nineteenth-century Jerusalemite R. Moshe Hazzan in his responsa *Keraḥ shel Romi* (Livorno, 1876). See A. Shiloah, 'The Music of the Jewish Communities in Greece and Turkey and its Relation to Byzantine Music', *Musica Antiqua*, 7 (Bydgoszcz, 1985), pp. 242–56.

tenacity of those who created the tradition, maintained and adapted it through the centuries — and knew when to express their creativity in other ways: some genres may have diminished or disappeared, but not the exuberant musical energy which produced them.[21]

The decay or eventual disappearance of the traditional Sephardic music has been a pivotal concern underlying its research in the past four decades. However, the adaptability of the Sephardic Jews to changing social circumstances in the past seems to be a cushioning factor that calls into question the apocalyptic forecasts of some scholars. Indeed, powerful processes, such as the dismantling of traditional social frameworks, the exposure to western culture, and the fall into disuse of the Judaeo-Spanish language drove many Sephardic music repertories to oblivion. However, entire liturgical repertories remain functional because they are practised by observant Sephardic Jews. Other Sephardic musical genres have been revitalized, in more or less arranged versions, for several reasons: as a vehicle for expressing Sephardic ethnic identity;[22] as a sub-genre of popular music with ethnic connotations, propelled in Israel by politicians of Sephardic background to appeal to latent public emotions;[23] as a manifestation of a supposedly archaic layer of Spanish music in the eyes of Jewish and non-Jewish artists; or as raw material for original artistic endeavours.

Future research in Sephardic music will have to address two main issues outlined in this study. First, it will need to re-evaluate historical processes by scrutinizing the extant sources (written and oral) with a critical approach that considers the context in which these sources were assembled. Biases of editors and field collectors, problems of music notation, and the

21. J.R. Cohen, 'The Impact of Mass Media and Acculturation on the Judeo-Spanish Song Tradition in Montreal', in S. Frith (ed.), *World Music, Politics and Social Change* (Manchester–New York, 1989), pp. 90–97. For the effect of the mass media on traditional Sephardic genres, see also I.J. Katz, 'Stylized Performances of a Judeo-Spanish Traditional Ballad: La mujer engañada', in F. Talmage (ed.), *Studies in Jewish Folklore* (Cambridge, MA, 1980), pp. 181–200.
22. For the role of folklore in the identity of a Sephardic community, see S. Stern, *The Sephardic Jewish Community of Los Angeles: A Study of Folklore and Ethnic Identity* (New York, 1980).
23. Such was the case of the activities of Isaac Levy in Kol Israel in the 1960s and early 1970s. Levy worked under the auspicies of Itzhak Navon, a prominent Sephardic politician who was then general director of the Ministry of Education and Culture, and later President of the State of Israel. For an example of manipulation of the revival of a traditional music genre for political purposes, see E. Seroussi, 'Politics, Ethnic Identity, and Music in Israel: The Case of the Moroccan *Bakkashot*', *Asian Music*, 17 (1986), 32–45. For an overall appreciation of these processes see, A. Shiloah and E. Cohen, 'The Dynamics of Change in Jewish Oriental Ethnic Music in Israel', *Ethnomusicology*, 27 (1983), 227–52.

erosion of memory are some of the constraints influencing musical sources. Another step in the historical approach to the study of Sephardic music is to detach it from theories stemming from literature or liturgy research. Secondly, the contemporary state of Sephardic music does not have to be considered as a terminal stage in its five centuries of evolution. Rather, music represents, and it has to be studied as, one of the most dynamic and tenacious aspects in the expressive culture of Sephardic Jews today, as it was for those of yesterday.

The Social and Communal Organization of the Jewish Communities in Epirus during the Sixteenth Century

LEAH BORNSTEIN-MAKOVETSKY

Only two significant Jewish communities existed in Epirus during the sixteenth century: Ioannina and Arta. The history of Ioannina has been recounted in many research articles, especially in recent years, for this community represented a significant surviving remnant of the Romaniote (Byzantine) Jewish population throughout the entire Ottoman era, until its annihilation during the Holocaust.[1] However, the history of the neighbouring sister-community, Arta, has only recently interested scholars.[2]

In this article, I describe certain typical characteristics of the social and communal organization common to both of these Jewish communities during the sixteenth century — although there were also some differences between them. Most of the essential information is drawn from the immense responsa literature written in the Ottoman Empire during the sixteenth century, especially the responsa authored by the two sages of Arta: R. Benjamin b. Rav Mattatya and R. Samuel Kal'ai.[3] Recently, a new manuscript containing the responsa of another sage of Arta, R. Solomon b. Rav Baruch, was discovered.[4] This sage was active in the second half of the sixteenth century, and his responsa add illuminating details to those already extracted from others. At present, information from additional sources — whether Jewish, European, or Ottoman — is very limited. A recent research project has demonstrated that there is no known Ottoman documentation on the community of Ioannina during the first half of the

1. On the Byzantine period, see S. Bowman, *The Jews of Byzantium: 1204–1453* (Alabama, 1985), pp. 25ff., 43, 73ff., docs. 15b, 36, 43; a new general study of the community of Ioannina from the beginning until its annihilation, emphasizing the modern era, has been written by R. Dalven, *The Jews of Ioannina* (Philadelphia, 1990). On the community from 1787 to 1913, see also M. Schwartz, 'The Jews of Ioannina', Ph.D. thesis (University of Birmingham, 1976).
2. L. Bornstein-Makovetsky, 'Life and Society in the Community of Arta in the Sixteenth Century', *Pe'amim* 45 (1991), 126–55 (Hebrew).
3. R. Benjamin b. Mattatya, *Responsa: Binyamin Ze'ev* (Venice, 1539); R. Samuel Kal'ai, *Responsa: Mishpetei Shmuel* (Venice, 1599).
4. R. Solomon b. Rav Baruch, Responsa, MS. This manuscript is in the library of R. Moshe Amar, Israel.

sixteenth century.[5] Without any doubt, most of the essential material for our discussion originates in the above-mentioned responsa literature.[6] It is well known that this literature not only recorded every-day problems, but often dealt with extraordinary questions which arose in both public and private life.

It is important to begin with some demographic data. So far no data are available concerning the origins of the Jewish communities at the time of the Ottoman conquest, except for the testimony that some Jews left these cities as *sürgün* (expelled by force). They were exiled to Istanbul where they organized themselves into congregations which were named Arta or Narda, and Yanya.[7] Unfortunately, there are no sources of information on, for instance, the size of these communities on the eve of the Ottoman conquest, or on the numbers exiled to Istanbul or how many remained in these cities. The only testimony that we have has been gleaned from the state records for the years 1540 and 1545, in which there is some information concerning these congregations in Istanbul. In 1540 the Arta congregation of Istanbul numbered 24 heads of households who paid taxes, and in 1545 there were 23 tax-paying heads of households, who would have represented about 120 family members. The Ioannina congregation in Istanbul in 1540 numbered 25 tax-paying heads of households, while there were 26 in 1545.[8] From these figures, one may assume that the number of émigrés from Ioannina was similar to the number from Arta. Furthermore, these facts lead us to the conclusion that the two communities of Arta and Ioannina were not entirely abandoned by the original Romaniote Jewish inhabitants, for at the beginning of the sixteenth century there is evidence in the rabbinical responsa literature that a Romaniote population had dwelled there. It seems highly probable that even after the conquest of these two towns by the Ottoman armies most of the population was still comprised of the original inhabitants, although they received an influx of émigrés from other Greek cities.[9]

There is evidence of a large Romaniote population in Arta. It is clear that the demographic facts found in the responsa literature concerning the Jewish community of Arta are more reliable than any found in Ottoman documents. The facts as recorded in the responsa literature show that in the twenties and thirties of the sixteenth century there was a Jewish

5. These details were noted by Melek Delibasi in the discussion at the V. Milletlerarasi Türkiye Sosyal ve İktisat Tarihi Kongresi, Istanbul, 22 Aug. 1989.
6. Significant documents on Ioannina existed in the responsa literature written by Greek and Turkish rabbis during the sixteenth century. It is important to note that we do not know of any responsa book written at that time in Ioannina.
7. Bowman, *The Jews*, p. 75; M.A. Epstein, *The Ottoman Jewish Communities and their Role in the Fifteenth and Sixteenth Centuries* (Freiburg, 1980), pp. 178–9.
8. Epstein, *The Ottoman Jewish Communities*, 178–9.
9. See Bornstein-Makovetsky, 'Life and Society', 127–8.

population of more than 300 householders in Arta, namely a total population of more than 1,500.[10] If we assume an average of six for every household, we arrive at a population of around 1,800. In contrast, the Ottoman state documents of 1564 record only 214 Jews who paid taxes in Arta.[11] One can assume that the number of Jews grew during the century and reached about 2,000. Despite the lack of exact figures, it seems that the population was most probably stable. Periodic drops in population growth were generally the results of plagues or lapses of security due to invasion by Christian armies. These events, as well as marriages between Artan and Ioanninan Jews, influenced local population movement, with Jews from Arta settling in Ioannina and vice versa. During the Ottoman siege of Corfu in the year 1537, many Jews fled from the island to Ioannina.[12] Many émigrés arrived in Arta between the years 1492 and 1541, when the Jews who had been expelled from Spain, Portugal, Sicily, Calabria, Apulia, Naples, and a few from Provence in southern France settled in cities throughout the Ottoman Empire.[13] As already mentioned, most of the Jewish population of Arta was of Romaniote extraction. In the first half of the sixteenth century there were not even 30 families in the congregation established by the émigrés from Apulia.[14]

An examination of the sources on Ioannina reveals an interesting fact. We have already noted that there is a wealth of information in the responsa literature concerning Arta, whereas concerning sixteenth-century Ioannina sources are scant. Consequently, in my opinion, there is absolutely no basis for the assumption on the part of scholars that Ioannina was already a significant community at the beginning of the sixteenth century. Although a few famous rabbis dwelled in the city for short periods, they were usually Romaniotes who would eventually relocate to other Grecian Jewish communities.[15] In a similar vein, the famous R. Joseph Ta'itatzak, a victim of the Spanish Expulsion, mentions in one of his responsa that he had stayed in Ioannina, apparently his first stop in Greece before moving on to Salonica. According to his report, for two years he had lived in the same house as R. David Ha-Cohen, one of the most illustrious of Greek rabbis.[16]

10. *Binyamin Ze'ev*, No. 303.
11. Epstein, *The Ottoman Jewish Communities*, p. 251.
12. Bornstein-Makovetsky, 'Life and Society', 128–9; M. Benayahu, *Relations between Greek and Italian Jewry* (Tel Aviv, 1980), pp. 78–9 (Hebrew).
13. L. Bornstein, 'The Jewish Communal Leadership in the Near East from the End of the 15th through the 18th Century', Ph.D. thesis (Bar Ilan University, Ramat Gan, 1978), pp. 45–66; id., 'Life and Society', 129–32.
14. *Binyamin Ze'ev*, No. 303.
15. For example, R. David Ha-Cohen settled in Ioannina during the years 1510–12. See M. Benayahu, *The Responsa of R. David ben Rav Ḥayyim Ha-Cohen from Corfu* (Jerusalem, 1988), Introduction (Hebrew).
16. M. Benayahu, *The Responsa of R. Joseph Ta'itatzak* (Jerusalem, 1987), No. 19, p. 138 (Hebrew).

This report may possibly constitute evidence for the existence of a yeshiva in Ioannina, but there is no concrete proof that in the first half of the sixteenth century we are talking about a large organized community. On the contrary, in other responsa throughout the sixteenth century it is referred to as a 'small city'.[17] In one responsum it is stated explicitly that there is only a small number of Jews in the city, without being more specific. It appears to me that the statement that 'all the poor and rich inhabitants of this city [Ioannina]' make their livelihood from trading cloth and furs, 'and they have no profits from any sources of income other than these commodities'[18] signifies that we are dealing with a small community. In medium-sized or large communities, the spectrum of Jewish occupations was much broader.[19]

There is extensive evidence concerning the organization of the two communities, and especially on the existence of autonomous congregations within them — a phenomenon most prevalent in the Ottoman Empire during the sixteenth century.[20]

The following demonstrates the dominance of the Romaniotes in both communities and surely modifies the theory recently proposed by Joseph Hacker according to which the Sephardic émigrés set the tone of communal life in Ioannina and Arta.[21] It is clear that from the twenties of the sixteenth century there were four congregations in Arta.[22] The largest and most famous of these was the congregation of the citizens, called also 'Gregush' or 'Corfu', whose members were either authentic Romaniotes who had remained in the community after the exile of their colleagues to Istanbul, or Romaniote immigrants from other Greek cities who had come to Arta after the exile, including many immigrants from Corfu. The other congregations, including the Calabrian, Sicilian, and Apulian congregations, were all founded by émigrés. In the responsa literature the names of some 120 inhabitants of Arta are recorded, covering a period of some 50 years (from the 1520s until the 1570s). Among these, there are only a few Sephardic names. The influence of the Sephardim was not felt in the life of the people, and they joined existing congregations. Apparently, Sephardic customs did not take root in the community.

On account of the various disagreements that broke out, the rabbis

17. *Mishpetei Shmuel*, Nos. 49–50; R. Solomon b. Rav Baruch, Responsa, MS, pp. 209a–213b.
18. *Mishpetei Shmuel*, No. 48.
19. For example, the community of Patras in the Morea numbered 500 families during the same period. See L. Bornstein-Makovetsky, 'On the Power Struggle in the Jewish Community of Patras in the Sixteenth Century'; *Michael* 7 (1981, Hebrew).
20. See Bornstein, *The Jewish Communal Leadership*, pp. 49ff.
21. J.R. Hacker, 'The Ottoman System of Sürgün and its Influence on the Jewish Society in the Ottoman Empire', *Zion*, 55 (1990), 68 (Hebrew).
22. On the Arta congregations, see Bornstein-Makovetsky, 'Life and Society', 129–32.

described at length the whole fabric of inter-congregational relations. First and foremost was the bitter controversy surrounding their autonomy in the question of the regulation that prohibited a betrothed man from entering the home of his fiancée before the wedding. This regulation was vehemently opposed by the Apulian congregation during the 1530s.[23] Controversies such as these were discussed by the sages of the sixteenth century. The conclusions reached by these rabbis established many significant precedents in halachic literature with regard to the autonomous organization of the congregations in every community throughout the Ottoman Empire. From the responsa we learn of the changes that these congregations underwent: often complete disintegration, including cases of members leaving one congregation and joining another. This instability was a wide-spread phenomenon in the communities — in spite of the fact that the congregations attempted to prevent this through legislation. These regulations, which attempted to stem the tide of disintegration, were instigated because the congregations feared losing their cultural identity or desired to preserve their ethnic framework.[24] Besides these social factors, there was an economic reason: the congregational leaders in Arta feared that if each individual changed his congregation there would be permanent strife over the payment of taxes levied by the Ottoman authorities on the community, which in the eyes of the imperial authorities constituted a single unit.[25]

The existence of separate congregations in Arta continued throughout the whole of the sixteenth century, and Romaniote hegemony became an established fact. Indeed, the minority, Italian émigrés, wanted to preserve some traditional social mores, but they generally accepted most of the Romaniote customs.[26]

In contrast to Arta, Ioannina had only one congregation, which was Romaniote, in the first half of the sixteenth century.[27] In the second half it split into two, with the two congregations existing side by side, and probably the second congregation was also largely Romaniote.[28] It appears that this Ioanninan tendency to secession already existed in the first half of the sixteenth century, but due to staunch opposition to the division of the congregation by the local Romaniote sage, R. Benjamin b. Rav Joseph Ibn Menachem, and the bar that he decreed against a split, the

23. Ibid., 131, 144–5.
24. Ibid., 130–32.
25. *Binyamin Ze'ev*, Nos. 286, 293–5, 302; *Mishpetei Shmuel*, No. 97; Bornstein-Makovetsky, 'Life and Society', 132.
26. Bornstein-Makovetsky, 'Life and Society', 129–31, 143ff.
27. *Mishpetei Shmuel*, No. 40.
28. R. Samuel de Medina, *Rashdam: Responsa*, 'Yoreh De'ah' (Salonica, 1594), Nos. 137, 147; *Rashdam*, 'Hoshen Mishpat' (Salonica, 1595), No. 442; R. Isaac Adrabi, *Divrei Rivot: Responsa* (Venice, 1587), No. 318; *Mishpetei Shmuel*, Nos. 28, 40.

congregation was not divided.[29] However, already in the first half of the sixteenth century, the separatist tendencies in the community found expression in the establishment of two different burial societies.[30] Notwithstanding all the efforts to preserve the unity of the community, in the second half of the century a new congregation was established by some individuals, who claimed that the old synagogue was situated within the city walls while their homes were outside the city.[31] Between the two congregations there was continual strife, with each passing bans and regulations aimed against the other. Peace, together with the abolition and public burning of the bans and regulations, was achieved through the good offices of the men of Arta. None the less, between the two congregations there was constant competition for membership.[32]

As we have no record of any demographic estimates of the two congregations, we can only attempt to deduce their relative sizes. At first the older congregation paid 80 per cent of the taxes levied on the community by the Ottoman authorites, while the younger congregation paid only 20 per cent. As the years passed, more individuals moved to the newer congregation and the tax division was two-thirds for the older congregation and one-third for the younger one. Later on the two congregations paid an equal sum. There was also a similar number of individuals who were not registered in the tax-payers' records of either community.[33] From the sources we can deduce that both congregations were Romaniote and that those Spanish or Italian émigrés who arrived at Ioannina — and indeed their names appear in all contexts — did not establish a congregation for themselves. From an examination of more than 50 names of members of the community mentioned again and again in the responsa literature during the sixteenth century, it is clear that the majority was Romaniote, whereas the minority was either Spanish or Italian.[34] It also appears that, contrary to the situation in Arta, the new inhabitants in Ioannina did not attempt to preserve their customs.

It should be noted that the spiritual leadership in Arta and Ioannina,

29. This regulation was confirmed by R. Samuel Kal'ai; *Mishpetei Shmuel*, No. 40.
30. *Binyamin Ze'ev*, No. 291; *Mishpetei Shmuel*, No. 53; *Rashdam*, 'Yoreh De'ah', No. 137; R. Solomon b. Rav Baruch, Responsa, MS, pp. 209a–215a.
31. *Mishpetei Shmuel*, No. 40.
32. R. Solomon b. Rav Baruch, Responsa, MS, pp. 209a–212a; *Rashdam*, 'Yoreh De'ah', Nos. 137, 147; *Mishpetei Shmuel*, Nos. 24, 28, 40, 53; R. Solomon Ha-Cohen, *Responsa*, Vol. I (Salonica, 1586), No. 45; *Divrei Rivot*, No. 318; R. Meir Melamed, *Mishpat Tzedek*, II (Salonica, 1799), No. 68.
33. *Rashdam*, 'Hoshen Mishpat', No. 442; *Divrei Rivot*, No. 318; *Mishpetei Shmuel*, No. 24. In the second half of the sixteenth century the two congregations together paid a tax of 90 aspers to the Ottoman authorities.
34. For example, Abraham de Boton, *Leḥem Rav: Responsa* (Izmir, 1740), No. 19; *Mishpetei Shmuel*, Nos. 13, 28, 34, 36, 40, 68, 80–82, 88, 100, 102; *Divrei Rivot*, No. 330; R. Solomon b. Rav Baruch, Responsa, MS, pp. 228a–234b; *Rashdam*, 'Yoreh De'ah', No. 137.

both the rabbis and the judges, were for the most part Romaniote rabbis while only a minority were rabbis of Spanish or Italian extraction.[35] In most cases there was co-operation between the rabbis of Arta and sometimes also between them and the rabbis of Ioannina.[36] In both communities there is evidence of a lack of respect for some rabbis, and of informants and slanderers who reported their alleged wrongdoings to the Ottoman authorities. These unfortunate incidents usually occurred when congregations wanted to rid themselves of their spiritual leadership.[37]

In both communities the lay leaders saw to all internal political, administrative, and financial matters.[38] Furthermore, like other communities in Greece and the Ottoman Empire, it was accepted practice in Arta and Ioannina to grant judicial authority to the lay leaders.[39] The élite of the congregations — the wealthy and the nobility — aided the lay leaders in the running of public affairs, and it was from this class that the lay leaders were elected.[40]

We have evidence of joint regulations passed by the four congregations of Arta on questions of taxation, tax evaders, and informers who were to report to the Ottoman authorities about internal interests of the community.[41] For the purpose of recording tax assessment, the community of Arta had a special register (*pinkas*).[42] Unfortunately, we have no comparable information concerning Ioannina, but it is quite likely that the same practice existed in that community. In one of the regulations passed in Ioannina it was decided to appoint nine leaders from the older congregation and three from the newer one.[43]

The responsa literature contains some of the Artan and Ioanninan congregational regulations, and the joint regulations for the congregations

35. Many sages had settled in Arta: see Bornstein-Makovetsky, 'Life and Society', 132–5. In the years 1510–12, R. David Ha-Cohen lived in Ioannina — see n. 15 above; in the mid-and later sixteenth century the rabbis Benjamin b. Rav Joseph Ibn Menachem, Samuel Kal'ai and Joshua Ibn Ezra were thriving in Ioannina. See *Mishpetei Shmuel*, Nos. 82, 107, 113; Rashdam, *Responsa*, 'Even Ha-Ezer' (Salonica, 1596), No. 220; 'Yoreh De'ah', No. 137.
36. Bornstein-Makovetsky, 'Life and Society', 134.
37. On Arta see ibid., 137ff; M.Benayahu, *Wolves that Savage Benjamin: The Book Beauty and Bands by Rabbi Eliyahu Capsali* (Tel-Aviv, 1990), pp. 53–66 (Hebrew); id., *An Introduction to the Responsa Binyamin Ze'ev by Rabbi Binyamin ben Mattatya of Arta* (Jerusalem, 1989), pp. 144–56 (Hebrew). On Ioannina see *Mishpetei Shmuel*, Nos. 37–40, 59; *Rashdam*, 'Yoreh De'ah', No. 137.
38. Bornstein-Makovetsky, 'Life and Society', pp. 135–6; *Divrei Rivot*, No. 232; *Mishpetei Shmuel*, No. 40.
39. Bornstein-Makovetsky, 'Life and Society', pp. 135–6; *Divrei Rivot*, No. 232.
40. On Arta see Bornstein-Makovetsky, 'Life and Society', 136. It is logical to assume that the situation was the same in Ioannina.
41. *Binyamin Ze'ev*, Nos. 249, 286, 416–20; *Ta'itatzak*, No. 10; Bornstein-Makovetsky, 'Life and Society', 136.
42. R. Solomon Ha-Cohen, *Responsa*, IV (Salonica, 1652), No. 49.
43. *Mishpetei Shmuel*, No. 40.

in each city. Few of the ordinances have survived due to the fact that the congregational or court records were lost, and these deal at length with daily life of the society and all aspects of its organization, family affairs, inheritance, social and religious morality, and economic regulations. The majority of surviving ordinances are from Arta, with a few from Ioannina.[44]

A few of these regulations should be mentioned. Either in 1520 or 1521, an important ruling that was passed by all the rabbis of Arta concerned the retroactive abrogation of any marriage not enacted in the presence of ten men including a rabbi (*minyan*). The rabbis nullified the consummation of such a marriage in advance, without necessitating any divorce proceedings.[45] A similar regulation was rejected in Ioannina although there were also many cases of marriage libels and marriages of doubtful validity.[46]

During the sixteenth century, the inheritance customs of the community of Corfu, based upon the ancient Romaniote customs, were accepted practice in both communities of Epirus. In Arta a special clause concerning the inheritance due to the wife according to custom of Corfu was added to the marriage contract (*ketubba*).[47] This was accepted by the majority of the inhabitants of the city throughout the sixteenth century and was formulated as follows:

> Should the wife die without children, her husband would inherit a third of her dowry, and her heirs (from her father's family) would inherit two thirds. Should she die and leave children, they would get her full dowry. And should her husband die before the wife, whether they had children or not, the wife would receive from the husband's estate an additional sum equal to the value of a third of her dowry together with the original dowry and possessions.[48]

The regulations of Ioannina were also based on the customs of Corfu, as they appear in various marriage contracts drawn up in Ioannina.[49] Only in

44. On Arta regulations see Bornstein-Makovetsky, 'Life and Society', 143–9. On Ioannina regulations see *Mishpetei Shmuel*, Nos. 40, 58; *Leḥem Rav*, No. 49; *Rashdam*, 'Hoshen Mishpat', No. 407.
45. *Binyamin Ze'ev*, No. 308; *Rashdam*, 'Even Ha-Ezer', No. 50; *Mishpetei Shmuel*, Nos. 49, 83; Bornstein-Makovetsky, 'Life and Society', 146.
46. R. Joseph Ibn Lev, *Responsa*, Vol. II (Amsterdam, 1725), No. 2; *Divrei Rivot*, No. 330; *Mishpetei Shmuel*, No. 68; *Rashdam*, 'Even Ha-Ezer', No. 220; R. Solomon Ha-Cohen, *Responsa*, Vol. I, No. 155; II, No. 127 (Venice, 1592); *Leḥem Rav*, No. 19; A.H. Freimann, *The Procedure of Marriage* (Jerusalem, 1945), pp. 199–201 (Hebrew).
47. *Mishpetei Shmuel*, No. 90; R. Elijah Ibn Ḥayyim, *Ra'anaḥ, Responsa* (Venice, 1610); R. Solomon b. Rav Baruch, Responsa, MS, pp. 185a–187a; R. Ḥayyim Sabetai, *Responsa* (Salonica, 1651), No. 24; *Mishpat Tzedek*, III (Salonica, 1795), No. 18; Bornstein-Makovetsky, 'Life and Society', 146–9.
48. R. Solomon b. Rav Baruch, Responsa, MS, pp. 79a–82a.
49. *Leḥem Rav*, Nos. 19, 49; *Mishpetei Shmuel*, No. 34; R. Solomon b. Rav Baruch,

the month of May 1585 did the leaders there abolish the Corfu-based custom of the city concerning inheritance and institute new regulations.[50]

With regard to the stringent observance of eating kosher meat, there is the well-known disagreement over the question whether it was permitted to eat a ritually slaughtered animal in which there was a *sirḥa* or adhesion of the lobes of the lung. According to some halachic opinions, the animal could be declared kosher after inflation of the lungs in order to examine the nature of the adhesion. Throughout the entire Ottoman period, the inhabitants of Ioannina, however, scrupulously observed the Romaniote practice, refusing meat declared kosher on this basis. This was in contradiction to the lenient Sephardic practice of permitting such meat to be eaten.[51] Although there is no indication of which custom was observed in Arta, it is logical to assume that there the Romaniote restriction was also observed, due to the preponderance of a Romaniote population.

Common to both communities is the almost complete lack of evidence of any sanctions against those who transgressed the various regulations. However, one cannot ignore the fact that, through the threat of a ban (*ḥerem*), the leaders of Arta and Ioannina were successful in enforcing the community's authority on troublesome or criminal elements. The effectiveness of this was without doubt limited; nevertheless, it was a cogent weapon.[52] On the other hand, one must note the many cases of disobeying the rabbinical court decisions, especially in Arta. There were even a few cases of litigants refusing to be judged by the rabbinical courts. As has been proved elsewhere, it was acceptable among the Jews of Arta to appeal to Gentile courts, and this was probably also the case in Ioannina.[53]

The evidence shows that, from the point of view of the internal organization of the community, the communities of Epirus were not different from the others in the Ottoman Empire in the sixteenth century. These communities were unique due to the predominance of the Romaniotes in both cities throughout the sixteenth century. Similarly, evidence points to the conclusion that Arta was the more significant of the two communities of Epirus. The above-mentioned sources show that both communities for the most part observed Romaniote regulations, and that in both the Romaniotes set the tone by their control of Jewish community affairs, even after the Spanish, Italian, and Sicilian émigrés had settled in Epirus.

Responsa, MS, pp. 197a–201b; *Rashdam*, 'Even Ha-'Ezer', No. 231; R. Ḥayyim Nissim Raphael Muziri, *Be'er Mayyim Ḥayyim: Responsa*, Vol. I (Salonica, 1594); 'Hoshen Mishpat', No. 26.

50. Muziri, *Be'er Mayyim Ḥayyim*, Vol. I; 'Hoshen Mishpat', No. 26.
51. R. Joseph Molcho, *Responsa Ohel Yoseph* (Salonica, 1756); 'Yoreh De'ah', No. 9.
52. See Bornstein-Makovetsky, 'Life and Society', 149–50.
53. Ibid.

The Rise and Fall of Salonica Woollens, 1500–1650: Technology Transfer and Western Competition

BENJAMIN BRAUDE

Among the often cited accomplishments of the exiles from Iberia has been the establishment early in the sixteenth century of a major woollen cloth manufacturing centre in Salonica. Woollen cloth was a prime initial reason why Salonica became the most long-lived and most Jewish city not only within the Ottoman realm and the Mediterranean world, but also for all of modern Jewish history. The creation of Jewish Salonica — Jews remained the largest single group within the city, generally the majority, until after the First World War — was one of the most striking consequences of the expulsion of the Jews from practically all of the Christian Mediterranean. Their swift transformation of the city after 1492 into a bustling industrial and commercial centre so impressed the Venetian *bailo* Jacopo Contarini, that already in 1507 he was moved to comment on the amazing fact of '10,000 [clearly an exaggeration] Jewish homes' newly established and on 'the astonishing quantity of woollens they produced.'[1] The explosive growth of Jewish Salonica, coupled with the rapid expansion of the Ottoman Empire wherein it flourished, helped to fix the image of the Jew as an economic force intimately allied with the Orient. It was an image with far-reaching consequences for the subsequent evolution of Jewish history.

THE FALL OF BYZANTINE THESSALONIKI

Salonica had been the second city of Byzantium in its heyday. It had historically represented a bastion of Greek Christianity for its Slavic hinterland. How did so dramatic and rapid a transformation occur?

Acknowledgements: I thank Aron Rodrigue and Shmuel Bolozky who commented on earlier drafts of this essay and John Munro who furnished very useful bibliography on textile technology. Work on this article was begun during my tenure as Visiting Associate Professor of Judaic Studies and Esther Levi Sosland Scholar in Residence at Smith College, and final revision was completed at the Institute for Advanced Studies, the Hebrew University, Jerusalem. I am grateful to both institutions for their support and hospitality.

1. Marino Sanuto, *I Diarii*, Vol. VII, ed. R. Fullin (Venice, 1882), p.19.

The developments which made this change possible had in fact occurred gradually over the previous century and a half. In the Byzantine era, during the struggle between the rival Cantacuzenus and Palaeologus dynasties, the lower-class poor, the so-called Zealots, controlled the city from 1342. They offered it to Stefan Duşan for his Serbian realm in 1350, only to have it wrested back by the Byzantine emperor, John Cantacuzenus. For his own success, however, he depended to a great degree upon Ottoman military support. Not surprisingly, the Byzantines could not long hold it. From 1387 to 1403 it was generally directly under Ottoman control, a dominance only ended by an invasion from the steppes of central Asia, Timur's defeat of Sultan Bayezit. From 1403–23 Byzantine rule was restored but it was undermined by continuous Ottoman harassment, particularly during the last of these two decades. From 1423 to 1430 the Venetians were given ownership so that it would not fall into the hands of the Turks; none the less, in 1430 it was conquered for the final time by the Ottomans whose rule lasted, effectively, until 1912. The near-century from 1342 to 1430 had witnessed some eight changes in power.

The result was a demoralized and devastated settlement, a faint echo of the city's past glories. From a population claimed to be 40,000 it had declined nearly four-fold to perhaps 10–13,000. Those few thousand left to resist the Ottoman invasion were overwhelmed by an army which — according to the contemporary chronicle, the *Diegesis* of John Anagnostes — slew some 6,000, 'emptied the city of people' and wealth, taking some 7,000 men, women, and children as captives and then returned to ensure 'that everything in the city was taken away and overturned, and nothing, neither holy church nor monastery nor house of even the humblest was left undisturbed'. Eventually, in 1432–33, Sultan Murat II repopulated the city, but only a fraction of its original inhabitants was resettled, for in that year its population consisted of some 1,000 restored Thessalonians and some 1,000 newly settled Muslims. It took nearly a half-century for the city to begin to approach its already lowly pre-conquest state and even then its total population was little more than some 10,000, of whom fewer than 7,000 were Christian. The economic, political, and spiritual poverty can be seen in the fact that these souls were served, according to the 1478 Ottoman fiscal records *(tahrir defterleri)* studied by H. Lowry, by only nine priests![2]

The devastation of Greek Thessaloniki weakened the ability of the remaining indigenous population to resist the western Jewish influx. There were few magnates, no entrenched urban patriciate, at best a decimated clergy, and only a small demoralized and struggling band of Thessalonians.

2. S. Vryonis, Jr. 'The Ottoman Conquest of Thessaloniki in 1430', in A. Bryer and H. Lowry (eds.), *Continuity and Change in Late Byzantium and Early Ottoman Society* (Birmingham, England–Washington D.C., 1986), pp. 281–321.

This weakness applied to the indigenous Jewish population as well which had suffered the same fate as Thessalonian Christians. Whatever few Jews were able to return to the city after 1430, as is clear from both the Turkish and Hebrew sources, were deported to Istanbul two decades later, long before the exiles left the Christian Mediterranean.[3]

THE RISE OF JEWISH SALONICA

It was the fortuitous conjuncture of this tale of decline and devastation in the eastern Mediterranean with an equally tragic tale of devastation and expulsion in the western Mediterranean that ultimately reversed the fortunes of Salonica. In 1492 after the victory of the most Catholic rulers of Aragon and Castile over the Muslim ruler of Granada, Ferdinand and Isabella announced the expulsion of the other major irritant to the Christian tranquility of their realm, the Jews. Like dominos falling neatly in a row, one after another of the other lands of the Christian Mediterranean took similar steps: Portugal in 1496–97, Navarre in 1498, Naples (with characteristic southern Italian diligence) expelled its Jews thrice — in 1497, 1510, and finally in 1540, while Provence needed only two tries at finishing the job, in 1493 and 1501. By the end of the sixteenth century only isolated parts of Italy, of all the Christian Mediterranean, remained officially open to Jewish settlement. This systematic expulsion destroyed long-established Jewries rich with vibrant cultural traditions marked by high intellectual attainments and supported by varied and successful economic enterprise. The oldest had been founded before either Christianity or Islam had reached their shores and most were intimately integrated with every aspect of the life of their lands.

It is not easy to assess the human cost of this tragedy. Whatever the estimate, the total number of refugees was huge for early modern times. Some sought haven in nearby North Africa, and gradually a few, much later in the sixteenth century, settled in the relatively open atmosphere of Venice and Livorno. But most eventually made their way to a more distant and less familiar point in the Mediterranean, the Ottoman Empire, and perhaps 15–20,000 established themselves in the renewed city of Ottoman Salonica, one of the largest concentrations of refugees. This was indeed a fortunate conjucture of a people in need of a land in search of a land in need of people.

As yet we have no contemporary evidence concerning the actual steps taken by the Ottoman government to direct the Jews to Salonica. But it would have been unneccesary for the state to have used the *sürgün* system

3. J. Hacker, 'Ottoman Policy toward the Jews and Jewish Attitudes toward the Ottomans during the Fifteenth Century', in B. Braude and B. Lewis (eds.),

(the system of forced migration used elsewhere to resettle urban areas), since, in the event, as the major and closest Ottoman port on the Aegean, Salonica would have been a natural destination for all sea-borne refugees. Thus the government did not need to take active measures to bring them there, rather they had to develop policies which would make it attractive for them to stay. Although the remark about these new subjects attributed to Bayezit II — 'You call Ferdinand a wise king, he who impoverished his own country and enriched mine!' — is undoubtedly apocryphal, it clearly reflected Ottoman awareness of the utility of these immigrants, particularly since their urban skills suited long-established Ottoman policies and met long- recognized Ottoman needs. As important as Ottoman policy towards the Jews and Jewish perception of this policy was the creation under the *Pax Ottomanica* of an effective Balkan common market which made Salonica economically strong enough to sustain so large a population. Salonica had great advantages. The economic geography of the Balkans made it the natural entrepôt. Situated near the mouths of the Vardar and Gallikos rivers, it controlled through these river valleys the easiest entry from the eastern Mediterranean to Central Europe. From Roman times merchants from Novgorod would pass through the city on their journey to Alexandria. It was closer to Belgrade than was Venice. The Egnatian Way from Istanbul to Durres, which was the only passable route across Macedonia connecting the Aegean to the Adriatic formed its high street. Not only was it well-located on land routes; it was also convenient for the sea lanes of international commerce. The city was closer to Alexandria than Venice or Istanbul, or any other port on the Adriatic or near the northern Aegean. Most importantly, it possessed an ideal harbour. In addition to these advantages it also had ready access to a large and potentially prosperous hinterland capable of producing food, minerals, and other raw materials for the trade and industry of the day. The fact of access during the heyday of Ottoman power, when practically all the Balkans came under Muslim sway, was crucial to the prosperity of the city. The political unity of the Balkans, something more easily achieved by imperial might than by factious national self-determination, gave the city great economic strength. The centres of production, consumption, and distribution were thus allied and their focus was Salonica.

Incidentally, although Ottoman policies helped revive Salonica, it never achieved the lofty status it had under the Byzantines as the empire's second city. In the Balkans, Skopje and Edirne were roughly comparable and at times even larger. At a further distance, Bursa, and later Aleppo,

Christians and Jews in the Ottoman Empire, 2 vols. (New York, 1982), Vol. I, pp. 117–18, and H.W. Lowry, 'Portrait of a City: the Population and Topography of Ottoman Selânik (Thessaloniki) in the Year 1478', *Diptycha*, 2 (1980–81), 261–3.

Damascus, and Cairo were all far more populous. At best Salonica was the Ottomans' sixth city.[4]

WOOLLEN MANUFACTURE IN SALONICA

The major employer and polluter of the Ottoman city was the textile industry, particularly the manufacture of woollen broadcloth. In a letter written around 1540, Salonican Jews described the textile industry as the sole support for most of the population.[5] In the same period a rabbinic jurist noted, in a ruling concerned with the noxious odour of urine which was used to process raw wool, that the majority of the inhabitants of Salonica were employed in cloth manufacture.[6] Later in the century another rabbi was called upon to resolve a similar complaint — demonstrating how widespread was the practice of soaking wool in urine and how ubiquitous were textile activities.[7] Its maximum production capacity was — according to my calculations — approximately 40,000 pieces of woollen cloth per year which placed it among the major producers in the Mediterranean world.[8]

Here again, the ready access of the city to steady supplies of raw materials made this a natural development. From the Balkan hinterland the city had supplies of fine wool. Near the city were the excellent grazing lands on the slopes of Mount Corthiat, an area called Calamari, but the entire region provided pasture. In the summer, sheep would feed in the Balkan highlands, but in the winter shepherds from all over the peninsula, but particularly from Albania, would lead their flocks to the open plain of Salonica, and other low-lying areas. From October to May the sheep

4. Ömer Lûtfi Barkan, 'Tarihî Demoğrafî Araşıtirmalrı [*sic*] ve 'Osmanlı Tarihı' [Studies in historical demography and Ottoman history], *Türkiyat Mecmuası*, 10 (1952–53), 22.
5. *Yosef da'at*, 1 (1888), p. 120.
6. Joseph b. Leb, ***Responsa*** (Amsterdam, 1726), section III, question 33, especially p. 16, column b.
7. Samuel b. Isaac Ḥayyon, ***Beney Shmuel*** (Salonica, 1613), question 14.
8. B. Braude, 'Community and Conflict in the Economy of the Ottoman Balkans, 1500–1650', unpublished Ph.D. thesis (Harvard University, 1977), pp. 32–9. For these purposes a cloth was approximately 24 yds. by 2 yds. in size. My estimate was calculated on the basis of the work force available for production as well as the amount of wool available in the Salonica region. By English statute the width of a broadcloth was set at 2 yds. Since looms were fairly standard in size, that width was probably maintained for Mediterranean production as well. The actual length of cloth varied from 18 to 46 yds., but the normal length was between 22 and 26 yds. My notional estimate of 24 yds. is thus based on an average length. Lengths were normally given by the pick (*pico*, *zira*, cubit) which is approximately 27 inches or three-quarters of a yard. Shemuel Avitsur estimated the production at 500,000 m. of broadcloth, which, by my cloth standard, would be about 20,000 cloths. See his, 'The Woollen Textile Industry in Saloniki [*sic*]', *Sefunot*, 12/2 (1971–78) (*The Book of Greek Jewry*), 159 (Hebrew).

would spend the cold months in the lowlands, but before their return, perhaps in late March or April, they would be shorn of their wool. The spring shearings in the Salonica plain were not the only source of wool. More distant sources of supply included Sofia, Plovdiv, Skopje, Stromnitza, Tatar Pazarı, Ipsala, Tırkala, Larissa, Jannitza, Lepanto, Karli, Scutari, Elbasan, Çatalca, Büyük Çeşme, Midia, and Livadia. Other raw materials readily available were such natural dyes as madder-root, kermes, and yellow-seed. However the most popular dye, indigo, which yielded the characteristic blue of Salonica cloth had to be imported from further east. An essential ingredient in the dyeing process was alum, which was used as a fixative to stabilize the colour of the cloth. It was mined in the town of Ipsala, three to four days' journey from Salonica on the road to Istanbul.

Salonica textiles have often been studied, most notably by Isaac Samuel Emmanuel.[9] However, there are still a surprising number of basic questions which have either yet to be addressed or remain open to debate. Recently scholars have differed over the period of both the industry's actual foundation and its effective demise. Moreover the precise circumstances which made the new Jewish contribution to the Ottoman textile industry technologically distinctive are insufficiently recognized. Since the foundation of the industry is linked to its innovative technology, both must be studied in order to resolve this part of the debate.

ORIGINS OF THE TEXTILE INDUSTRY

An important question for scholars interested in the continuity of Byzantine and Ottoman social and economic history concerns the circumstances of the industry's foundation. Was it a new Ottoman development or merely the continuation of a long-standing Greek enterprise? Although Jewish historians take for granted the assumption that Salonica textile manufacture was initiated early in the sixteenth century by Jews from Spain, specialists in Ottoman and Greek history working from Turkish sources have claimed that the industry was founded earlier. In fact, given the natural advantages which Salonica had as a textile producing centre, it is reasonable to ask if such an industry antedated the arrival of the exiles.

9. I.S. Emmanuel, *Histoire de l'industrie des tissus des Israélites de Salonique* (Paris, 1935). For a survey of the bibliography see S. Schwarzfuchs, 'Quand commença le déclin de l'industrie textile des Juifs de Salonique?', in A. Toaff and Simon Schwarzfuchs (eds.), *The Mediterranean and the Jews: Banking, Finance and International Trade (XVI–XVIII Centuries)* (Ramat Gan, 1989), pp. 215–34. See also S. Faroqhi, 'Textile Production in Rumeli and the Arab Provinces: Geographical Distribution and Internal Trade (1560–1650)', *Journal of Ottoman Studies*, 1 (1980), 61–83.

İsmail Hakkı Uzunçarşılı, drawing on an early seventeenth- century manuscript, claimed that Salonica Jews produced woollen cloth for the Janissary corps as early as the era of Murat II (1430–51).[10] Following him, Apostolos Vacalopoulos made the same claim.[11] More recently Heath Lowry in two different articles, drawing upon his research in the 1478 Ottoman *tahrir defterleri*, has argued that at that time Salonican Turks had already established a significant textile industry, decades before the Jews arrived.[12] Thus according to these scholars, rather than founding a new industry the Iberian exiles merely built upon an existing one.

The claims for the existence of a Jewish industry during the first half of the fifteenth century are not confirmed by other sources. Given the lateness of Uzunçarşılı's evidence this is not surprising, but Lowry's evidence cannot be as easily dismissed. However, a closer examination suggests some significant differences between early Turkish cloth production and the later industry developed by the Jews.

Lowry's study of the Ottoman fiscal records reveals details of the nature of Salonica cloth in the late fifteenth century. He notes that a total of 135 were involved in some aspect of textile production. Of these 125 were Muslims: 6 cloak-makers (*abacı*), 8 dyers (*boyacı*), 73 weavers (*culah*), and 38 tailors (*hayyat*). In addition there were 10 Christians employed in this same craft: 4 dyers, 3 tailors, and 3 linen weavers (*kettancı*). All told this totals roughly 25 per cent of the non-government occupations listed in the records, a sizeable percentage.

Given the importance that textile manufacture might be expected to play in any Ottoman urban centre it is worth asking how representative this one-quarter proportion was. We are fortunate in having data for comparison from the neighbouring Balkan city of Skopje, which was similar in size.[13] There textile workers constituted only 14 per cent of the total non-governmental work-force. A detailed analysis of the labour forces in Skopje and Salonica reveals some of the differences. Skopje had a total of 79 employed in various aspects of textile production. Of these there were: 5 cloak-makers, 2 weavers, 12 felt-makers (8 *kebeci* and 4 *kececi*), 10

10. İsmail Hakkı Uzunçarşılı, *Osmanlı Devleti Teskilatından Kapkulu Ocakları* [The Imperial Janissary Corps in the organization of the Ottoman state] (Ankara, 1943), Vol. I, p. 269. According to V.L. Menage, this manuscript seems to be identical to a work composed under Aḥmed I described in *Ist. Kit. Tarih-Coğrafya Yazmaları Kataloğları*, i/10, 813 (MS Esad Ef. 2068), and to a work contained in MS Revan 1320. See s.v. *Devshirme*, *Encyclopedia of Islam*, 2nd edn. (Leiden, 1965).
11. Apostolos E. Vacalopoulos, *History of Macedonia, 1354–1833*, trans. P. Megann (Salonica, 1973), p. 540.
12. Lowry, 'Portrait of a City', 290–91, and '"From Lesser Wars to the Mightiest War": The Ottoman Conquest and Transformation of Byzantine Urban Centres in the Fifteenth Century', in Bryer and Lowry, *Continuity and Change*, pp. 337–8.
13. Barkan, 'Tarihī Demoğrafī', 26.

spinners of goat-hair (*mutaf*), 42 tailors (*terzi* — perhaps dressmaker?), 4 yarn-makers (*iplikci*), 3 cotton fluffers (*hallac*), and 1 felt-cap maker (*arakıyeci*).

Statistically the most significant difference between the two, in addition to Salonica's 25 per cent as opposed to Skopje's 14 per cent, was the number of weavers. Thus Salonica had a total of 76 weavers, while Skopje had only 2. Clearly Salonica had a disproportionately large number of weavers late in the fifteenth century, but the figures by no means reach the hundreds and thousands which must have been the case in the sixteenth century. This is striking evidence that the city had a small but comparatively significant weaving craft decades before the arrival of the exiles. Clearly it was not a major industrial centre. Whatever they were weaving in Salonica, they probably were not finishing it there. There were only 12 dyers. Furthermore, other important crafts necessary for producing the finished product are completely missing from the list of occupations that Lowry provides.

But what were they weaving in Salonica? In all likelihood, given the nature of the raw materials readily available, it was woollen cloth of some kind. However the only woollen cloth which appears in these records is *aba*, a very stout kind using a somewhat coarse home-spun yarn which was woven in many different parts of the empire. The woollen cloth which does not appear and which was to be the distinctive product of the new Jewish industry was *çuha*, or *çuka*, the equivalent of English broadcloth, a much stronger felted fabric, the product of a particular mechanical technology which the Jews brought with them from the West.[14]

That this was in fact a new technology is confirmed by a number of different sources. One is the letter of the Greek Cardinal Bessarion to the Despot of the Morea, sent around 1444, in which he decries the technological backwardness of his land, and proposes a programme for the training of Greek youth in Italy. Specifically they were to learn new techniques in the use of water-powered mills and other labour-saving devices, as well as in the manufacture and dyeing of silk and woollen cloth.[15]

This was the area of technological innovation introduced by the new Jewish industry. They were making a cloth stronger than *aba* at a cost lower than in the past through the application of water-powered mills to the fulling process. Fulling itself was known in the Middle East. For

14. For an example of late sixteenth-century broadcloth, this manufactured in England, see British Library, Lansdowne Manuscripts, No. 114, fols. 92b–93a.
15. S. Lambros, 'Hypomnēma tou Kardinaliou Bessariōnos eis Konstantinon ton Palaiologon', *Neos Hellenomnemon*, 3(1906), 25–7, and A.G. Keller, 'A Byzantine Admirer of Western Progress: Cardinal Bessarion', *Cambridge Historical Journal*, 11 (1953–55), 343–8.

instance, the fuller (*qaṣṣar*) appears in Geniza documents dating from the eleventh century.[16] This craft continued into Ottoman times. The Ottoman Turkish term *kısaret*, derived from the Arabic, meaning 'the occupation, calling of a bleacher or fuller' and the related verbs, *kassarlatmak* (causative), *kassarlamak* (active) and *kassarlanmak* (passive), all appear in the authoritative nineteenth-century Turkish-English dictionary of Sir James Redhouse.[17]

Fulling is a process in which woollen cloth is beaten, scoured, and cleansed with a detergent. As a result the cloth is degreased of lanolin and shrunken in width and length by anything from a fifth to a half. Indeed it is the fact that fulling shortens the cloth which explains the derivation of the Arabic and Ottoman Turkish terms which have as their root *QṢR*, meaning 'to be or become short.' The process also serves to felt the cloth, inextricably entangling the fibres to make the weave pattern almost invisible. This produces a stronger fabric with a smoother and softer surface. The resulting increased thickness and weight give the cloth greater resistance to wear and weather. In medieval Egypt the fuller would use fuller's earth, *tufal*, to take care of the heavier textiles.

Originally the beating of the cloth was done by hand-held club or by foot, but in medieval western Europe a mechanical method of fulling was invented. To quote Eleanora Carus-Wilson, 'An Industrial Revolution of the Thirteenth Century': 'The invention was a two-fold one.... The action of the two feet was replaced by that of two wooden hammers, alternately raised and dropped on the cloth as it lay in the trough, and controlled probably by a revolving drum.... This revolving drum was attached to the spindle of a water-wheel, and this supplied the motive power.'[18] Mechanical fulling did not completely replace the older method. Particularly for fine luxury cloths the manual method was preferred, but the new technology did allow a saving in time and money which has been estimated as a productivity gain of between three- and fifty-fold.[19]

It is this process of mechanical fulling which the Jewish exiles transferred to the Ottoman Empire. Use of the water-driven fulling mill now spread. In the Salonica region nearly 300 people were employed in the fulling mills set up on the banks of the Vardar (Axios) River and its tributaries. A bit further away there was an important mill in Kara-Ferya (Beroea or Veroia). Other sites were Pleven, Tırkala, Larissa, and

16. S.D. Goitein, *A Mediterranean Society* (Berkeley, 1971, 1978), Vol. II, p. 438; Vol. III, pp. 341, 503.
17. Sir James Redhouse, *A Turkish and English Lexicon* (Istanbul, 1890), pp. 1455–6.
18. E. Carus-Wilson, 'An Industrial Revolution of the Thirteenth Century', *Economic History Review*, 11 (1941), 40, 43.
19. J. Munro, 'Textile Technology', in J.R. Strayer *et al.* (eds.), *The Dictionary of the Middle Ages*, 11 (NY, 1988), pp. 705–7.

Rhodes.[20] A smaller textile centre emerged in Safed in the Galilee region of Palestine, though it suffered from two disadvantages. It had to import its raw wool from the Balkans and its fulling mills could only operate in the winter rainy season.[21]

The mechanical fulling mill was called *batan* in the Hebrew sources of the sixteenth century. He who performed this function was called a *batañero*. A contemporary Jewish source of the sixteenth century details the technical processors involved in preparing the finished broadcloth. The absence of such craftsmen from the occupation lists published by Lowry suggests the far-reaching transformation of the industry which had occurred in the course of nearly a century since 1478. The following passage is taken from a rabbinic responsum of the leading Jewish legal authority in Salonica, Samuel de Medina:

> I will preface with an introduction to cloth manufacture in Salonica. After the cloth has been woven it is given to the fuller (*batañero*) who fulls it. Then it is given to the comber (*perharo*) to perform the task which is called combing (*bettildar*). Then it is given to shearing (*tundir*). After this task is done they see if the cloth is pure and clean of oil. If it is not as clean as it should be it is given to another man to degrease it and this task is called bleaching (*desteñir*). Nitre and water are placed upon the cloth. After it has been degreased the comber takes it to comb (*perhar*) it a second time. Afterwards the tenter-man (*tendirdor*) takes it to tenter it in the sun of the courtyard as is the custom.[22]

The Spanish origin of the terminology employed in the industry suggests the western origin of its technology. This is confirmed in an Ottoman fiscal document concerning the Salonica region in the late seventeenth century which mentions the most distinctively innovative process of cloth production, mechanical fulling, and uses the word, *batancıyan*, a Turkish version of the Spanish *batañero*. Avitsur's work indicates that the term entered local Arabic usage in Safed as well.[23]

20. Uzunçarşılı, *Kapkulu Ocakları*, Vol. I, p. 276; Aaron b. Ḥayyim Hacohen Peraḥyah, *Peraḥ mateh aharon* (Amsterdam, 1703), sect. II, question 62; sect. I, question 31; Samuel de Medina, *Responsa* (Lemberg, 1862), sect. II, question 123, question 81; sect. III, question 18; sect. II, question 259; Levi b. Ḥabīb, *Responsa* (NY, 1962), questions 31 and 99; Ḥayyim Shabbetay, *Torat Ḥayyim* (Salonica, 1651), sect. II, question 34.
21. S. Avitsur, 'Safed — Center of the Manufacture of Woven Woollens in the Fifteenth [*sic*, should be Sixteenth] Century', *Sefunot*, 6 (1962), 41–69 (Hebrew), and 'The Batan, a Water-Powered Fulling Mill in Naḥal 'Ammud — Relic of the Wool-Textile Industry in Safed', *Israel, Land and Nature*, 7 (1981), 18–21. I thank Joseph Shadur for furnishing a copy of the second article.
22. Samuel de Medina, *Responsa*. sect. IV, question 45.
23. V. Demetriades, 'Phorologikes kategories tōn chōriōn tēs Thessalonikēs kata tēn

There remains, however, one issue which should be addressed before we can firmly establish that this was a new technology with no origins in the Middle East. The term *batan* itself has a Semitic ring to it which raises the question as to whether or not both the term and the fulling-mill itself might have originated in the Arab-Muslim industry. Perhaps it was a technology known to or even introduced by the Muslims of Spain and subsequently reintroduced to the Muslim East by the Jews? Indeed *BṬN* is an Arabic root which among its many usages can mean 'to penetrate', certainly an element involved in the process of mechanical fulling. Furthermore, some words derived from this root concern textiles. However, with one exception, none of the dictionaries I have consulted on this point suggest that the term, fulling mill, was defined by a word derived from *BṬN*. Thus the Arabic origin of the term and the invention seem unlikely. The one exception constitutes an elegant confirmation of my argument. Hans Wehr's lexicon does so define the term, but he lists its root as a separate entry distinct from the other root.[24] Furthermore, he notes that this *BṬN*, meaning to full, has a distinctly regional usage limited to Tunisia. The related term *batan* he defines as fulling mill, but he derives both words from the Spanish! Thus it is reasonable to conclude that the distinctive term and technology spread from the Maghrib into the Mashriq, and into the Ottoman realm, as a result of the events of 1492. Aside from areas specifically influenced by Iberian Jews, such as Safed and Salonica, the term *batan* does not seem to appear in the Arabic or Turkish vocabulary of the eastern Mediterranean.

The evidence gathered here suggests that the claim that woollen cloth had been woven in Salonica before Jewish exiles reached the city at the turn of the century is correct. However there is little evidence to suggest that this weaving was part of a larger textile processing enterprise. Moreover, the technological innovation the exiles introduced and the scale of the manufacture they established were quantitatively and qualitatively so different that one is justified in claiming that indeed this was a new industry.

There remains another basic question concerning the foundation of the Salonica industry which has hardly been raised, let alone answered. Was this in fact an Iberian Jewish enterprise? After all, as has been clear from the outset of this discussion, the Jewish exiles who settled in Salonica did not all come from Spain and Portugal. The Ottoman records, among other sources, also testify to their origins in Sicily, and other parts of Italy, as well as the Provence, although the Iberian element was the largest single

Tourkokratia' [Tax categories of villages in the Thessaloniki district during Ottoman rule], *Makedonika*, 20 (1980), 412–13.

24. H. Wehr, *A Dictionary of Modern Written Arabic*, J. Milton Cowan (ed.) (Ithaca, NY, 1976), p. 64.

group.[25] Is it possible that the source of this new technology has been misattributed? Joseph Nehama in fact has suggested that it was established by Sicilian immigrants.[26] Recently scholars have argued that the economic role of the Jews in fifteenth-century Spain has been much exaggerated.[27] In the secondary literature there is a tradition of Sephardic contributions to Ottoman industry — notably in the manufacture of 'canons, bows, crudeshooting engines, swords, shields, and spears' — which dates back at least to the chronicle attributed to the sixteenth-century historian, Joseph Ha-Cohen, and which has been continuously rpeated, for example recently by Fernand Braudel.[28] However — beyond the often unreliable accounts of Christian travellers in the Ottoman realm — there seems little evidence to corroborate the claim that Spanish Jews had special expertise in military technology. Is the claim about textile technology equally false?

How to resolve this question? It is clear that by the mid-sixteenth century the work-force and the terminology of the Salonica industry were, respectively, mostly Iberian and Spanish. The origins of its technology are less clear. As has been argued, it is very unlikely that it was indigenous to the Near East. The etymological evidence alone suggests, but does not prove a Spanish origin. Local textile industries existed throughout Spain and Jews could easily have acquired the skills there.[29] Water-powered fulling mills while apparently unknown in the East had been extant in the West for centuries. This was by no means an advanced technology to be found only in the most developed economic centres. By virtue of its dependence on water power, fulling mills were often found in rural areas and small towns and villages, precisely in the areas whence many of the exiled Jews came. Thus a Spanish origin, whatever the precise distribution of the pre-Expulsion Jewish population, seems quite reasonable. Since the technology was so widespread in the West, it is also possible that Nehama's claim about Sicily is not incorrect. Thus there may have been a number of different western sources for this textile innovation in the Ottoman

25. B. Lewis, *Notes and Documents from the Turkish Archives* (Jerusalem, 1952), 25–7.
26. J. Nehama, 'The Jews of Salonika [*sic*] in the Ottoman Period', in R.D. Barnett and W.M. Schwab (eds.), *The Sephardi Heritage*, Vol. II, *Western Sephardim* (Grendon, England, 1989), p. 208.
27. H. Kamen, 'The Mediterranean and the Expulsion of Spanish Jews in 1492', *Past and Present*, No. 119 (May 1988), 30–55, and M. Kriegel, *Les Juifs à la fin du Moyen Age dans l'Europe méditerranéenne* (Paris, 1979).
28. Joseph Ha-Cohen and The Anonymous Corrector, *The Vale of Tears*, trans. Harry S. May (The Hague, 1971), p. 115; F. Braudel, *The Mediterranean and the Mediterranean World in the Age of Philip II*, trans. S. Reynolds (NY, 1976), Vol. II, pp. 808–9.
29. For example, P. Iradiel Murugarren, *Evolucion de la industria textil castellaña en los siglos xii–svi* (Salamanca, 1974), p. 111. F. Baer, *Die Juden im christlichen Spanien*, Pt. I, *Urkunden und Regesten*, Pt. II, *Kastilien/Inquisitionsakten* (Berlin, 1936), contains abundant evidence for the involvement of the Jews in the details of processing cloth. See also *Aragonen und Navarra* (Berlin, 1929), Vol. 1, p. 876.

Empire. However, after a few decades the terminology of this technology, like much else in Ottoman Jewry that developed in the aftermath of the Expulsion from Spain, became Hispanified and its simultaneous Sicilian as well as perhaps Italian and Provencal origins were lost.

THE DECLINE OF THE TEXTILE INDUSTRY

The other issue which has provoked discussion is the timing of its effective demise. Most recently Simon Schwarzfuchs has concluded that the beginning of the decline of the industry should be set much earlier than others — myself included — have suggested, perhaps as early as the 1530s.[30] Drawing largely upon the abundant responsa literature, he has demonstrated the problems that Salonica manufacturers had during that period in obtaining an adequate supply of raw wool. He concludes that the high price and scarcity of its essential raw material started the unravelling of the industry. With the help of a statistical analysis conducted by the Responsa on Computer Project of Bar-Ilan University, he further notes that in the subsequent decades of the century leather goods increasingly replaced textiles as the subject of the legal opinions of Ottoman rabbis which have been preserved in print.

His contribution raises a number of important methodological and substantive questions. Legal sources in general and responsa literature in particular, especially when they are used on their own without the control of other kinds of documents, present certain difficulties as historical sources. When they involve litigation they may or may not be representative of society as a whole. In Ottoman historiography the *sicillat*, the records of the courts of Islamic law, have proved to be an invaluable source for social and economic history. Their strength lies in their abundance and their nature. They contain the daily record of centuries. Moreover the Islamic court was not only a forum for litigation and conflict resolution, but also had a notarial function. It was a repository for all issues of real estate, inheritance, and other questions of the possession of property not only among Muslims, but also for Christians and Jews. Such sources have helped scholars reconstruct the life of Cairo, Jerusalem, and other Ottoman cities. Unfortunately these documents cannot help us reconstruct the first two centuries of Ottoman Salonica since they have not survived. The Salonica *sicillat* which we do have date only from the very end of the seventeenth century: thus we do not have the best sources.

Rabbinic responsa do represent a valuable resource. Until the mid-sixteenth century they represented the only sources — in addition to Jewish sermons — printed within the Ottoman Empire, for printing, like the fulling mill, was another European technological innovation introduced

30. Schwarztuchs, 'Le déclin de l'industrie'.

in the aftermath of 1492. Since printing in Arabic script was allowed only in the eighteenth century and since very little was printed in either Armenian or Greek prior to that century, for all intents and purposes these books are the most widely available internal record we have for what was arguably the most important period of Ottoman history.

However, like any source, they do present problems. While the comprehensiveness of the domain of the *sicillat* reduces the question of the representativeness of that legal material, no such statement can be made for responsa. The process of selection and elimination which was essential for the preservation and publication of the rabbinic sources which have survived introduced a number of considerations which have militated against their value as historical documents. Censorship, printers' errors, and various editorial considerations have at times altered the original text in significant ways. Responsa were written, preserved, and published primarily because of their insight into halacha (Jewish law) and only secondarily because of their intrinsic contemporary social and economic significance, and certainly not at all for their value as a historical source. In other words it was the halachic answer and not the social and economic question which determined whether or not a given datum of local realia survived. It was possible that the more routine and commonplace a problem and therefore the more subject to a boilerplate legal response, the less likely that it would have prompted the steps necessary to seek an authoritative legal ruling and the less likely that it would have been preserved in a published collection. Furthermore, after the initial flurry of legal attention to a new social and economic phenomenon — such as the textile industry during the first decades of its foundation — it is quite possible that the engagement of legal minds might turn in different, less representative, and more novel directions. Moreover the accident of genius plays a role as well. It so happened that during the mid-sixteenth century two giants of Jewish law, Samuel de Medina and Joseph ben Lev, dominated the legal discourse of Salonica. None of their successors, though certainly worthy, merited the same degree of respect in their own time or in subsequent generations. Consequently their opinions, quantitatively and qualitatively, are disproportionately represented in the surviving corpus of printed responsa. Thus this body of material contains an inherent bias which tends to over-represent cases from approximately the period which is identified as the beginning of the end.

None of this is to deny the value of the evidence which Simon Schwarzfuchs has gathered. Nor should his conclusion be dismissed. He presents a strong case that the 1530s witnessed an early crisis for the industry and that it very likely was the beginning of one of the problems which ultimately brought it down, but I would argue that it was only one of a series of problematic decades which slowly and ultimately weakened it.

He may be correct in identifying that decade as the beginning of the decline, but the decline itself was not precipitous or immediate, but rather slow and lingering. The path of decline of the Salonica textile industry — and for that matter of any human enterprise — was never consistent and straightforward. There were repeated ups and downs before the ultimate and final demise occurred.

Here is not the place to repeat the detailed evidence I have published elsewhere, but it would be useful to summarize these and related arguments, particularly since one of the articles has appeared only in Hebrew.[31]

Obtaining an adequate supply of wool for the newly emerging woollen textile industry was certainly a major challenge. I have estimated that the combined needs of the new woollen centres of manufacture established in the aftermath of 1492 represented a significant demand on the flocks of Balkan sheep, requiring as much as a 12 per cent increase, if pre-existing needs were to continue to be met.[32] Moreover these were not the only new pressures on this rural sector of the economy. After the mid- sixteenth century the flocks of the Spanish Mesta, which had long been the principal source for the Italian woollen textile industry, began their long decline.[33] At the same time the Venetian textile industry underwent a tremendous expansion which required an increasing supply of raw wool from the Ottoman markets. The empire was not the only source of supply; wool continued to be shipped from Spain, but wool from the Balkans was the closest to Venetian looms.[34] The result of this market pressure was a steady rise in the price of wool throughout the sixteenth and seventeenth centuries.[35]

Foreign textile producers significantly affected not only the local market in raw wool, but also competed successfully with the new industries. Ottoman markets were readily permeable to foreign trade in any commodity. This was particularly true for the Balkans whose Adriatic inlets beckoned to traders from across the narrow sea. Such Venetian possessions as Zakynthos (Zante), Ithaca, Corfu, and Crete lay even closer

31. 'International Competition and Domestic Cloth in the Ottoman Empire, 1500–1650: A Study in Undevelopment', *Review of the Fernand Braudel Center*, 2 (1979), 437–51, and 'The Manufacture of Salonica Cloth in the Economy of the Eastern Mediterranean', *Pe'amim, Studies in the Cultural Heritage of Oriental Jewry*, 15 (Jerusalem, 1983), 82–95 (Hebrew).
32. Braude, 'Community and Conflict', 40–42.
33. J. Klein, *The Mesta: A Study in Spanish Economic History, 1273–1836* (Cambridge, MA, 1920), p.27.
34. J. Gentil da Silva, *Stratégie des affaires à Lisbonne entre 1595 et 1607* (Paris, 1956), pp. 34, 65.
35. L. Berov, *Dvizhenieto na tsenite na Balkanite prez XVI–XIX b. i Evropeiskata revolutsiia na tsenite* [Movement of prices in the Balkans during the sixteenth to nineteenth centuries and the European price revolution] (Sofia, 1976), pp. 243–4.

to the shores of the empire. But even the inland towns of the peninsula had their resident or travelling Frank merchants whose stock-in-trade was often cloth.[36]

During the sixteenth century Italy was the most important source of foreign textiles in Ottoman markets. The major Italian penetration in all likelihood did not occur until the second half of the sixteenth century when Venetian production took off. As Italian production increased a division of the Balkan market evolved. The Ottoman industry produced kersies and less expensive broadcloth while the Italians produced luxury cloths. Both Italian and Ottoman producers operated within similar price constraints since both depended on Ottoman raw wool, the cheapest Mediterranean source. Although English cloth was occasionally to be found, it did not make a major impact until after the 1580s with the founding of the Levant Company and the beginning of direct export to the eastern Mediterranean. Towards the end of the sixteenth century, the English started to flood the Levant with their textiles. The ability of the English to sell large quantities of their cloth to the Ottoman Empire stemmed from three advantages they had over their Mediterranean competitors. First, the size of their industry allowed economies of scale. Second, the price of their wool rose at a lower rate. Third, the structure of the Levant Company facilitated the dumping of English cloth.[37]

The difficulties confronting Salonica's textile manufacturers clearly grew in the course of the sixteenth and seventeenth centuries. In response they developed a number of strategies which involved the ever greater intervention of the Ottoman government and progressively reduced the commercial viability of the industry.

Initially the cloth produced in Salonica was marketed throughout the Ottoman Empire, but the single most important customer was the government itself. From the beginning the Abode of Felicity had a clear interest in the production of good quality woollens to clothe the élite of the Ottoman military, the Janissary corps. Previously it had to depend on the alien and hostile producers of Christendom. Now the corps had a dependable supplier of winter garments. One of the garments made from Salonica broadcloth was a winter cloak whose thick and felted surface protected the Janissary from wind and rain. It may very well have been that this well-documented connection with the Ottoman military became the

36. D. Shopova, *Makedonija vo xvi i xvii vek, Dokumenti od Carigradskite Arhivi* [Macedonia during the sixteenth and seventeenth centuries: Documents from the Constantinople Archives] (Skopje, 1955), pp. 11–12, 28, 29.

37. For a parallel case from Bursa see M. Cizakca, 'Price History and the Bursa Silk Industry: A Study in Ottoman Industrial Decline, 1550–1650', *Journal of Economic History*, 40 (1980), 534, 545–6; repr. in Islamoğlu-İnan (ed.), *The Ottoman Empire and the World Economy* (Cambridge–Paris, 1987), pp. 248, 256, 430, 433, 434.

basis for the unsubstantiated claims that Spanish Jews were making all sorts of weapons for them as well.

Despite this interest however, the Ottoman government did not initally maintain a complex bureaucracy to supervise the cloth industry. Certain policies suggest that it treated the industry as an independent commercial enterprise, not as an adjunct to the military supply system. Thus in 1510–11 they paid for the cloth outright.[38] The producers functioned as independent entrepeneurs. Prior to the mid-sixteenth century, Ottoman contact with the cloth producers was one among the many miscellaneous duties entrusted to the factotum of Ottoman administration, the local kadi. But in the middle of the century, in response to the very real economic pressures which the producers did experience, particularly in the supply of raw wool and in the supply of cloth to the government, a military official was appointed as superintendent of the cloth and this responsibility was removed from the local kadi. This marked the beginning of a process whereby the cloth industry ceased to be an independent commercial enterprise and became instead part and parcel of the military supply system. A century later this process culminated in a dramatic change in the organization of cloth production. In 1664, with the approval of the head of the Janissary corps, the imperial treasurer, the grand vezir, and the sultan himself, all of the individual artisans and workshops were gathered together into one work-house in the Haghia Sofia district of the city. Not only was their place of work to be determined by the corps but also their place of residence. The Jews who made cloth for the Janissaries were to live in the Citadel. No longer were they subject to the normal civil authorities: now they were directly under the authority of the Janissary corps itself.[39]

Of course the Janissary corps and the Ottoman government as a whole, which now acted as the protector of the industry, had also played a role in helping to bring it down. In response to the crisis of the mid-sixteenth century, the Salonica textile producers arranged to provide cloth not for sale to the corps on the open market but as cloth levy for the corps in lieu of certain taxes. Initially this arrangement was advantageous to both sides, but as the corps expanded in size, Ottoman demands on the industry rose. During the decade 1580 to 1590 there was a doubling of the cloth needs of the corps and a tripling of the price of raw wool. Although there is evidence that the Ottoman government at least partially compensated Salonica producers for this cloth, it is clear that the compensation did not keep pace with price rises.

38. Halil Sahillioğlu, 'Yeniçeri Çuhası ve II. Bayezid'in son yıllarında Yeniçeri Çuha muhasebesi' [Janissary broadcloth and janissary broadcloth accounts during the last years of Bayezit II], *Güney-Doğu Avrupa Araştırmaları Dergisi*, 2–3 (1973–74), 425.

39. Uzunçarşılı, *Osmanlı Devleti*, I, pp. 273, 276.

The climax of the crisis came in 1637 on the occasion of the delivery of Salonica cloth to Constantinople. The Ottoman government ruled that the quality was unacceptable. Rabbi Judah Covo, the leader of the delegation, negotiated with officials to effect a compromise. After several months of fruitless discussion the government decided to act. Judah Covo was executed for his economic crime. As the deed was carried out the public crier proclaimed, 'This is one who has adulterated cloth.'[40]

Thus in the first half of the seventeenth century the Salonica industry confronted three major challenges: 1) continuing steep increases in the price of wool; 2) increasingly stiff competition from English cloth imports; and 3) the special demands of the Janissary corps. It is clear that while the last was the most dramatic and tragic, in economic terms it was probably the least important source of difficulty. Other textile-producing centres in the Balkans, which did not have to face that demand do not seem to have been any more successful.

Evidence that Salonica was no longer a commercially significant producer of cloth for the Ottoman market by the 1630s emerges from another source as well. Although we have no *sicillat* for Salonica in this period, the records for Edirne have survived. One important category of court records is particularly valuable for our purpose: the *tereke* of the *askeri* class, which are the inquisitions post-mortem compiled by the local kadi for the bureaucratic-military class. For Edirne they have been surveyed by the distinguished Turkish historian, Ömer Lûtfi Barkan, for the period 1549–1659.[41] The following table, based on his published research, illustrates the process by which European (and particularly English) cloth replaced Salonica cloth in the wardrobes of the wealthier classes of the city, those who in fact constituted the major market for the product.

Cloth in the Wardrobes of the Edirne *Askeri* Class

Year	*Origin*	*Type*
1548	Salonica	Broadcloth
1553	Salonica	Broadcloth London-style
1553	Salonica	Kersey
1553	?	Red broadcloth
1554	Salonica	Kersey
1604	?	Broadcloth

40. Hayyim Shabbetay, *Torat Ḥayyim*, sect. I, question 94.

41. Ömer Lûtfi Barkan, 'Edirne Askeri Kassamına âit Tereke Defterleri, 1549–1659' [Inquisitions post-mortem of the Edirne military-bureaucratic class, 1549–1659], *Belgeler*, 3 (1966), 91, 120, 127, 129, 225, 265, 270, 325, 331, 375, 382, 427–8, 430–31.

1605	?	Navy-blue broadcloth
1624	Salonica	Broadcloth
1624	Salonica	Broadcloth
1624	?	Violet broadcloth
1624	?	Violet broadcloth
1624	London	Broadcloth
1624	London	Broadcloth
1637	Venice	Broadcloth
1637	?	Violet-purple-brown broadcloth
1649	London	Blue broadcloth
1649	London	Broadcloth
1655	?	Blue broadcloth
1656	Edirne	Broadcloth

What is remarkable about this evidence is that it demonstrates that Salonica cloth was still significant until the 1620s, suggesting that despite the challenges the industry confronted it remained commercially viable. This conclusion is generally confirmed by an entirely different contemporary source, *The Merchants Mappe of Commerce; wherein the Vniversell Manner and Matter of Trade is compendiously handled*, a practical treatise on international trade published in London in 1638, making it one of the earliest such works. The distinction it garnered insured that it was reprinted at least twice. Its author, Lewis Roberts, was both an experienced merchant in the Ottoman Empire and one of the pioneers in the study of foreign commerce. In 1617, at the age of 21, he entered the service of the East India Company and later the Levant Company and remained continuously active in the pursuit and study of trade until his death in 1640. His family connections mirrored this interest. His wife was the daughter of a cloth merchant. His eldest son became a high official of the Africa Company and his eldest daughter married an English merchant in the Levant. This was not an ill-informed traveller eager to regale an audience with sensational half-truths from the lands of the infidel. Clearly he was a man who lived and breathed England's trade with the East. Thus his judgements deserve attention not only because of his knowledge and experience but also because of the respect that his work continued to command for decades after his death. Unfortunately, despite its value, the book is now neglected by students of Mediterranean commerce.

Roberts devoted a brief chapter to Salonica, describing it in the following terms (note that I have preserved the original spelling and punctuation):

> ... now there rests a word of the trade of this ancient and famous citie of *Salonica*, anciently called *Thessalonica*; to the inhabitants whereof

> Saint *Paul* write one of his Epistles: it is yet a rich and large Citie, and the residence of the *Sangiac* of *Macedonia* under the *Grand Signiour*. The present inhabitants are *Greekes*, *Turkes*, and principally *Iewes*; who are here found to be very rich and eminent Merchants, 80 Synagogues of them being accounted to bee in this towns employing themselves in severall Arts and Merchandising. It is seated in the bottome of a gulphe called by the Cities name; and by the demurre that happened in the *English* trade to *Turkie* some years past, these *Iewes* and inhabitants, and some *Moores* banished out of Spaine, have here set up some *Loomes*, and made cloth, in imitation of our *English* Suffolke clothes, which hath proved a great detriment to the sale of *Hampshire* kersies, once vented in great quantities generally throughout *Turkie* and especially in these parts; besides which sort of cloth now here made, and hence taking name, the place affoordeth the generall commodities of the *Morea* as *pouder of Berry Orgrame* for Diers, *Wools*, *Cottons*, *Wax*, *Hony*, *Cordovants*, *Aniseeds*, and the like.[42]

Thus Roberts, who had every reason to be well-informed on the subject, claimed that in the period some years before 1638, the Salonica industry was still strong enough to effect adversely the sale of English kersies. Furthermore he adds a fact mentioned nowhere else, namely that among the Salonica weavers were not only Jews but also 'Moores banished out of Spain', presumably the *moriscos* expelled in 1609. Unfortunately it is impossible to date with precision the period to which Roberts is alluding, but it must have been some point after the establishment of the Levant Company in the 1580s, when major English cloth exports began, and before 1638 when the book appeared. I suggest that this was probably the second decade of the seventeenth century. During the last decades of the sixteenth century English kersey exports to the Levant were still quite strong, but during the 1620s broadcloth became the major commodity, replacing the smaller and cheaper kersies. Presumably it was because of the Salonican competition that this shift took place. Roberts' reference to the expulsion of the *moriscos* is another indication that the decade of their arrival in the East was the period to which he was alluding. However it is also clear that this 'demurre' was something 'that happened ... some years past'. Clearly, by the 1630s Roberts and his fellow English merchants no longer feared the competition of their Salonican rivals.

* * * * * * * * * * * * * *

42. Lewis Roberts, *The Merchants Mappe of Commerce; wherein the Vniversell Manner and Matter of Trade is compendiously handled* (London, 1638), Ch. CCXLVI, p. 181 [should be 189] — Please note pagination is not continuous, both chapter and page number are necessary.

The woollen textile industry in Salonica, established by exiles from the Christian Mediterranean, for more than a century with skill and ingenuity struggled to overcome a variety of internal and external challenges. Although it managed to achieve a number of remarkable successes, it failed to survive as a commercially viable enterprise after the advent of the Levant Company.

Spanish Ladies from the Cairo Geniza

JOEL L. KRAEMER

In his lively portrayal of sixteenth-century Jerusalem, the Portuguese traveller Fr. Pantaleão de Aveiro mentions a visit to the area of the Siloam pool.[1] He describes a large edifice, having ample water, where women — Christian, Muslim, and Jewish — do their laundry and bathe. There, he tells, he and two other monks met two women, one old and the other young, accompanied by a little girl. De Aveiro and his companions disregarded them and went to the building to see what was there. The women inside began to shout angrily in Arabic (which the monks did not understand) not to enter. Realizing that women were inside, the monks became flustered, and returned to those on the outside. When the young lady noticed their predicament, she addressed them in speech 'half Italian and half Spanish', from which de Aveiro gathered that she was Jewish. He replied in Castilian — this delighted her, he says — asking if she was Jewish and where she was born. She answered that she was a Jewess and had been born in Rosetta, Egypt, where Jewish refugees from Castile and Portugal lived, mostly merchants, and that her parents were originally from Toledo.[2] The young lady enjoyed their conversation, as de Aveiro understood Castilian well — a language which he says she had studied and knew best.

See p. 267 for manuscripts cited in the article.

1. António Baião (ed.), *Itinerario da Terra Sancta e suas Particularidades* (Coimbra, 1927), pp. 284–6; see M. Ish-Shalom, *Christian Travels in the Holy Land* (Tel Aviv, 1979), pp. 293–4 (Hebrew). Pantaleão de Aveiro, a Portuguese traveller, was a Franciscan priest from the Algarves. He accompained Bonificio de Aragosa on his Journey to Palestine at the head of a group of 60 monks, at the time of Pope Pius IV, dedicating his book to Archbishop Don Miguel de Castro of Lisbon. His visit took place some time around 1565. See N. Slousch, 'The Population of Palestine and Syria in the 16th Century (according to a Portugal Source)'. *Bulletin of the Jewish Palestine Exploration Society* (1934–35), 323–8 (Hebrew); J. Braslavsky, *Studies in Our Country: Its Past and Remains* (Tel Aviv, 1954), p. 198 (Hebrew); Ish Shalom, *Christian Travels*, pp. 25–6.
2. The Jewish population of Rosetta (Rashîd) was greatly increased by the influx of Iberian refugees in the sixteenth century, and it ranked just after Cairo and

She tried to convince de Aveiro to convert to Judaism, uttering blasphemies against the Christian religion, 'exactly as she had learned in the house[3] of her parents'. For Jews, he says, 'teach nothing else to their children but to curse the dulcet name of Jesus and to hate the name Christian'. She then used courteous language, encouraging him to convert by promising that the Jews would marry him to a beautiful and wealthy Jewess and give him money. In response, de Aveiro pointed out her poverty and misery. Though genteel, he said, she washed laundry at the spring, barefoot like a maid *(escrava)*; but if she were a Christian lady, in Spain, she could live as a princess like other New Christians.

Upon hearing this, she began to weep, and admitted that her parents told her these and similar things, though not regretting that they left their birthplace. When the old woman noticed that the conversation was going on and on, without the young woman seeming to object, she intervened as she heard her weeping. She shouted at Pantaleão de Aveiro and his friends to leave the area.

This poignant account gives us an *aperçu* into the mind and predicament of a Spanish lady, a daughter of exiles, in the mid-sixteenth century. Though born in Egypt, she knows Castilian, a language in which she had, in fact, been instructed.[4] Her attachment to Castilian is touching, a fine example of the language (and cultural) loyalty of Spanish Jews.[5] She expresses the colossal loss and tragedy of post-Expulsion Jews and their profound ambivalence towards their former homeland. She detests the Christian religion, but she loves the Castilian language, and recalls her parents' wistful nostalgia for Spain and quiet despair in their new home. Her attempt to convert de Aveiro is quite remarkable and signifies not only a powerful commitment to Judaism but also a sense of safety and security in her environment, which was predominantly Muslim.

Alexandria in renown. It was then a major port for commerce between Egypt and Turkey. On the settlement of Spanish exiles in Rosetta, see Joseph Sambari, *Sefer Divre Yosef*, MS Paris AIU H 130A, fol. 129b (in the offset edition of S. Shtober [Jerusalem, 1981], 254) (=ed. Ad. Neubauer, *Medieval Jewish Chronicles* [Oxford, 1887], Vol. I, p. 156); E. Strauss-Ashtor, *History of the Jews in Egypt and Syria under the Rule of the Mamlūks* (Jerusalem, 1944–70), Vol. II, p. 445 (Hebrew); A. David, 'Egyptian Jewish Settlements from the 16th Century to the 18th Century', in J. Landau (ed.), *The Jews in Ottoman Egypt (1517–1914)* (Jerusalem, 1988), p. 26 (Hebrew).

3. The word is *escola*, lit. 'school'.
4. She must have known Arabic and perhaps some Hebrew. And she knew Italian, if the author's description of her first remark is accurate.
5. See Y.H. Yerushalmi, 'Castillian, Portuguese, Ladino: The non-Hebrew Literatures of Sephardi Jewry', in Z. Ankori (ed.), *Then and Now: Annual Lectures on the Jews of Greece* (1977–1983), Tel Aviv, 1984, pp. 35–53 (Hebrew). Yerushalmi stresses that Spanish Jews who remained in the West spoke Castilian and those in the East spoke Judaeo-Spanish.

Pantaleão de Aveiro reports similar feelings of despair among Jewesses, witnessed during his visit to Safed.[6] There, he and his colleagues lodged with a Jew from de Aveiro's home town, a man with whom he had gone to school as a youth. The school had been administered by still another Jew. His host treated him with great courtesy, proud to have him as a guest. That night, some Jews came to argue with him about their faith, but he declined on the pretext of exhaustion. The same night, he says, some Portuguese Jewesses came to visit him. They had tears in their eyes, lamenting that their sins had caused them to leave Portugal, not for the Promised Land, however, as they had thought, but to a land of despair *(terra da desesperação)*, as they witnessed with their eyes and experienced in their misery.[7]

Pantaleão de Aveiro's visit to Tiberias features another Iberian lady — this time a quite famous one — Doña Gracia, who had escaped from Portugal with great wealth.[8] As he puts it, she had purchased the town of Tiberias from the Sultan *(grao turco)* for a yearly payment and was about to settle there with her family, along with all Jews who wished to come with her. The news overwhelmed the Jews living then in Palestine, who believed that her settling there augured the advent of the Messiah. He tells that Doña Gracia escaped from Portugal along with her sister, both widows, emigrating to Venice, where they lived for some years, and from there she travelled to Istanbul with all her wealth. The family name, de Aveiro notes, was Luna. In Istanbul, Doña Gracia engaged in commerce, especially overseas trade and ship building, through the intermediacy of her agents.[9] Doña Gracia's sister lent out money on interest to the Republic of Venice. The sister, lacking faith in the advent of the Messiah, converted to Christianity.

De Aveiro goes on to tell that Doña Gracia's sister had a very beautiful daughter who inherited her wealth, and had been secretly married in Venice to a New Christian, who served the Marquis de Villareale.[10] This New Christian brought her stealthily at night, with her consent, from the house where she lived to Ancona, and from there to Rome, where he got the patronage of the Ottoman ambassador. The author adds that Doña Gracia was so powerful that the Jews called her simply 'Seniora'.

6. Baião (ed.), *Itinerario*, pp. 482–4; Ish Shalom, *Christian Travels*, pp. 299–300.
7. On the despair of the exiles, see R. Hacker, 'Pride and Depression — Polarity of the Spiritual and Social Experience of the Iberian Exiles in the Ottoman Empire', in M. Ben-Sasson *et al.* (ed.), *Culture and Society in Medieval Jewry: Studies Dedicated to the Memory of Haim Hillel Ben Sasson* (Jerusalem, 1989), pp. 541–86 (Hebrew).
8. Baião (ed.), *Itinerario*, pp. 471–4; Ish Shalom, *Christian Travels*, pp. 296–8.
9. Here the author tells us that his informant was a Portuguese Jew from Lisbon, a close friend of Doña Gracia, who had come from Istanbul to Portugal to visit his relatives.
10. This was, of course, Doña Gracia's nephew, the famous Joseph Nasi.

The story of Doña Gracia Mendes, or Beatrice Benveniste (or de Luna), and Don Joseph Nasi, illustrious members of the 'Spanish-Marrano diaspora', has been often told.[11] Pantaleão de Aveiro gives the general lines fairly accurately.

De Aveiro often portrays Jews, in particular those of Spanish and Portuguese origin. He was one of a number of Portuguese travellers in the sixteenth century who took a keen interest in them, and particularly in *conversos* of Jewish descent. Some of these travellers were themselves *conversos* or descendants of *conversos* who tried to bring salvation to Jews who held steadfast to their faith.[12] It is possible that de Aveiro was himself of *converso* descent.[13] If so, his posture as a devout and militant Catholic and his anti-Jewish motifs take on a different hue.

Of particular interest to us are the numerous accounts of exiled Iberian Jewesses he met, not only in Palestine, but also in his travels in the eastern Mediterranean. In Paphos (Cyprus), he met Jewish refugees from Portugal, and describes their miraculous escape and their mood.[14] Their ship had been wrecked near Cyprus; Greeks fell upon those who survived, killing and pillaging them mercilessly. Among the survivors was a Jewess, formerly a New Christian from Coimbra, with her two young children. She had fled so as to re-embrace her ancestral faith. In dire straits, aboard the ship, she had taken an oath to return to the Catholic religion, 'and the Lord (Jesus) answered her prayers so she was saved'. But she was ungrateful to

11. For a good summary and sources, see S. W. Baron, *A Social and Religious History of the Jews*, Vol. 18 (New York, 1983), pp. 74–86; and see C. Roth, *The House of Nasi: The Duke of Naxos* (Philadelphia, 1948); P. Grunebaum-Ballin, *Joseph Naci: Duc de Naxos* (Paris–The Hague, 1968); and B. Ravid, 'Money, Love and Power Politics in Sixteenth Century Venice: The Perpetual Banishment and Subsequent Pardon of Joseph Nasi', in *Italia Judaica* (Atti del I Convegno internazionale, Bari, 18–22 maggio 1981) (Rome, 1983), pp. 159–81, especially p. 159, n. 1, for bibliography (my thanks to Benjamin Arbel for orientation). It should be noted that Doña Gracia was very active in the Sultan's court, which explains the lease of land in Tiberias and its vicinity, and the 'proto-Zionist' experiment of Jewish settlement there. A good account is given by Baron, in *Social and Religious History*, Vol. 18, pp. 109ff. The experiment rested partly on the burgeoning textile industry in the area. Joseph Nasi imported Spanish wool and perhaps Merino sheep. The synagogue Della Señora in Istanbul was named after her. Doña Gracia was also a patron of scholars. Samuel Usque dedicated his famous *Consolations for the Tribulations of Israel* to her. See the translation from the Portuguese by M.A. Cohen (Philadelphia, 1977), p. 37
12. Slousch, 'The Population of Palestine', 323–4, mentions among these Portuguese travellers Gorge Henriquez, who wrote *Itinerario de veagem a Terra Santa* (1561–62); Fr. Francisco Guerreiro, *Itinerario* (Lisbon, 1774) (1586); and Juan di Jesus Christo, who wrote a travel book on Jerusalem in 1580, published in Lisbon in 1831.
13. The suggestion was made by Slousch, 'The Population of Palestine', 324.
14. Baião (ed.), *Itinerario*, pp. 54–5.

God, de Aveiro adds, for on his return trip with his companions, he came across her in Tripoli and found that she had remained a Jewess. When he rebuked her, she replied that God had saved her because she had been a good Jewess her whole life.

At Salines (modern Larnaca in Cyprus) he met a Portuguese Jewish woman married to a Jewish physician named Don Joseph, with whom she escaped from Lisbon. He was her second husband (her first may have converted to Christianity). She asked Pantaleão de Aveiro to inquire after her brother in Portugal — a wealthy merchant and a Christian — whom she loved greatly. She wanted him to send notice of her brother's welfare with a Venetian ship from Lisbon.[15]

Pantaleão de Aveiro's *Itinerario* is a rich source for knowledge of Iberian Jewesses in the post-Expulsion era. His vivid eyewitness descriptions provide valuable information about the mood and mentality of these women, exiled from their original homes. Though we need not accept his literary narrative at face value, and may ascribe some features to the sentiments and artifice of the author, his accounts may still be taken as precious evidence that may be used (with due caution) in recreating the socio-cultural ambience of these ladies.

The loyalty of Jewish women to their native religion, which Pantaleão de Aveiro took the trouble to portray more than once, is a well-attested phenomenon during the Spanish persecutions in 1391 and throughout the fifteenth century up to the time of the Expulsion in 1492.[16]

Y. Baer suggests that the reason for the greater loyalty of women, was that they were people of simple faith, not cosmopolitan and sophisticated intellectuals. The simple of heart rather than the philosophers (the Jewish Averroists), according to Baer's thesis, preserved the tradition at the time of the Expulsion.[17]

15. Ibid., pp. 64–5. On the Jews in Cyprus during this period, see B. Arbel, 'The Jews in Cyprus: New Evidence from the Venetian Period', *Jewish Social Studies*, 41 (1978), 23–40.
16. See Y. Baer, *A History of the Jews in Christian Spain*, trans. L. Schoffman (New York, 1978), Vol. II, p. 103. About Jewish women's loyalty to their religious creed, see also ibid., pp. 131–2.
17. See Baer, *History*, Vol. II, p. 509. He cites a well-known text from R. Joseph Jabetz, *Or ha-Ḥayyim*, Ch. 5: 'For it was the Spanish women who persuades their husbands to martyr themselves for God, blessed be He, while those who prided themselves on their wisdom exchanged their glory on the bitter day; and this is mighty proof that had they not sought after wisdom, but remained among the simple, their simplicity would have saved them, for God preserveth the simple'. It is significant that Baer returns to this passage on the last page of his chapter on the Expulsion (which is, in fact, the last in the book before the appendix), stressing the verse from Psalms 116:6, 'God preserveth the simple'. Baer's thesis that the Jewish rationalists had weakened the hold of religion on the faithful cannot be discussed here.

Another explanation for the loyalty of women is offered by Mark Meyerson in his 'Aragonese and Catalan Jewish Converts at the Time of the Expulsion'.[18] He gives three cases from the petitions of recently converted Jewish men. Their wives and children had proved unwilling to convert. The petitioners were ready to embrace Christianity, it is suggested, because they wished to preserve their property while they took refuge in Navarre. The conversion of some but not all members of families, writes Meyerson, had a powerful impact on them; for the older and younger generations, siblings, and marriage partners were split by the baptism of some and the allegiance of others. These cases 'suggest a pattern of resistance to conversion on the part of Jewish women at the time of the Expulsion', according to Meyerson, who notes that this pattern corresponds well with the prominent educational role played by *converso* women in crypto-Jewish society. For women were mainly responsible for education and the transmission of religious values in the home, and as such they became even more active in this role in *converso* society when formal institutions of learning broke down.[19] And Meyerson concludes that Jewish women were reluctant to convert in 1492 because of their intense feeling of responsibility for preserving tradition by educating their children, so that baptism meant for them both leaving Judaism and forsaking their maternal role. As the maternal bond was strong, we find in the cases cited that the children emulated their mothers. In sum, these women contributed heavily, as mothers and educators, to the maintenance of group identity.

18. The article was submitted for publication in the *Frank Talmage Memorial Volume* (Haifa University Press). Meyerson kindly sent me a typescript copy of his splendid study.
19. Meyerson cites R. Levine Melammed, 'The Ultimate Challenge: Safeguarding the Crypto-Judaic Heritage', *Proceedings of the American Academy of Jewish Research*, 53 (1986), 91–109. The author used records of trials of the Inquisitorial Court of Toledo, preserved in the Archivo Histórico Nacional in Madrid, and covered the trials of 111 women accused of judaizing between 1492 and 1520. Levine Melammed actually states that 'more women, namely, *conversas*, were brought before the Inquisition as *judaizantes* than were men' (92, n. 1). Surveying Haim Beinart's *Records of the Trials of the Spanish Inquisition in Ciudad Real* (Jerusalem, 1974–1985), one has the distinct impression that women appear in disproportionately great numbers before the tribunal. See also his *Conversos on Trial: The Inquisition in Ciudad Real* (Jerusalem, 1981), pp. 226, 266–8, 275. Cf. A. Meyuhas Ginio, 'Women in the *Converso* Community of Fifteenth Century Spain, *Pe'amim*, 46–7(1991), pp. 170–89(Hebrew). It may be added that a similar phenomenon existed among *morisco* women and for kindred reasons; they preserved Muslim tradition in the realm of the home. See, for example, J. Fournel-Guérin, 'La femme morisque en Aragon', in *Les Morisques et leur temps* (Table Ronde Internationale, 4-7, 1981, Montpellier) (Paris, 1983), pp. 523–36. I am grateful to Dr. Reuven Shaffer for this reference.

Our knowledge of Spanish Jewesses in the post-Expulsion period, in the area of Egypt and Palestine, is fairly limited. It is based largely upon rabbinic responsa and rescripts *(taqannot)*. There is also the *sijill* of the Jerusalem court, which has been studied and intensively utilized by Amnon Cohen.[20]

In this study I propose to use an additional source of information on Iberian Jewish ladies of this period, namely, the Cairo Geniza. The manuscripts of the Geniza have been used extensively for the study of Mediterranean Jewish history in what S.D. Goitein has called the classical period, which extends from c. 1000 to 1250. The vast bulk of surviving documents are from this period. And there are also documents from the Mamlūk age (1250–1516), used by E. Strauss-Ashtor in his *History of the Jews in Egypt and Syria under the Rule of the Mamlūks*. Far fewer Geniza manuscripts are from the early Ottoman period, which will occupy us here.[21] From this fount, Ch. Turniansky published seven sixteenth-century letters in Yiddish, six written by a woman, Rachel Sussman of Prague, and one by her son to her.[22] These letters, contemporary to the women's letters to be discussed below, contain splendid comparative data. Avraham David has published several women's letters from the Geniza, some of Spanish background, and has very generously given me access to many others from his corpus of Geniza documents from this period, to be published in a volume in the series Oriens Judaicus.[23] Eleazar Gutwirth has treated a number of women's letters from the Geniza which were written in Judaeo-Spanish, and is continuing to study this material.[24]

20. A. Cohen, *Jewish Life under Islam: Jerusalem in the Sixteenth Century* (Cambridge, MA–London, 1984). See also M. Rozen, *The Jewish Community of Jerusalem in the Seventeenth Century* (Tel Aviv, 1985) (Hebrew), which deals with a later period but is also of value for the previous century. For details on women, see Cohen, *Jewish Life*, pp. 127ff.; Rozen, *The Jewish Community*, pp. 239ff. The responsa of R. David Ibn Abi Zimra (Radbaz) are particularly valuable for the cultural, religious, and socio-economic life in general in Egypt and Palestine in the sixteenth century; see, for example, the material assembled by I.M. Goldman, *The Life and Times of Rabbi David Ibn Abi Zimra* (New York, 1970).
21. The dating of the documents to the sixteenth century (unless otherwise noted) is based on handwriting, names of personages mentioned and *realia* alluded to (such as coins, garments, etc.).
22. Ch. Turniansky, 'A Correspondence in Yiddish from Jerusalem, Dating from the 1550s', *Shalem*, 4 (1984), 130–208 (Hebrew).
23. See especially, A. David, 'New Geniza Documents on the History of Relations between the Jews of Eretz Israel and the Jews of Egypt in the 16th and 17th Centuries', *Cathedra*, 59 (1991), 19–55 (Hebrew). Avraham David kindly read an earlier draft of this paper and made helpful comments.
24. See 'A Judeo-Spanish Letter from the Genizah', *Judeo-Romance Languages*, ed. I. Benabu and J. Sermoneta (Jerusalem, 1985), pp. 127–38, and 'The Family in Judeo-Spanish Genizah Letters of Cairo (XVIth–XVIIIth C.)', *Vierteljahrschrift für Sozial- und Wirtschaftsgeschichte*, 73(1986), 210–15. Ruth Lamdan is working on this material for her Tel Aviv Ph.D. dissertation.

These Geniza documents are of great value, as they are the only ones in which women actually speak for themselves. Of course, they often dictated their letters to scribes or to relatives or friends, depending on their level of literacy, and in such cases there may have been interference and stylization.[25] However, scribal copies may be of great value as when they preserve the actual speech of the woman as she dictated, mistakes and all.

These letters of Spanish ladies are included in a corpus I am preparing consisting of some 180 women's letters from the Cairo Geniza. The letters in this corpus are from the tenth to the seventeenth century, from all parts of the Mediterranean; and they are written in Judaeo-Arabic, Hebrew, Judaeo-Spanish, and Yiddish. They were sent by and to women of all social levels — wealthy merchant women and poor beggars; orphans, widows, mothers, sisters, daughters, grandmothers, and so on. There are love letters and business letters, letters of complaint, invitations to weddings, petitions, appeals, and wills.

The letters considered here are mainly in Hebrew; some are in Judaeo-Spanish; one is in Judaeo-Arabic. The most dramatic linguistic change from the classical and the Mamlūk period of the Geniza to the sixteenth century is the increasing predominance of Hebrew, the steady decline of Arabic, and the use of Judaeo-Spanish. This is obviously due to the large-scale immigration of Spanish Jews and an increase in the Ashkenazi community.

All documents, unless otherwise noted, are from the sixteenth century, that is, the period following the Expulsions from Spain and Portugal.

The main value of these letters is their contribution to understanding the socio-economic status of women, and more broadly the family structure, household, and demography of Iberian Jews in this period. They also contain rich information about cultural and social *realia*. And they give information on the onomasticon of Jewish women of the time.

These letters complement our knowledge drawn from the sources noted above. They do not change the picture appreciably but supplement and augment our perspective. They also bring to life the women who left these documents for posterity and are mentioned in them.

25. When women are recorded as giving testimony in court, their actual language was often distorted, as we know from the responsa of Radbaz: 'I have seen scribes who, being wise in their own eyes, would either add, subtract or interpret of their own accord. I have seen it many times in instances where woman testified and where they gave their testimony in the vernacular [Spanish] or in Arabic, and often the translation of that testimony changed the complexion of the whole case. Be thou among the careful ones' — cited with slight changes from Goldman, *Life and Times*, p. 11, quoting Radbaz, *Responsa*, Vol. I, No. 98.

As a general thesis, if one is sought, it may be proposed that under the conditions that prevailed in the eastern Mediterranean following the Expulsion in 1492, with the breakdown and dispersion of families and the emerging role of Spanish women as educators and preservers of tradition, they come into their own. They achieve more of an individual identity and play a stronger role in the economic life of the society.

These ladies belonged to the communities of refugees from the Iberian Peninsula who arrived in waves on the shores of Egypt and Syro-Palestine from the fourteenth to the sixteenth century, many of them of *converso* origin.[26] Even before 1492, in the 1470s, *conversos* began to flee from Spain for Mediterranean countries; and by 1487 there were Spanish sages in Jerusalem, Spanish widows, and a number of repentants who were descendants of *conversos*. In 1492, soon after the Expulsion, Spanish exiles appeared in the ports of Egypt and Syro-Palestine, and from the end of the fifteenth century the fugitives became a steady stream. In the earlier centuries, from the thirteenth to the fourteenth, Spanish emigration to Palestine was scant. After the persecutions of 1391 emigration became a mass movement, which swelled to major proportions after 1492. Then, after the Ottoman conquest in 1516, Jewish settlement in Palestine expanded significantly. These waves of immigration were often accompanied by Messianic hopes, as in 1391, in the fifteenth century, and later after the Ottoman conquest in 1516. Many came to Jerusalem to repent their former conversion to Christianity.

The Spanish Jews who emigrated to Palestine found there local, native Jews, called *Musta'rabs*, and also Jews from North Africa *(Ma'aravīs)*.[27]

26. On the waves of emigration from the Iberian Peninsula to Egypt and Syro-Palestine, see especially Strauss-Ashtor, *History of the Jews*, Vol. II, pp. 441ff.; Baer, *History*, Vol. II, pp. 103, 158–9; M.Benayahu, 'A Document from the First Generation of Spanish Exiles in Safed', in M.D. Cassutto *et al.* (eds.), *Sefer Assaf* (Jerusalem, 1953), pp. 109–25 (Hebrew); B. Dinur, 'The Emigration from Spain to Eretz Yisrael after the Disorders of 1391', *Zion*, 32 (1967), pp. 161–74 (Hebrew); J. Hacker, 'The Connections of Spanish Jewry with Eretz-Israel between 1391 and 1492', *Shalem*, 1 (1974), 105–54 (Hebrew); 'Some Letters on the Expulsion of the Jews from Spain and Sicily', in E. Etkes and Y. Salmon (eds.), *Studies in the History of Jewish Society... Presented to Professor Jacob Katz* (Jerusalem, 1980), pp. 64–97 (Hebrew); A. David, 'Two 16th Century Emissaries' Letters from Jerusalem', *Shalem*, 3 (1981), 325–32 (Hebrew); J. Hacker, 'On the Intellectual Character and Self-Perception of Spanish Jewry in the Late Fifteenth Century', *Sefunot*, 2 (1983), 21–95 (Hebrew); 'Spiritual and Material Links between Egypt and Palestinian Jewry in the Sixteenth Century', in A. Cohen and G. Baer (eds.), *Egypt and Palestine: A Millennium of Association (864–1948)* (Jerusalem–New York, 1984), pp. 241–50.
27. On the three communities see Strauss-Ashtor, *History of the Jews*, Vol. II, pp. 493–9; A. Cohen and B. Lewis, *Population and Revenue in the Towns of Palestine in the 16th Century* (Princeton, 1978), pp. 156, 159; M. Rozen, 'The Position of

To be sure, there were also a number of Ashkenazi Jews. Very soon the Spanish newcomers came to dominate, their scholars gaining ascendancy in the community. By the beginning of the sixteenth century there were in Safed separate congregations for Spanish Jews, *Musta'rabs* and *Ma'aravīs*. Spanish Jews continued to speak Spanish, especially merchants, but gradually Spanish exiles learned Arabic.

FAMILY LETTERS

Most of the women's letters from the Geniza in the post-Expulsion era — as, in fact, previously — are family letters, and I shall treat them under the headings of family relationships.

A common theme in these letters, as in those of what Goitein calls the classical period of the Geniza, is the absence of the husband, the father of the family, and the related motifs of abandonment and of yearning by his wife and children.[28] This theme of separation and wistful longing continues in our period, which witnessed incredibly intensive travel, despite the manifold dangers of shipwreck, piracy, and other perils.

In the classical period, we often find that the *pater familias* has absconded to Syria; occasionally he has sailed as far away as India. The motives for leaving home and hearth were myriad: pursuing a livelihood, or simply escaping from family, a nagging wife, tormenting children, poverty, humility, and hardship. The constant threat of abandonment, widowhood, and divorce forced women to depend on their children, especially their sons, for security and support. The mother-son relationship was emotionally intense.[29] With lack of security (divorce, polygyny, abandonment or death of husband), women came to rely on their sons for assistance. For the same reason, women depended upon their brothers. Husbands come and go — they die, they divorce, they marry second wives — but a brother is always a brother.[30] But we also find that a brother

Musta'arabs [*sic*] in the Inter-Community Relationships in Eretz Israel from the End of the 15th Century to the End of the 17th Century', *Cathedra*, 17 (1981), 73–101 (Hebrew). And see A. David, 'The Jewish Settlement in Palestine at the Beginning of the Ottoman Empire (1517–1599)', in A. Carmel *et al.* (eds.), *The Jewish Settlement in Palestine 634–1881* (Wiesbaden, 1990), pp. 86–141, especially pp. 90–117.

28. Lovers of Shalom Aleichem will think of Menachem Mendel and his exchange of letters with Sheina Shendel
29. The dynamics of the mother-son relationship in Arab-Muslim society are similar. See A. Bouhdiba, *Sexuality in Islam*, trans. A. Sheridan (London, 1985) from *La Sexualité in Islam* (Paris, 1975); id, 'The Child and the Mother in Arab-Muslim Society', in L. Carl Brown and N. Itzkowitz (eds.), *Psychological Dimensions of Near Eastern Studies* (Princeton, 1977), pp. 126–41.
30. S.D. Goitein, *Mediterranean Society* (Berkeley–Los Angeles, 1967–88), Vol. III, p. 21. A Jewish woman from Yemen is cited: 'A husband — I can get; children — I can bear; but a noble brother — from where shall I get him?'

appeals for help to his more fortunate sister. And a younger brother shows deference to an older sister.[31] Many of these themes familiar from the classical period of the Geniza, which ended in the thirteenth century, recur in our letters of Spanish ladies in the sixteenth century, but the autonomy and self-reliance of these women surpassed that of women in earlier centuries who were more dependent and subservient, more vulnerable to their male relatives' domination and whims.

Husband–Wife

A certain Doña Sarah (written Soro) writes to her husband, a scribe named Solomon. He has left home ostensibly because of tax arrears. She implores him to return.[32] Ashtor dates our letter from the fifteenth century, from prior to the Expulsion. But there is no compelling reason for this dating; the letter may have been written at the end of the fifteenth century or at the beginning of the sixteenth. The fact that it was found in Fusṭāṭ indicates, according to Ashtor, that R. Solomon returned to Egypt. But there are other possibilities, namely, that it is a draft copy, or perhaps it was never sent.

The entire family, in fact, had come together to write the letter — the wife, daughters (Reina, Rachel, Rebecca), and a son-in-law Moses (husband of Rebecca), but it is written in the name of the wife. The family had not heard from Solomon, though they had written several times asking him to return home. They express their hope that he succeed with a certain physician, also named R. Solomon, in getting an exemption from taxes. They entreat the *pater familias* not to travel to Turkey as they heard he planned. 'By the life of the Lord of Hosts, if you do this thing, do not speak to us any longer. For you wish to do something that will cause all of society (*ha-'olam*) to look down upon us.' And the letter continues:

> You will provoke a quarrel between your son-in-law and your daughter, who is pregnant, and she will become upset and perchance suffer a miscarriage. You will also destroy the good fortune of Rachel, who has become a mature young lady, beautiful, good natured and modest. Everyone will say: 'Look at this man who is a decent and elderly scribe, who has abandoned his wife and daughters, and has

31. Goitein, *Mediterranean Society*, Vol. III, pp. 20–21.
32. The document is preserved in MS T-S 13 J 21.10, published by J. Mann, *The Jews in Egypt and in Palestine* (Oxford, 1920–22, repr. New York, 1970), Preface and Reader's Guide by S.D. Goitein), Vol. I, p. 242; Vol. II, p. 309; and translated by Kobler, *Letters of Jews throughout the Ages* (London, 1953), Vol. I, pp. 233–4, on the basis of the (somewhat erroneous) text in Mann; and see the comments of Strauss-Ashtor, *History of the Jews*, Vol. II, pp. 83–4.

> been absent for a number of years! Perhaps he has become deranged, for he has gone to a distant country'.

And his wife quotes Prov. 17:24: 'The eyes of a fool are on the ends of the earth.' This allusion to his defective mental capacity may be contrasted with the formal respect in the salutation: 'Great peace and speedy salvation to my teacher and lord, light of my eyes, crown of my head, my lord and husband, his eminence R. Solomon the scribe, may he live forever'.[33]

In another husband–wife letter, the father has taken off for Mocha in South Arabia, evidently on a business trip, and a daughter writes in the name of her mother as her clerk.[34] She is one of three sisters of the Perdonel family.[35] The father's name, Moses Perdonel, appears in the mail address on the reverse side of the document. A Jew of Granada named Isaac Perdonel is known to have received special permission from King Ferdinand and Queen Isabella to leave Spain with his immediate and extended family and possessions, by intervention of the former Muslim king of Granada. The Perdonel family was active in both Egypt and Palestine.

The girl's father, Goitein suggests, had presumably taught her the Hebrew script and Arabic phrases used in letters, as there was no son. She writes in Judaeo-Arabic but with errors caused by the influence of her (Spanish) mother-tongue. She had trouble pronouncing Arabic, and confused *tā'* and *ṭā'*, *kāf* and *qāf*, *alif* and *'ayn*, *sīn* and *ṣād*, and muddled long and short vowels. The letter also contains colloquialisms.[36] The sociolect of women's letters in general exhibits relatively more vernacular speech than men's letters. The letter serves as an example of the acquired Arabic speech of first or second generation Spanish exiles in Egypt, and wavers between concern for the father's health and requests for South Arabian textiles, gold threads, and other commodities. After each order the father is assured that his health is most important.

33. The combination of conventional, stylistic deference and frank ridicule is a motif found in Shalom Aleichem's portrayal of Sheina Shendel's letters to her absentee husband Menachem Mendel. So indeed is the fear of social embarrassment — what the '*olam* would say.
34. MS T-S 13 J 24.22. Goitein, *Mediterranean Society*, Vol. V, p. 222, dates the letter to the sixteenth century, when Mocha became prominent. It is included in his (projected) *India Book*. See also *Mediterranean Society*, Vol. III, pp. 240, 478. Mocha (Mukha) is a well-known port on the Red Sea in south-west Yemen, from where coffee (mocca) was shipped. A Jewish community existed there in the sixteenth century; see Zechariah al-Ẓahrī, *Sefer ha-Musar*, ed. Y. Ratzaby (Jerusalem, 1965), pp. 40, 285.
35 See Goitein, Vol. V, p. 568, n. 17, citing F. Baer, *Die Juden im christlichen Spanien*, Vol. II (Berlin, 1936), p. 413.
36. For instance the adverb *qawī* for 'very', *kitīr* ('much', 'often'), *takul* (for *ta'kul*).

'This letter to you is from the one who kisses your hands and feet, your daughter Najma (Star) ...from here in Fusṭāṭ'. She then alludes to a letter which had arrived from the father: 'We rejoiced in it and it took the place of seeing your dear face for us'.[37] After expressing concern about his health — he apparently wrote that he had been ill — she says, 'We only wish from you either the hyacinth fabric or whatever fine fabric you find for us in Mocha'. Again his health is mentioned.

She then delivers the message from her mother: 'My lady, my mother sends her best regards to you. And she says to you that you should hurry and come, for she has entered her month'.[38] The mother has heard that there is in Mocha very fine white cloth. Again she seeks news of his health, as he promised to send, to gladden her heart. One sister, named Clara, could not eat or drink because of her loneliness for the father — a common motif. Again, the writer implores the father to return as soon as possible: 'We hope for you only and for your well-being and health. Peace'. Then in a postscript she mentions her five pieces of fine cloth (or linen) he had sent. She has given three to Raḥma, presumably her sister. And she adds a request for another purchase.

The father was evidently a merchant. He did not trade in the eastern Mediterranean area, as was more common, but rather in far-flung South Arabia. His wife and daughters were probably not ordering gifts but were involved (along with the father) in textile manufacturing on a family scale.

Absentee husbands are the rule, but we also have cases in which the wife is away. Wives also left home! In a Judaeo-Spanish letter a certain R. Jacob writes, complaining that he is blind and has no one to look after him, so it is easy to deceive him. 'I am afraid that what happened to me will happen again'. And he asks his wife to come to Cairo, 'for it is better that a woman who has sight take care of me'. 'Twice I decided to take a ship', he says, 'but did not find a companion'. He wishes that his wife would sell her clothing and jewellery to sustain him.[39]

Mother–Son

The dependence of mothers upon their sons, mentioned above, is attested in women's letters from the sixteenth century. But we also find active

37. As Goitein notes, these were Arabic phrases that were in current use in the classical period.
38. She wants to say that her mother is pregnant. She did not know, or at least did not write, which month of pregnancy it was.
39. This is a Judaeo-Spanish letter preserved in MS T-S Ar. 29.105, cited by Gutwirth, 'The Family', 214. Gutwirth observes that letters written by husbands to wives are not very common in this era, not even in Hebrew. He suggests that the wife lived in Palestine. If so, how did the letter wind up in the Geniza in Fusṭāṭ? Did she return home?

involvement of mothers in their sons' business affairs, as in the next letter.

The mother opens with felicities and then continues with an item of business.[40] The son was away from home, and his wife had remained with his mother. He was presumably in Cairo; the mother may have been in Palestine. The document is torn at the top.

She first sends greetings from his dear sister Reina,[41] and also from a woman named Sol, 'most blessed of women in tents';[42] and the mother adds that 'your wife[43] kisses your hands, as do Ḥabīb and Esther, Isaac and Rivah'. She then sends regards to 'dear Doña Miriam' and to her son and grandchildren.

Then the mother comes to the main point of the letter. She tells her son that a certain R. Abba had delivered some merchandise, including Candia wine.[44] She ends with typical maternal concern: 'I entreat you not to forget to write to us a letter with anyone who comes. Best wishes.[45] Your mother Esther'.

This is a family letter, but as in other cases, as the letter to Mocha, family letters often treated business matters. Here, the son went to Egypt for trade, and his mother reports on a commercial transaction that took place in his absence.

40. MS T-S NS 338.51.
41. The name Reina — common in this period — is written Rana (R'nh).
42. The phrase is from Judges 5:24, where it refers to Jael, wife of Heber the Kenite. It is almost *de rigueur* in the epistolary etiquette of the milieu, mostly abbreviated as here.
43. The expression is *newat betkha* (literally 'housewife' or the like); see Psalms 68:13. It is used consistently as an expression for 'wife' in these texts. In the classical period the word *ohel* ('tent', etc.) or *bayit* ('house', etc.) was used. It was considered discourteous to inquire about the welfare (or send greetings) to someone's wife, or even to mention her by name. See, for example, Maimonides' letter to Judah b. Joseph in D.H. Baneth (ed.), *Epistles* (Jerusalem, 1946; 2nd edn., 1985), p. 71; trans. *Ethical Writings of Maimonides*, ed. R. L. Weiss and C. E. Butterworth (New York, 1975), p. 123, where the sage says that although it is forbidden to ask about the welfare of a woman (according to Bab. Tal. *Qiddushin* 70b), a prayer is not prohibited, quoting I Sam. 25:6: 'Greetings to you and to your household and to all that is yours' (JPS). Note that the wife is alluded to indirectly as 'household'. The same diffidence is of course exercised in Arab-Muslim society.
44. Candia was the name of Venetian Crete as well as at its capital. The Jews, most of whom were Venetian subjects, traded in wine and other products (sugar, indigo, etc.). See E. Ashtor, 'Jews in the Mediterranean Trade in the Later Middle Ages', *Hebrew Union College Annual*, 55 (1984), 176. See also J. Starr, 'Jewish Life in Crete under the Rule of Venice', *Proceedings of The American Academy for Jewish Research*, 12 (1942), p. 88; Hacker, 'The Connections of Spanish Jewry with Eretz-Israel', 126. And on Candia wine, see also E. Bashan, 'Economic Life from the 16th to the 18th Century', in M. Landau (ed.), *The Jews in Ottoman Egypt (1517–1914)* (Jerusalem, 1988), p. 83 (Hebrew).
45. *Sa' shalom*, abbreviated as usual.

Typical of the complex emotional ties between mother and son is the following letter.[46] It also reflects the mother's involvement in her son's affairs. Doña Jamīla, a widow, writes from Safed to her son Abraham Shalom in Cairo.[47] The letter was definitely written by a scribe.

Gutwirth sees the letter as witness to the intimate family ties in letters of this period;[48] the close relations between Safed and Egypt (Cairo and Alexandria); and the relative poverty of Safed as compared with Egypt.[49] And he presents it as a good example of Judaeo-Spanish vernacular usage in sixteenth-century Safed.

The reverse side has the mail address. The letter was sent from Safed to Cairo, and from there R. Judah Castellaz(zo)[50] is to send it on to Alexandria. We see from this that rabbis, who travelled on scholarly, fund-raising, and commercial missions, were employed as convenient mail-carriers.

The mother addresses her son as 'my dear and beloved son, strings of my heart' *(telas de mi koracon)*. But she immediately launches into a complaint about his absence and neglect, in particular his failure to show up at his sister's wedding.

> Know, as it pleases you, that Esther married. May you marry your sons with greater advantage, and may they never be orphans.[51] ... And more tears flowed from my eyes than from a fountain, for I found myself with no other relative or support than the Lord, blessed be He,

46. Preserved in MS T-S NS 298.11, published and translated by E. Gutwirth, 'A Judeo-Spanish letter from the *Genizah*', 127–38. Gutwirth published a facsimile, a copy in the original Hebrew script and a reconstituted transcription into present-day Judaeo-Spanish and a translation.
47. For the identity of Abraham Shalom, see Gutwirth, 'A Judeo-Spanish Letter'. 129. He suggests that this is the Abraham Shalom who was born in Salonica, settled in Safed c. 1530, and died c. 1557. Abraham Shalom was a disciple and colleague of R. Jacob Berav and received ordination from him. Doña Jamīla is the widow of Yom Tov Shalom. On the latter see Joseph Sambari, MS Paris, 73a (Shtober, 142) (=Neubauer, I, 140); David Conforte, *Qore' ha-Dorot*, ed. D. Cassel (Berlin, 1846), 36a, 49b. And see M. Benayahu, 'The Revival of Ordination in Safed', in S.W. Baron *et al.* (eds.), *Yitzhak F. Baer Jubilee Volume* (Jerusalem, 1960), p. 249 (Hebrew); H.Z. Dimitrovsky, 'Rabbi Yaakob Berab's Academy', *Sefunot* 6 (1962), 93, 111; 7 (1963), 49, 66 (Hebrew); H. Bentov, 'The Autobiographical and Historical Register of Rabbi Joseph Trani', *Shalem*, 1 (1974), 227ff. However, as this was a common name, it is possible that a different Abraham Shalom is intended.
48. The same could be maintained, of course, for the classical period of the Geniza.
49. Citing E. Bashan, 'The Freedom of Trade and the Imposition of Taxes and Custom Duties on Foreign Jewish Traders in the Ottoman Empire', in H.Z. Hirshberg and E. Bashan (eds.), *East and Maghreb* (Ramat Gan, 1974), pp. 105–66.
50. Gutwirth: Kasteliano. He read the final letter *ṣadi* as a *yod* and a *nun*.
51. As the mother was a widow, her daughter was considered an orphan. I have followed Gutwirth's translation (p. 135), with some minor adjustments.
52. Not 'between you', as Gutwirth adds in parentheses. This was a common term for frequent travellers.

> and your brother-in-law, who came. But you did not even send me two lines with him — or, if he did not call on you, then with those who come and go.[52]
>
> You should have written, not for me but for your sister, who tears out her heart and does nothing but weep because of your lack of affection. If not for your aunt I would not have had enough for a dowry. May she be rewarded for what she did for this orphan of the Lord....

The mother adds that she knows why her son failed to show up. 'I know full well that you are angry because of the *sīnī*.[53] She reminds him that the year he pawned it she redeemed it, and he pawned something else which she redeemed. "And I sold everything on account of hunger, and nothing was left. And you, all your anger is on account of the *sīnī*. May I never see you, nor your wife or the children if I have ever laid eyes on any of it.'

Despite the harsh tone of the reprimand, the mother ends with prayers for the lives of her son, his sister and wife, and requests that he send 'two lines' (the conventional expression) so that she should know what is happening with him.

She ends with prayers for the scholar R. Judah, solicitation of health for all, and sends regards to her son's wife and kisses for the children. She adds a request by the recipient's brother- in-law for some paper (which was expensive in Safed and more available in Egypt). He also awaits the shipment of pepper and other products. She repeats her request for two lines: 'I ask you again, do not forget us, by the life of the children. Send us two lines'. And she asks him, now that 'your sister is wed', to concern himself with her own shroud.[54]

There is a postscript in which the scribe/sender, the young R. Khalifiya Aguman ('who kisses your hands'), sends greetings from himself and from Doña Jamīla,[55] widow of the perfect scholar, R. Yom Tov Shalom.

We thus find the mother as caretaker of her son's possessions in his absence, a role she played badly in his view. She conveys a business request in the name of her son's brother-in-law, but she was not necessarily involved. The main point is that she needs her son, and he should be with her. The trying conditions certainly exacerbated her outrage.

The custodian capacity of the mother in the son's absence is remarkably illustrated in the next case. Again the mother takes care of the son's property.[56] The letter is definitely from the Ottoman period; the writing

53. *Sīnī* is 'china' or 'porcelain'. See Goitein, *Mediterranean Society*, Vol. IV, p. 145.
54. The text after this point is unclear.
55. The name is written with a long *a* after the initial letter.
56. Cf. the case of Moses b. Eliezer, in the Ashkenazi correspondence published by Turniansky, 'Letters in Yiddish', 153. Moses goes with his wife to Egypt, where her family lives. His books, candlesticks, etc., remain with his mother.

and other features indicate that it may be from as late as the seventeenth century. Certain evidence not cited here indicates a Spanish provenance. The beginning of the letter reads as follows:[57]

In the name of the Lord

Following the opening greetings, etc.[58] You know how your letters reached us and the letter of the excellent son of R. Abraham, your brother-in-law. And thus we were very delighted at your health and the well-being of all the sons, and that of my mother, my lady, may God save you from every trouble and distress. You know, my son, praise God,[59] that we are in good health. As far as what you write concerning (...) the books, I have already carried out what you requested,[60] and have removed the old books from the box and placed them in the house; and I sent the new books to you by way of R. Shalom b. Phineas, as you requested. There is also one old book, a Commentary on Job, which was in the house of Abraham the Judge, which you forgot.[61] I placed it in the bookcase. And also the 500 *levanim* [=Turkish *akčes*], which you left in the possession of Joḥanan Ḥeteẓ — he has already given them to the Jews, as you requested.

And as for the matter of the clothes that are in the possession of R. Sulaymān b. Qamrah, which you requested that I receive from him and give him 500 *levanim* — I have spoken with R. Shalom b. Phineas, and he told me that he no longer needs Sulaymān b. Qamrah. He (R. Shalom) wishes to go to Egypt, and he will bring the clothes with him and (...) is taking them from him. And another matter you requested from me was that I care for your books and also the clothing by airing them. I am just about to do this, as you instructed me.

Note that in this letter the mother not only takes care of the son's clothing and books; she also arranges for finances and business affairs, which he had entrusted to her. She was probably a widow but was quite competent to perform on her own.

57. The letter is preserved in MS T-S Misc. 35.23. A note in the volume in the Cambridge University Library containing the letter says that this is a 'letter by a lady. 1 piece papar [*sic*] Dr. S[chechter]'. A small excerpt, the passage concerning books (see n. 61 below), is given by A. Scheiber, 'Books in the Private Letters of the Geniza', *Israel Oriental Studies*, 2 (1972) (Stern Memorial Volume), 411–12 (repr. in his *Geniza Studies* [Hildesheim, 1981]), pp. 409–12, on pp. 392–5.

58. Cf. Arabic *ammā baʿdu*.

59. Cf. Arabic *al-ḥamdu li-llāh*, or *ṣubḥāna llāh*.

60. Lit. 'what vou wrote' — also below.

61 The son's book had been on loan to Abraham. Which commentary on Job is referred to is not stated. Scheiber mentions as possible candidates the commentaries of Sa'adya Gaon, Ḥai Gaon, Judah b. Bal'am, Moses b. Chikitilla, and an anonymous commentary, ed. A. Grün (Budapest, 1931).

In the category of letters from mother to son I include a family letter in which a woman wrote to her son-in-law, sending many greetings, in a message obviously intended also for her daughter, if not primarily so.[62] The letter is written in very unidiomatic Hebrew. It contains abundant greetings and many names. The letter is torn at the top. The mother's hesitant allusion to her poor health is touching.

She entreats the recipients to write to her and send the letters with any of those who come and go. Aware that griping will only alienate her children, she ends touchingly: 'My son, I do not wish to write all that is in my heart so as not to trouble you, and so that you write about your (pl.) health and peace, which will console me....'[63]

A straight business relation between mother and son is indicated in a court declaration in which a mother and son appear together as lender and borrower. Accordingly, the esteemed, elegant, and learned bridegroom R. Joseph Aroyas, son of the elder R. Jacob Aroyas Toldedano, acknowledged that he owes the lady, his dear mother, Doña Mas'ūda, forty *peraḥim* (florins) of gold. He attested further that 'if she needs them, then whenever she demands them, he is obliged to deliver them to her within thirty days'. The document was witnessed and signed on Wednesday 10 Shevat 5321 (28 January 1561), in Cairo.[64]

Brother–Sister

Close ties between brother and sister, attested for the classical period, appear also in our letters.[65] But we find that the heavy reliance of sister on brother, witnessed in the previous period, has been shifted, at least according to the evidence we have, so that the emotional and material dependence has become more mutual. This may be an effect of the greater vitality and more active participation of women in economic life indicated in these documents.

Doña Jamīla writes from Jerusalem to her brothers in Egypt, Joshua and Jacob Zalmati.[66] She was the widow of the Kabbalist R. Judah Ḥallēwa. He had emigrated from Fez with other noted Kabbalists in the sixteenth century, and lived in Safed and then in Jerusalem.[67]

62. The document is preserved in MS T-S NS 32.53b.
63. The document is torn at the end.
64. The document is preserved in MS Mosseri VII, 41.1; see *Catalogue of the Jack Mosseri Collection* (Jerusalem, 1990), p. 193.
65. See Goitein, *Mediterranean Society*, Vol. I, pp. 15–26.
66. Preserved in MS Cambridge Or. 1080 J 161, published by A. David, 'New Genizah Documents', 35–6.
67. Her husband was the author of *Ṣafenat Paneaḥ*, completed in 1545. See M. Idel, 'R. Yehudah Ḥallēwa and his Ẓafenat Pa'aneah', *Shalem*, 4 (1984), 119–48. For her two brothers, see David, 'New Genizah Documents', 35, n. 86.

Doña Jamīla requires her brothers' help to get the support of a certain wealthy patron who had not sent her what he had promised. Avraham David suggests that this was Solomon al-Ashqar, who had previously supported her husband.[68]

Doña Jamīla had received a letter from her brothers. She asks for their advice, for she says that there is no one in Jerusalem to depend upon. One of the brothers had written advising her to do whatever was proper in her eyes. This disappointed her, for the matter, she says, is grave. She had also written about certain funds she had, inquiring whether to send them or not. However her brother had not given her advice. 'But the right counsel is from God, the exalted', she consoles herself.

Doña Jamīla explains that the reason she did not come with the present caravan — the one which carried her letter — was that Jews from Damascus came to her and remained in her house, with their wives and children, having escaped from an epidemic that raged there.[69] She would take another caravan. The writer expresses her consternation at the death of Esther and Samuel, 'may God the exalted make them an atonement'. She signs off with greetings to her brothers' families.

We thus find the widow Doña Jamīla left to her own devices in Jerusalem following her husband's death. She sought her brothers' advice concerning financial matters, but from their distant perspective they were reluctant to give counsel beyond urging her to rely on her own good sense. The fugitives from the epidemic in Damascus took refuge in her house, although her husband, the rabbi, was no longer living. We may gather from the letter that the widow was a woman of strength and independence, traits noted several times in these letters.

In another letter a brother in Tripoli (Lebanon) writes to his sister in Cairo begging for her help in a tone of great desperation.[70] The document is unfortunately badly torn in several places. The letter is lengthy and repetitive, and is written in a mood of extreme emergency.

The writer addresses the recipient as 'my sister, my darling, my faultless dove'.[71] He tells her of a calamity he suffered while he was in Alexandria. Police had arrested him, and he was apparently — the text is unclear — detained for two months. For this and other reasons he became impoverished. He had, it seems, borrowed money from a certain Sulaymān for rent and food. 'I therefore entreat[72] you to forgive me for the love of God,

68. See also A. David, 'Sages in Jerusalem in the 16th Century', *Shalem*, 5 (1987), 246–7 (Hebrew).
69. On contacts, particularly after 1516, between Damascus and Jerusalem, see A. Cohen, 'Damascus and Jerusalem', *Sefunot*, 2 (1983), 97–104 (Hebrew).
70. MS T-S 16.265.
71. Song 5:2.
72. This is in the plural, out of respect.

for He knows how distressed I was at my separation from you. Perhaps God will give me the opportunity for the sake of his mercy to return to you, for the desire is engraved in my heart'.

The brother sends regards to a certain Solomon[73] and requests that his sister chide him for forgetting ties of friendship ('the loaf of bread and salt') that had existed between them. The writer had entrusted money to Solomon and requests that he send it to him with the first one who comes: 'For I am dying of starvation and thirst and lack everything. I am in a foreign country, with no one to take pity upon me'. He says that he intends to go to his uncle on the island of Chios,[74] and thus requests that his sister tell Solomon to send the money there, or to deliver it to the perfect sage, R. Samuel Ḥaqan;[75] for he, the brother, is returning once again to Istanbul. 'And by your good life, my sister, darling, and beloved as the pupil of my eye, I entreat you to make every effort for me in this matter and remember the loaf of bread and salt....'

The brother goes on to implore his sister to fulfil his wishes and send the money through his uncle or by means of R. Samuel Ḥaqan 'if you wish my love'. He also appeals to a relative named Doña Hannah, most blessed of women in tents, and requests that she accompany her sister on a mission to the delinquent Solomon[76] to ask that the money be sent immediately 'so that I can return home in peace, for I am tossed from one sorrow to another in a foreign land, in hunger and thirst'. Regards are also sent to a R. Jacob, who should also be enlisted to speak, for love of him, to Solomon and to a certain R. Manṣūr. There follows a reference to his mother, who was living with the sister, or nearby. 'I kiss the hands and feet of my lady, my mother and entreat her to pray to God for me and so that I see your faces'. After sending greetings once more to 'my sister, my darling', he begs her, 'for the sake of God, do not neglect the matter, to speak with Solomon to send the money with the first traveller'. If he has spent the money, she should send 20 *bunduqī* (Venetian gold ducats).

The letter is given the mail address: 'my sister, my darling, my faultless one, Hannah, most blessed of women in tents, wife of Joseph al-Butaynī, may his Rock and Redeemer protect him, to Cairo from Tripoli, may the letter's deliverer be blessed'.

73. He is not to be confused with the Sulaymān mentioned above.
74. For Jews in the commerce of Chios, see E. Ashtor, 'Jews in the Mediterranean Trade', 159–78. See also D. Jacoby, 'The Jews in Chios at the Time of the Rule of Genoa (1346–1566)', *Zion*, 26 (1961), 180–97 (Hebrew), and B. Z. Kedar, 'Jerusalemite Jews in Genoese Famagusta, 1388', in B.Z. Kedar (ed.), *Jerusalem in the Middle Ages*, (Jerusalem, 1979), pp. 198, 200–201 (Hebrew).
75. Samuel b. Moses Ha-Levi Ibn Ḥakim died some time after 1547; see *Encyclopaedia Judaica*, s. v., Vol. VII, 1151–2. He served as rabbi in Egypt, and then moved to Istanbul probably after 1527. See also Ashtor, *History*, Vol. II, p. 481.
76. But actually spelled Sulaymān here.

The forlorn brother, then, relies on his sister to step into the breach and recover his money, detailing to her how to go about it and in what ways she could send the money to him. Her husband is still living, but for some reason her brother does not appeal for his help. The letter arrived at its destination; whether the hapless brother ever got his money we do not know.

Mutual recrimination is the theme of a letter which a woman named Miriam writes to her brother in a Judaeo-Spanish letter from the first half of the sixteenth century.[77] She mentions R. Abraham b. Shoshan, who is included by Joseph Sambari among the exiles from Spain.[78] Her brother had carped about her failure to write. She defends herself: 'Your blaming me for not having written to you is very strange to me.' She turns the tables on him, going on to complain how disappointed she was that he did not visit at Passover.[79] She was informed of this on the eve of the holiday by 'my lady' *(señora)*, namely, her mother: 'You know how I stood with my eyes on the road, waiting to hear an answer from you. Now that it has come, but in a disappointing way,[80] consider what I felt'.

She gives her brother advice about hastening the marriage of his son, though, it is implied, his mother was not co-operating.

A certain Abraham Palieche writes to his sister (Elul 1564).[81] He is in Cairo; she is apparently in Palestine. He complains of his fate *(fortuna)* since she departed and expresses his wish to be reunited with her. He wants her to board the first ship and come: 'Here we shall make a good living... you on your part, I on mine'. He thus hints that she is economically independent, and tells her what to bring. Your brother R. Isaac, he says, will guide you in all this.[82] 'Most important — bring money! Take a

77. MS Cambridge Or. 1080 J 194; see Gutwirth in 'The Family', pp. 212–13.
78. Sambari, *Sefer Divre Yosef*, MS Paris, 73a (Shtober, 142)(=Neubauer, I, 140), cited by Gutwirth. For the famous Ibn Shoshan family, communal leaders in Toledo from the twelfth century, see *Encyclopaedia Judaica*, Vol. VIII, 1199–1201.
79. Passover was a time of pilgrimages by Egyptian Jews to Palestine; see Hacker, 'Spiritual and Material Links, 246.
80. Lit. 'a different way'.
81. Elul began on 9 Aug. that year. See MS T-S 13 J 25.23; Gutwirth, 'The Family', 211–12. On the family, see *Encyclopaedia Judaica*, s.v. Palache, Vol. XIII, pp. 16–17. The name appears in Spain as Palyāj. See Ibn Daud, *Sefer ha-Qabbalah*, ed. G. D. Cohen (Philadelphia, 1967), p. 66, and n. 64 (Ibn Falija [Palyāj]). The variants Pallache, Palacio, de Palatio, Pallas, Palaggi, Balyash are also attested. Abraham Palieche is signed to a Ladino document from Elul 5324 (began 9 Aug. 1564), preserved in MS T-S 13 J. 25.23a (as pointed out to me by Avraham David). On the other side is a letter to Isaac Palieche, presumably a relative of Abraham and perhaps identical with the Isaac mentioned in our document as is his brother. An Abraham Palieche of Safed, who lived at the beginning of the nineteenth century, is mentioned by Benayahu, 'The Revival of Ordination in Safed', 249.
82. In letters, relatives were often designated by their kinship to the recipient; thus

companion, either a respectable man *(hombre nekhbad)* or a woman. Debark in Alexandria'.

This letter, then, shows that brothers were occasionally dependent on their sisters. After the sister left Cairo his luck turned for the worse, and he wants her back ('bring money!'). Here we have further evidence of the enfranchisement of women as active participants in business enterprises.

Our last example of letters between a brother and sister concerns a woman named Qamar of Jerusalem who writes to her brother, R. Moses Parikh, living in Cairo.[83] Moses may be, according to Avraham David, Moses Ha-Levi Parikh, who signed a Prosbul contract in 1616, acting as scribe and witness, along with R. Judah Castro.[84]

The writer begins her letter with greetings to her brother, his wife, called 'his household, most blessed of women, the dear one', and to his sons. Qamar then goes on to complain that her brother had forgotten to write to her. If he was prevented from corresponding, then at least one of his sons could have written. She yearns for her brother and his family, and prays that God will save them from distress and that her brother will come soon to Jerusalem to bury her there.

She then gets to the main part of the letter. Do not forsake me, she pleads, and explains that her travel expenses were heavy, almost 20 *peraḥim* of gold. She had obviously borrowed money for the voyage from her brother. She notifies him that a certain Qamar Castro will bring a deposit. She is also sending three pieces of *'aqrias*[85] for her brother and his sons, a black fabric for Marḥaba, who was her brother's daughter-in-law, and other items, which she will send at the earliest opportunity.

She sends greetings to members of the family, to Marḥaba and to their mother, who both receive condolences, to the sage Eleazar Parikh,[86] to R. Abraham and R. Raphael, and to her sister-in-law Qamar, and she sends a kiss to the student Ezra Parikh Levi, who may have been a son of the recipient of the letter. She also sends greetings to family members ('your relatives') living in Alexandria.

Isaac may be Abraham's brother as well. But it is also possible in this case that he was only a half-brother of the writer.

83. The document is preserved in MS Bodleian Heb. e.94.29, and was published by A. David, 'New Documents from the Geniza', 33–4.

84. See David, 'New Documents', 33, n. 75. The contract is preserved in MS ENA NS 54.7. Moses Parikh is also mentioned, according to David, in MS ENA 4101. If the identification is correct, then our letter takes us to the beginning of the seventeenth century.

85. The word is written *'qryy's*. David cites (34, n. 81) its appearance in MS T-S 16.295a (see above) and in MS T-S AS 153.214, where it is written *'qrysynys*. He takes it as referring to garments or textiles.

86. For the identification, see David, 'New Documents', 33, n. 76. He is mentioned by Joseph Sambari, MS Paris 129b (254) (=Neubauer, I, 156) as one of the sages of Alexandria. And see further references cited by David. He was also called Shuqayr.

Qamar ends by reminding her brother not to forget to write to her, for she is in distress. She wishes him to send with dispatch a certain Abraham who is to bring some items. She is alone in Jerusalem and depressed.

A Letter to In-Laws

We have seen all along that women were involved in family businesses, often in the textile industry. This is well illustrated in our next letter. A woman named Esther, who had gone with her husband from Safed to Egypt, where her parents lived, writes to her in-laws in Safed, where she had resided previously.[87]

She greets her in-laws and expresses her desire to see them, kisses their hands, and reports on the welfare of her own family and their arrival in Egypt without damage or mishap.

Her father sends greetings and eight cubits of cloth[88] for her mother-in-law, Thea, and also a new vest[89] of wool. She assures them that their son Prima (that is, her husband) misses them. He kisses their hands and conveys his greetings; and he worries about them. He sends a gift to his mother *(adonato)*. The letter is unfortunately damaged here along its left side.

In a marginal note Esther thanks her mother-in-law, whom she calls *señora* (*adonati* at the beginning) for the artisanship she learned[90] in Safed. She had already made two or three headcoverings,[91] she says. She expresses her gratitude to Doña Sarah who taught her this art, and promises to send a kerchief *(mandīl)* when it is available.

It was not unusual for women to learn weaving and sewing, and to work at home, and even in small workshops away from home. Women were occasionally employed to teach textile manufacture, as we find here. And they also taught in schools. In our document, weaving and sewing appear to have been a family business on both sides. Safed was a centre of the textile industry.[92]

87. MS T-S 16.295a, published by David, 'New Documents', 52–3. On living with in-laws, see Bashan, 'Economic Life', 232.
88. The word is *kumasi*; see David, 'New Genizah Documents', 52, n. 187.
89. The word is *bandolita*.
90. She wrote this in Arabic, and immediately translated the words into Hebrew. This was evidently a slip. She began the marginal note automatically in Arabic, her own native language — she was Egyptian — although the letter to this point was in Hebrew. She caught herself and translated the words into Hebrew.
91. The word is written *'rqyysh*; see n. 87 above. Ruth Lamdan takes it here and elsewhere to mean 'headcloth'.
92. On the textile business in Safed during this period, see Y. Kena'ani, 'Economic Life in Safed and its Vicinity in the 16th Century and the First Half of the 17th Century', *Zion*, 6 (1934), 172–217 (Hebrew); S. Avitsur, 'Safed — Center of the Manufacture of Woven Woolens in the Sixteenth Century', *Sefunot*, 6 (1962), 41–69 (Hebrew); David, 'The Jewish Settlement in Palestine at the Beginning of the Ottoman Empire', 89, 101–3.

NON-FAMILY LETTERS

There are also non-family women's letters, which are of interest. Occasionally a woman was addressed in a letter along with her husband, indicating a certain level of personal status. Thus a certain Shalom Shemesh writes from Cairo to a man named Barukh Toqo and also to his wife Doña Miriam.[93] They lived in Safed.

The wife is addressed as 'his wife (*newat* [*beto*]),[94] the precious elder *(zeqena)* Doña Miriam, most blessed of women in tents'. The writer wishes them to pray for him 'to return to our land, our household, and our children in safety'. He tells that he is a friend of their son who is studying in Egypt, and reports on his academic progress and his piety ('fasting and good deeds').

A woman writes a Judaeo-Spanish letter from Palestine — she swears by this holy land — appealing for help.[95] She calls herself *ena esklava*, as women called themselves *mamlūka*, etcetera, in the Geniza classical period, and the recipient is addressed as 'the mighty woman of valour like unto Esther and Abigail'. She mentions her poverty — 'the difficulty which I am in because of my sins' — and asks for the addressee's charity *(vuestra sedaqa)*.

As Gutwirth notes, this letter points to the economic security of the woman in Egypt addressed by the writer.

Women Mentioned in Letters

Apart from letters written by and to women, there are also several in which women, some of them notable figures, are mentioned in letters written by men. These documents are also pertinent for understanding the circumstances and status of the women we are studying here. And in view of the active role they played in the socio-economic life of the time, it is not surprising that they show up in court documents of the sixteenth century, preserved in the Cairo Geniza, as we have also seen above.

A certain Blanca, wife of Moses Franco, is mentioned in a letter from someone in Egypt to R. Moses Franco in Jerusalem.[96] The Egyptian first

93. MS T-S 16.295b, published by David, 'New Genizah Documents', 54–5. This is on the reverse side of the previous letter. As both messages were sent to Safed, the senders availed themselves of the same piece of paper. Or perhaps the document is a draft copy, and both senders used the same piece of paper. In any case, there remains the question of how the paper got into the Geniza in Fusṭāṭ if it had been sent to Safed.
94. He omitted the word *beto*.
95. MS T-S Ar. 218.151, mentioned by Gutwirth, 'The Family', 214.
96. The letter is preserved in MS Jewish Theological Seminary, ENA 2591.16, and was published by David in 'New Genizah Documents', 36–57. The large Franco family was originally from Spain, and dispersed over the Ottoman Empire after the Expulsion.

apologizes for not writing. Among the reasons he gives is his wife's illness for 17 months. He had emigrated to Egypt from Salonica. He teaches children — a low income job — and plans to stay a short while in Egypt and then go to Jerusalem. For our writer, as for many others, Egypt was a transit station on the way to the Holy Land.

The letter is addressed both to Moses and to his wife *(newat betkha)*, most blessed of women in tents. The writer explains that from the day he left Salonica he suffered one trouble after another. He had been ill for six months. Whenever he spots a caravan arriving from Safed or Damascus, or anywhere else where the recipient Moses Franco may be located, he inquires about him: 'You have written to me that your wife is pregnant, may God bring her out to the light. Amen. And may he give you proper sons, a permanent seed who will take the place of your ancestors'. The writer then relates that he teaches boys, and his wish is to stay but a short while in Egypt.

He assures Moses that from the day he came to Egypt he inquired about him and about his wife Doña Blanca from all those who came from Jerusalem, and learned that the addressee was fine except for some illnesses. He adds that he had been sick and that his wife had been very ill.

We may readily assume that the intention of the letter, never expressed explicitly, was to get support from Moses Franco and Doña Blanca for the move to Jerusalem. For our purposes, the apparently equal status of Doña Blanca to her husband Moses is worthy of attention.

Esther, the wife of Moses Castro, is mentioned in a letter sent by her husband to R. Abraham Ibn Sanji.[97] Moses Castro, of the famous Castro family, which had as one of its members the renowned Abraham Castro (who lived in Egypt and Jerusalem), was apparently a pupil of Jacob Berav (according to Avraham David). His step-father was R. Abraham Ha-Levi, a victim of the Expulsion from Spain and active in the Jerusalem community in 1520s and 1530s. And he was apparently a nephew of Abraham Zacuto, the illustrious astronomer and historian. Moses Castro, notes David, went to the same places as the step-father: he was born in Spain, and then went to Portugal, Italy, Greece, Istanbul, and Egypt — a good example of the itinerant Spanish exile. His step-father and uncle, says Moses in the letter, were extremely poor. He sends regards from his teachers, the prominent R. Levi Ibn Ḥabīb and R. David Ibn Shoshan.[98]

97. See MS Cambridge Or. 1080 J. 193, published by A. David, 'On the History of Sages in Jerusalem in the 16th Century', in *Shalem*, 5 (1987), 236–43. See also David, ' Settlement', 109. For Sanji, see Bashan, 'Economic Life', 80.

98. For David Ibn Shoshan, see A. David, 'A Letter from Jerusalem from the Early Ottoman Period in Eretz-Israel', in A. Cohen (ed.), *Jerusalem in the Early Ottoman Period* (Jerusalem, 1979), p. 51 (Hebrew).

At the end of the letter we find Moses sending his wife's greetings to the wife of the recipient: 'The children kiss your hands, and Esther, most blessed of women in tents, sends regards to the mistress, the wife *(newat bet)* of the lord'.

Esther Castro is also mentioned in a court deposition, dated 5 Shevat 5307 (= 27 December 1546), which actually relates to the letter just cited and throws light on the relations between Moses Castro and Abraham Ibn Sanji.[99] Accordingly, Esther brought a monetary claim against two men in Egypt following her husband's death. One was Samuel, son of Abraham Ibn Sanji, the recipient of the letter from her husband Moses.

According to the document, the sage R. Moses Castro was owed money by the sage R. Abraham Ibn Sanji (both were deceased) and by R. Abraham Ibn Zur. And when Samuel b. Abraham (Ibn Sanji) was in Jerusalem, Esther, the widow of Moses, claimed the money from him as payment for her *ketubba.*[100] Samuel, according to the declaration, undertook to raise the money and send it to her. The witnesses attested that they had seen this written and signed in a court deposition brought by R. Samuel from Jerusalem.

Again we find a widow functioning on her own in financial matters, in this case putting in a legal claim for money owed to her deceased husband whose estate could not defray the money he owed her on the basis of her marriage contract.

The daughter of Isaac Sholel Ha-Nagid is mentioned in a letter which her brother Abraham Sholel sent from Jerusalem to Egypt, to his brother-in-law, on Tuesday 4 Iyar 5320 (=30 April, 1560).[101] The wife of Isaac Sholel was Qamar, daughter of R. Abraham b. Ḥayyim, who died in 1545. This daughter was not a natural daughter of Isaac, according to Avraham David, but the offspring of another marriage of his wife. The letter has to do with financial support by Palestinians for individuals and for a congregation in an Egyptian community. Abraham Sholel apparently sent the letter with his sister Simḥa. She had been visiting in Jerusalem, and there is a reference to her haste to return to Egypt. Abraham sends regards to several women and to his daughter-in-law Rachel and her family.[102]

99. MS Oxford Bodl. Heb. e 98.57a, published by A. David in *Te'udah*, 7 (1991).

100 That is, the money due her upon her husband's demise on the basis of her marriage contract. Those indebted to her husband were thus responsible for a share of the sum to be defrayed.

101. MS T-S 8 J 15.30, published by A. David, 'On the History of the Family of Sholel in Egypt and Eretz-Israel at the End of the Mamlūk Period and the Beginning of the Ottoman Period', in A. Mirsky *et al.* (eds.), *Galut ahar Golah: Studies in the History of the Jewish People Dedicated to Professor Haim Beinart* (Jerusalem, 1988), pp. 392–3, 410–11 (Hebrew).

102. It is possible that her husband was not of Spanish origin.

A woman named Qamar, wife of Joseph, is mentioned in a document of sale as having sold property to Doña Esther, widow of R. Abraham Tortos in the Karaite Quarter (Ḥārat al-Qarā'iyyīn) in Cairo, in the year 5321 (1561).[103]

CONCLUSION

The Spanish Jewish women who appear in our Geniza documents are vibrant and dynamic, and hardly fit the stereotype of the traditional Jewish female as ensconced in her house and under the protective wings of father, husband, and brothers. The Expulsion from Spain and the consequent dispersion of families throughout the eastern Mediterranean was a great watershed in Jewish history in general and, as we see here, had a forceful impact on the status of Jewish women in Egypt and Palestine. It is hard to determine with available evidence how the lives of non-Sephardic Jewish women were altered as a result of the Sephardic immigration. We do not have the extent of documentation from the Geniza for local *(Musta'rab)*, North African or even Ashkenazi women that we have for Spanish. I have not found letters of *Musta'rab* or North African women in the Geniza, and letters of Ashkenazi women are limited to the Rachel Sussman correspondence. If women's letters from the local and North African communities existed, they have not turned up, for whatever reason.

Of major significance is that fact that in the previous Mamlūk period (1250–1517) the status of Jewish women was virtually the same as it was in the Geniza classical period in matters of family and profession, and even women's names continued to be predominantly Arabic as formerly.[104]

For the Ottoman period, which begins shortly after the great wave of Spanish immigration, it has been observed, on the basis of other evidence, that women in general were legally and socially quite independent, certainly more so in practice than in legal theory. In fact, this gap existed in prior periods as well, but the hiatus between theory and practice seems to have become greater in later centuries. Women could own inherited property, bestow legacies, and grant pious foundations; they could buy and sell property, deal with guarantees and securities and post bonds. Some were economically independent, involved in business and real estate. A significant number of active women were widows, as noted in our

103. MS ENA 2562 (L 36).

104. See Strauss-Ashtor, *History of the Jews*, Vol. II, pp. 342ff. There are virtually no women's letters from the Mamlūk period. Strauss-Ashtor assigns the letter to Solomon the scribe (above, pp. 247–8) to the fifteenth century (II, 84). But, as noted above, this is not certain.

documents. Women in growing numbers refused to stay home and perform their traditional role.[105]

Pressing financial circumstances forced women into the work market. It has been seen that many were engaged in the textile business, mostly on a small family scale. This often took them outside the home. Interestingly, R. David Ibn Abî Zimra declares that a woman may work outside her home, even among Gentile employers, with her husband's permission.[106] Only arduous conditions would account for this latitude.

Women were still expected to be modest and not to rove around freely. There were few opportunities to meet other men. One was doing laundry at the Siloam pool, as shown above.[107] There were also holidays and family festivities. The last day of Passover was spent outdoors in nature, with men and women mixing. There were also visits *(ziyarāt)*, for instance, to the tomb of Samuel at Nebi Samuel on eve of his death, at which times women went unveiled.

A woman is addressed in our letters as 'dear and modest' *(yeqara we-senu'a)*.[108] The evidence shows that there was no longer a taboo, as in the the Geniza classical period, to refer to a man's wife by name, although the expression 'your household' *(nevat betkha)* for 'your wife' was widespread. This is true of Hebrew epistles. And Gutwirth observes for women's letters in Judaeo-Spanish that the taboo against mentioning the wife's name disappeared, and that he recalled the use of 'your house' *(su kaza)* as a euphemism for 'your wife' perhaps only once.[109] If so, Judaeo-Spanish letters are even more direct than Hebrew letters in this regard.

Women's names from the Spanish community are predominantly Hebrew. Esther was extremely popular, and the biblical names Miriam, Hannah, Rebecca, et cetera, recur. Esther was, of course, the redeemer of

105. Cohen, *Jewish Life under Islam*, pp. 127–8; Rozen, *The Jewish Community*, p. 239.
106. David Ibn Abi Zimra (*Responsa*, I, 67, no. 919) dealt with a situation going back to the Mamlūk period whereby indigent Jewish women who worked for Gentile employers in the weaving industry, along with strange men, would lodge there for three or four days and nights. The Nagids pronounced decrees against the practice, but it continued, and the Radbaz bowed to exigency, permitting the women to work but only if they were over forty and did not share worktables with men. See Strauss-Ashtor, *History*, II, 153; Goldman, *Life and Times*, p. 142; Bashan, 'Economic Life', 109.
107. For this and the following, see Rozen, *The Jewish Community*, pp. 239–40.
108. MS T-S 13 J 34.7.
109. Gutwirth, 'The Family', 214–15, citing MS NS 264.77. Gutwirth also mentions Misc. 25.76, from the late eighteenth century. The man is asking for charity or repayment: 'My wife is naked (*desnuda*), and the dress she wears on the Sabbath is the one she wears all week'. Gutwirth says that this represents freedom in speaking of the wife, although (as he observes) there is extreme need. It may be added, however, that in the classical period women freely spoke of themselves as naked, namely, poverty-stricken.

her people, and the predilection for this name cannot have been accidental. Reina (Reyna, Regina) was also very much in favour, and had been popular in Spain.[110] The name Jamīla was also fashionable: an Arabic name, it was already widespread.[111] To be sure, Romance names such as Blanca, Clara, Sol, and Thea were already in use in Spain. Arabic names such as Mas'ūda, Najma, Raḥma and Qamar (or Qamra) came into use in the Arabophone ambience of the Middle East. What is striking is that, as far as I know, names compounded with Sitt ('Lady'), such as Sitt al-Kull, Sitt al-Nas, prevalent in the classical and Mamlūk periods, do not show up.[112]

Increased mobility within a community was matched by great movement from town to town and from country to country. We see in these letters an immense dispersion of families and tremendous migratory activity, a phenomenon noted also for Ashkenazi families.[113] Transportation in the sixteenth century was relatively easy. There was heavy traffic between Jerusalem and Safed in Palestine and Egypt. Indeed this was a triangle forming a socio-cultural and economic unit. The passage of men and goods seems to have been an everyday matter. Travellers were a 'living bridge' between Egypt and Palestine, and brought goods, money, letters, and news. This migration created powerful family ties between Egypt and Palestine.[114] Amnon Cohen's sources indicated that there were more Jews travelling from Egypt to Jerusalem than the other way.[115] While this may have been the case, it is not indicated in the Geniza material. As Palestine in the early Ottoman period was economically less well off than Egypt, there was pressure to take refuge in Egypt. The dependence of Palestinian Jewish communities on their Egyptian confrères was tremendous in the early sixteenth century under Mamlūk rule, but this reliance declined under the Ottomans.[116] To be sure, Jerusalem was a place of pilgrimage, especially around holidays such as Passover. And people fleeing plagues

110. See Baer, *Die Juden im christlichen Spanien*, Vol. I, I, p. 1127; Vol. II, p. 566; J. Régné, *History of the Jews in Aragon, Regesta and Documents 1213–1327*, ed. Yom Tov Assis, with A. Gruzman (Jerusalem, 1978), p. 695.

111. See Baer, *Die Juden*, Vol. I, 1, p. 1117; and Vol. II, p. 559; Regne, *History of the Jews*, p. 687; Yom Tov Assis and J. Ramón Magdalena, *The Jews of Navarre in the Late Middle Ages* (Jerusalem, 1990), p. 231. It was apparently more common in Castille than in Aragon and Navarre.

112. A study of the onomasticon of Jewish women of Egypt and Palestine in the sixteenth century would be very useful.

113. Turniansky, 'A Correspondence in Yiddish', 153.

114. See Hacker, 'Spiritual and Material Links', 243.

115. A. Cohen, 'Sixteenth Century Egypt and Palestine: The Jewish Connection as Reflected in the *Sijill* of Jerusalem', in A. Cohen and G. Baer (eds), *Egypt and Palestine: A Millennium of Association (868–1948)* (Jerusalem–New York, 1984), p. 234.

116. Hacker, 'Spiritual and Material Links', 247.

had a tendency to come to Jerusalem under cover of the pilgrimage, as it was difficult to find refuge under other conditions.[117]

Letters often refer to sending mail with those who travel back and forth. There was also an organized mail service, at least references to messengers ('runners') who plied the route between Egypt and Palestine. We often find rabbis carrying letters, or performing services as intermediaries.

A main reason for migratory activity was economic, the relative prosperity of Egypt, for instance, as compared with the hard times often prevailing in Palestine. Another was epidemics and famine, which were very frequent and destructive.[118]

Families moved from place to place, seeking health, safety, and security. For Sephardic families, this was merely an extension of the wandering that accompanied the Expulsion itself and its aftermath.

Marriage was another reason for mobility. The communities of Cairo, Jerusalem, and Safed were composed of various sub-groups, as we have noted, and there was intermarriage among them, and among Egyptian and Palestinian Jews. We have seen that marriage was not always virilocal, or did not remain so. A wife returns to Egypt with her own family and her husband.

The dispersal of families due to mobility, and the plagues, famine, and high mortality rate, especially of children, created a situation of small family cells of one or two generations. This is observed for the seventeenth century,[119] but there are signs of this even in the sixteenth. For Sephardic families, the impact of the Expulsion and dispersion was no doubt a factor. There are no references to households with aunts, uncles, and cousins. The Sephardim did, however, practice polygyny, though our letters show no indication of it.

My conclusions are offered tentatively, as the material surveyed, while fascinating and significant, is not extensive. But it is hard to escape the conclusion that Sephardic ladies in Egypt and Palestine in the sixteenth century were looking ahead to new prospects.

117. Cohen, 'Sixteenth Century Egypt and Palestine', p. 239.

118. See Hacker, 'Pride and Depression', 564–65, citing plagues in Damascus (and probably Safed also) in 1513–14, and in Safed in 1533, 1540–43, 1547. And there were more.

119. Rozen, *The Jewish Community*, p. 241, who notes that only rarely do we find large families of several generations. The second generation often left town.

Manuscripts Cited

Cambridge University Library
T-S 8 J 15.30
13 J 21.10
13 J 24.22
13 J 25.23
Ar. 29.105
Ar. 218.151
Misc. 35.23
AS 153.214
NS 32.53b
NS 298.11
NS 338.51
Or. 1080 J 161
Or. 1080 J 193
Or. 1080 J 194

Jewish Theological Seminary
ENA 2562 (L 36)
ENA 2591.16
ENA 4101
ENA NS 54.7
Mosseri
VII 41.1

Oxford University, Bodleian Library
Heb.e.94.29

List of Contributors

María Fuencisla García Casar is Professor of Hebrew Language and Literature at the University of Salamanca. Her publications include *El Pasado judío de Salamanca*, and *El Pasado judío de ciudad Rodrigo*, as well as various articles on the history of the Castilian *conversos*.

Carlos Carrete Parrondo is Professor in the Faculty of Philology at the University of Salamanca. He is the author of numerous books and articles, including volumes in the series Fontes Judaeorum Regni Castellae (Salamanca, 1981–); 'Hebraistas Judeoconversos en la Universidad de Salamanca (Siglos XV–XVI)', *Simposio Biblico Español* (Madrid, 1984), pp. 723–39; 'Judería Soriana y moréna burgalesa; historia de amor', *Estudios Mirandeses*, 8 (1988), 57–62; 'Intervención de los judeo-conversos en la expulsion', in *Destierros Aragoneses* (Saragossa, 1988), pp. 61–85.

Sara Turel is completing her Ph.D. in the Department of Drama at Tel Aviv University and is a researcher at the Diaspora Museum. Her previous research was on surrealism in Spain, and on that topic she has published '*La Quimera* de García Lorca: expresión surrealista de un mito', *Cuadernos Hispanoamericanos*, 433–4 (1986), 351–9. Currently she is working on biblical themes in the theatre of the Golden Age in Spain.

Miguel-Angl Ladero Quesada is Professor of Medieval history at the University of Madrid. He is a member of the Royal Academy of History, and specializes in the history of the Kingdom of Castile in the thirteenth to fifteenth centuries. Among his publications are: *Los Reyes Católicos. La Corona y la unidad de España* (Madrid, 1989); *La Hacienda Real de Castilla en el siglo XV* (Tenerife, 1973); *Historia de Sevilla. La ciudad medieval* (Seville, 1988); *Granada. Historia de un país islámico (1232–1571)* (Madrid, 1989); *Los mudéjares de Castilla y otros estudios de historia medieval andaluza* (Granada, 1989); 'Publación, economía. Sociedad', in *Historia General de España y America*, Vol. 5 (1369–1517) (Madrid, 1982).

Elguja Khintibidze is Pro-Rector for Scientific Research at Tbilisi State University, and specializes in Church texts (chronicles and Hagiographa). Among his publications are *Links between Georgian and Byzantine Literature* (Tbilisi, 1989, Russian), and studies on the epos of Shota Rustaveli.

Felipe Maíllo Salgado is Professor of Arabic and Islamic Studies at the University of Salamanca. His many publications include: *Los arabismos del Castellano en la baja Edad Media (consideraciones históricos y filológicas)* (Salamanca, 1983; 2nd edn. 1991); *A través del Oriente* (Riḥla Ibn Yubayr) (Barcelona, 1990); *Crónica Anónima de los Reyes de Taifas* (Madrid, 1991), and he is co-author of *España--Al-Andalus--Sefarad: Síntesis y nuevas perspectivas* (Salamanca, 1988; repr. 1990).

Michal Oron is Senior Lecturer in Kabbalah and Hasidism at Tel Aviv University. Among her publications are *The Sefer Ha-Peli'ah and the Sefer Ha-Kanah: Their Kabbalistic Principles, Social and Religious Criticism and Literary Composition* (Jerusalem, 1980, Hebrew), and *The Mystical World of Rabbi Nahman of Bratslav* (Tel Aviv, 1986, Hebrew), as well as a critical edition in Hebrew of *Sha'ar Ha-Razim* by Todoros B. Joseph Abulafia (Jerusalem, 1989).

Renata Segre is director of the international project 'Sephardic Settlements of the Sixteenth Century in the Mediterranean Basin', sponsored by the Istituto dei beni culturali dell' Emilia-Romagna. She is author of numerous articles and publications, including 'Gli ebrei lombardi nell'età spagnola. Storia d'une espulsione', *Memorie d'ell'Accademia delle scienze di Torino*, 28 (1973), 123–4; *The Jews in Piedmont*, 3 vols. (Jerusalem, 1986–90); 'Documentary History of the Usque Printing-House, and especially of the Biblia de Ferrara', in *Simposio Sobre la Biblia de Ferrara* (Seville, 1991).

Benjamin Ravid is Jennie and Mayer Weisman Professor of Jewish History and Chair of the Department of Near Eastern and Judaic Studies at Brandeis University. He is the author of *Economics and Toleration in Seventeenth-Century Venice: The Background and Context of the* **Discorso of Simone Luzzatto** (Jerusalem, 1978), and has contributed historical notes, together with Howard Adelman, to Mark R. Cohen (ed. and trans.), *The Autobiography of a Seventeenth-Century Venetian Rabbi: Leon Modena's Life of Judah* (Princeton, 1988). He has written over 20 articles on the history of the Jews of Venice.

Elisabeth Crouzet-Pavan is Professor of Medieval History at the Charles de Gaulle University (Lille III). Her work deals with the social and cultural history of Italy during the last centuries of the Middle Ages, for example: *'Sopra la acque Salse'. Espaces, pouvoir et société à Venise à la fin du Moyen Age*, 1100 pp. (Rome-Paris, in press).

Ennio Concina is Professor of the History of Architecture at the School of Architecture, Venice, and at the Faculty of Engineering, Trieste. His various publications include: *Structure urbaine et functions de batiments, XVIe–XVIIIe siècles* (Venice, 1982); *La macchina territoriale* (Rome--Bari, 1983); *L'arsenale della repubblica di Venezia* (Milan, 1984); *Arsenale e città nell'occidente europeo* (Rome, 1987), and he is co-author of *La città degli Ebrei. Il Ghetto di Venezia, architettura e urbanistica* (Venice, 1991).

Ivana Burdelez specializes in Italian and English literature, and is archivist at the Historical Archives of Dubrovnik, working on the documents from the sixteenth to eighteenth centuries. Her publications include: 'Il ruolo degli Ebrei nel commercio marittimo di Ragusa e di Livorno', *Studi livornesi*, 3 (1988); *Commercial Relationships between Dubrovnik and Sofia through the Ladino Documents of the Historical Archives of Dubrovnik in the First Half of 17 C*, Resumés des communications de congrès international (Sofia, 1989); *La Repubblica di Ragusa, mediatrico tra Oriente e Occidente del 500; La lettura geografica: Studi in onore di Osvaldo Baldacci*, Vol. I (Bologna, 1991).

Edwin Seroussi is a faculty member of the Department of Musicology at Bar-Ilan University and researcher at the Jewish Music Research Centre of the Hebrew University of Jerusalem. He studies the musical culture of the Sephardic Jews in North Africa and the Ottoman Empire and has published several studies on this topic. His publications include *The Spanish-Portuguese Liturgical Repertory in Western Europe according to German Sources* (Jerusalem, forthcoming); 'R. Joseph Shalom Gallego: A Cantor from Saloniki in Seventeenth-Century Amsterdam', *Asufot* (Hebrew, in press); and 'Politics, Ethnic Identity, and Music in Israel: The Case of the Morrocan *Bakkashot*', *Asian Music*, 17 (1986).

Leah Bornstein-Makovetsky is a Lecturer in the Department of Jewish History at Bar-Ilan University, and specializes in the social and religious life of the Jews in the Ottoman Empire from the fifteenth to the nineteenth centuries. Her various articles include: 'Jewish Brokers in Constantinople during the 18th Century according to Hebrew Documents', in A. Toaff and S. Schwarzfuchs (eds.), *The Mediterranean and the Jews: Banking, Finance and International Trade* (XVIth–XVIIth Centuries) (Ramat Gan, 1989), pp. 75–104; 'The Relations between the Secular Leadership of the Jewish Communities in the Ottoman Empire and the Ottoman Government during the 16th and 17th Centuries', in A. Rodrigue (ed.), *Chapters in the History of the Jews in the Ottoman Empire* (Indiana, forthcoming).

Benjamin Braude is Associate Professor of History and Director of the Committee on Middle Eastern Studies at the University of Boston College. He has been visiting professor at Harvard University and Smith College, as well as research assistant in the Program in Near Eastern Studies at Princeton University. Among his publications are (co-edited with B. Lewis), *Christians and Jews in the Ottoman Empire: The Functioning of a Plural Society*, 2 vols. (New York, 1982).

Joel L. Kraemer is Professor in the Department of Middle Eastern and African History and an affiliated Senior Fellow of the Dayan Center, Tel Aviv University. He specializes in medieval Islamic and Jewish cultural and intellectual history. He is the author of *Humanism in the Renaissance of Islam* (Leiden, 1986); *Philosophy in the Renaissance of Islam* (Leiden, 1987); and *The History of Tabari: Incipient Decline* (Albany, NY, 1989). His latest book is *Perspectives on Maimonides* (Oxford, 1991).

The Editor: Alisa Meyuhas Ginio is a faculty member of the Department of History at Tel Aviv University. She specializes in the history of Judaeo-Christian polemics in medieval Spain.

Indices

Editor's Note:

Index *locorum* includes names of geographical sites or terms, according to the common transcription in the language of origin (for example, Livorno), except in cases of an English form adopted in current use (for example, Seville).

Index *personarum* includes individual names, listing last names (where known) in alphabetical order, with the exception of royalty, ecclesiastical dignitaries, and world-famous men of letters (for example, Judah Halevi).

Index *rerum* includes historical terms and concepts; names attributed to various social groups and denominations; legislative, judicial, administrative, and other offices and organizations; names of coins and weights; and titles of literary works cited in the text. Bibliographical references are included in the footnotes of each article.

Index locorum

Index personarum

Index rerum

www.ingramcontent.com/pod-product-compliance
Lightning Source LLC
LaVergne TN
LVHW102226300926
849160LV00005B/41